Praise for *Fundamentals of WiMAX*

This book is one of the most comprehensive books I have reviewed ... it is a must read for engineers and students planning to remain current or who plan to pursue a career in telecommunications. I have reviewed other publications on WiMAX and have been disappointed. This book is refreshing in that it is clear that the authors have the in-depth technical knowledge and communications skills to deliver a logically laid out publication that has substance to it.

—Ron Resnick, President, WiMAX Forum

This is the first book with a great introductory treatment of WiMAX technology. It should be essential reading for all engineers involved in WiMAX. The high-level overview is very useful for those with non-technical background. The introductory sections for OFDM and MIMO technologies are very useful for those with implementation background and some knowledge of communication theory. The chapters covering physical and MAC layers are at the appropriate level of detail. In short, I recommend this book to systems engineers and designers at different layers of the protocol, deployment engineers, and even students who are interested in practical applications of communication theory.

—Siavash M. Alamouti, Chief Technology Officer, Mobility Group, Intel

This is a very well-written, easy-to-follow, and comprehensive treatment of WiMAX. It should be of great interest.

—Dr. Reinaldo Valenzuela, Director of Wireless Research, Bell Labs

Fundamentals of WiMAX *is a comprehensive guide to WiMAX from both industry and academic viewpoints, which is an unusual accomplishment. I recommend it to anyone who is curious about this exciting new standard.*

—Dr. Teresa Meng, Professor, Stanford University,
Founder and Director, Atheros Communications

Andrews, Ghosh, and Muhamed have provided a clear, concise, and well-written text on 802.16e/WiMAX. The book provides both the breadth and depth to make sense of the highly complicated 802.16e standard. I would recommend this book to both development engineers and technical managers who want an understating of WiMAX and insight into 4G modems in general.

—Paul Struhsaker, VP of Engineering, Chipset platforms, Motorola Mobile Device Business Unit, former vice chair of IEEE 802.16 working group

Fundamentals of WiMAX is written in an easy-to-understand tutorial fashion. The chapter on multiple antenna techniques is a very clear summary of this important technology and nicely organizes the vast number of different proposed techniques into a simple-to-understand framework.

—Dr. Ender Ayanoglu, Professor, University of California, Irvine,
Editor-in-Chief, *IEEE Transactions on Communications*

Fundamentals of WiMAX is a comprehensive examination of the 802.16/WiMAX standard and discusses how to design, develop, and deploy equipment for this wireless communication standard. It provides both insightful overviews for those wanting to know what WiMAX is about and comprehensive, in-depth chapters on technical details of the standard, including the coding and modulation, signal processing methods, Multiple-Input Multiple-Output (MIMO) channels, medium access control, mobility issues, link-layer performance, and system-level performance.

—Dr. Mark C. Reed, Principle Researcher, National ICT Australia,
Adjunct Associate Professor, Australian National University

This book is an excellent resource for any engineer working on WiMAX systems. The authors have provided very useful introductory material on broadband wireless systems so that readers of all backgrounds can grasp the main challenges in WiMAX design. At the same time, the authors have also provided very thorough analysis and discussion of the multitudes of design options and engineering trade-offs, including those involved with multiple antenna communication, present in WiMax systems, making the book a must-read for even the most experienced wireless system designer.

—Dr. Nihar Jindal, Assistant Professor, University of Minnesota

This book is very well organized and comprehensive, covering all aspects of WiMAX from the physical layer to the network and service aspects. The book also includes insightful business perspectives. I would strongly recommend this book as a must-read theoretical and practical guide to any wireless engineer who intends to investigate the road to fourth generation wireless systems.

—Dr. Yoon Chae Cheong, Vice President, Communication Lab, Samsung

The authors strike a wonderful balance between theoretical concepts, simulation performance, and practical implementation, resulting in a complete and thorough exposition of the standard. The book is highly recommended for engineers and managers seeking to understand the standard.

—Dr. Shilpa Talwar, Senior Research Scientist, Intel

Fundamentals of WiMAX *is a comprehensive guide to WiMAX, the latest frontier in the communications revolution. It begins with a tutorial on 802.16 and the key technologies in the standard and finishes with a comprehensive look at the predicted performance of WiMAX networks. I believe readers will find this book invaluable whether they are designing or testing WiMAX systems.*

—Dr. James Truchard, President, CEO and Co-Founder, National Instruments

This book is a must-read for engineers who want to know WiMAX fundamentals and its performance. The concepts of OFDMA, multiple antenna techniques, and various diversity techniques—which are the backbone of WiMAX technology—are explained in a simple, clear, and concise way. This book is the first of its kind.

—Amitava Ghosh, Director and Fellow of Technical Staff, Motorola

Andrews, Ghosh, and Muhamed have written the definitive textbook and reference manual on WiMAX, and it is recommended reading for engineers and managers alike.

—Madan Jagernauth, Director of WiMAX Access Product Management, Nortel

Fundamentals of WiMAX

Prentice Hall Communications Engineering and Emerging Technologies Series

Theodore S. Rappaport, *Series Editor*

Fundamentals of WiMAX
Understanding Broadband Wireless Networking

Jeffrey G. Andrews, Ph.D.

Department of Electrical and Computer Engineering
The University of Texas at Austin

Arunabha Ghosh, Ph.D.

AT&T Labs Inc.

Rias Muhamed

AT&T Labs Inc.

PRENTICE
HALL

Upper Saddle River, NJ • Boston • Indianapolis • San Francisco
New York • Toronto • Montreal • London • Munich • Paris • Madrid
Capetown • Sydney • Tokyo • Singapore • Mexico City

Many of the designations used by manufacturers and sellers to distinguish their products are claimed as trademarks. Where those designations appear in this book, and the publisher was aware of a trademark claim, the designations have been printed with initial capital letters or in all capitals.

The authors and publisher have taken care in the preparation of this book, but make no expressed or implied warranty of any kind and assume no responsibility for errors or omissions. No liability is assumed for incidental or consequential damages in connection with or arising out of the use of the information or programs contained herein.

The publisher offers excellent discounts on this book when ordered in quantity for bulk purchases or special sales, which may include electronic versions and/or custom covers and content particular to your business, training goals, marketing focus, and branding interests. For more information, please contact:

U.S. Corporate and Government Sales
(800) 382-3419
corpsales@pearsontechgroup.com

For sales outside the United States, please contact:

International Sales
international@pearsoned.com

 This Book Is Safari Enabled

The Safari® Enabled icon on the cover of your favorite technology book means the book is available through Safari Bookshelf. When you buy this book, you get free access to the online edition for 45 days.

Safari Bookshelf is an electronic reference library that lets you easily search thousands of technical books, find code samples, download chapters, and access technical information whenever and wherever you need it.

To gain 45-day Safari Enabled access to this book:

- Go to http://www.prenhallprofessional.com/safarienabled
- Complete the brief registration form
- Enter the coupon code JRR8-NBAQ-SCDV-9ZQB-FNJU

If you have difficulty registering on Safari Bookshelf or accessing the online edition, please e-mail customer-service@safaribooksonline.com.

Visit us on the Web: www.prenhallprofessional.com

Library of Congress Cataloging-in-Publication Data

Andrews, Jeffrey G.

Fundamentals of WiMAX : understanding broadband wireless networking / Jeffrey G. Andrews, Arunabha Ghosh, Rias Muhamed.

p. cm.

Includes bibliographical references and index.

ISBN 0-13-222552-2 (hbk : alk. paper)

1. Wireless communication systems. 2. Broadband communication systems. I. Ghosh, Arunabha. II. Muhamed, Rias. III. Title.

TK5103.2.A56 2007

621.382—dc22

2006038505

ISBN 0-13-222552-2

Text printed in the United States on recycled paper at Courier in Westford, Massachusetts.
Second printing, June 2007

Dedicated to Catherine and my parents, Greg and Mary
—Jeff

Dedicated to Debolina and my parents, Amitabha and Meena
—Arunabha

Dedicated to Shalin, Tanaz, and my parents, Ahamed and Fathima
—Rias

Contents

Foreword

Within the last two decades, communication advances have reshaped the way we live our daily lives. Wireless communications has grown from an obscure, unknown service to an ubiquitous technology that serves almost half of the people on Earth. Whether we know it or not, computers now play a dominant role in our daily activities, and the Internet has completely reoriented the way people work, communicate, play, and learn.

However severe the changes in our lifestyle may seem to have been over the past few years, the convergence of wireless with the Internet is about to unleash a change so dramatic that soon wireless ubiquity will become as pervasive as paper and pen. WiMAX—which stands for Worldwide Interoperability for Microwave Access—is about to bring the wireless and Internet revolutions to portable devices across the globe. Just as broadcast television in the 1940's and 1950's changed the world of entertainment, advertising, and our social fabric, WiMAX is poised to broadcast the Internet throughout the world, and the changes in our lives will be dramatic. In a few years, WiMAX will provide the capabilities of the Internet, without any wires, to every living room, portable computer, phone, and handheld device.

In its simplest form, WiMAX promises to deliver the Internet throughout the globe, connecting the "last mile" of communications services for both developed and emerging nations. In this book, Andrews, Ghosh, and Muhamed have done an excellent job covering the technical, business, and political details of WiMAX. This unique trio of authors have done the reader a great service by bringing their first-hand industrial expertise together with the latest results in wireless research. The tutorials provided throughout the text are especially convenient for those new to WiMAX or the wireless field. I believe *Fundamentals of WiMAX* will stand out as the definitive WiMAX reference book for many years to come.

—Theodore S. Rappaport
Austin, Texas

Preface

Fundamentals of WiMAX was consciously written to appeal to a broad audience, and to be of value to anyone who is interested in the IEEE 802.16e standards or wireless broadband networks more generally. The book contains cutting-edge tutorials on the technical and theoretical underpinnings to WiMAX that are not available anywhere else, while also providing high-level overviews that will be informative to the casual reader. The entire book is written with a tutorial approach that should make most of the book accessible and useful to readers who do not wish to bother with equations and technical details, but the details are there for those who want a rigorous understanding. In short, we expect this book to be of great use to practicing engineers, managers and executives, graduate students who want to learn about WiMAX, undergraduates who want to learn about wireless communications, attorneys involved with regulations and patents pertaining to WiMAX, and members of the financial community who want to understand exactly what WiMAX promises.

Organization of the Book

The book is organized into three parts with a total of twelve chapters. Part I provides an introduction to broadband wireless and WiMAX. Part II presents a collection of rigorous tutorials covering the technical and theoretical foundations upon which WiMAX is built. In Part III we present a more detailed exposition of the WiMAX standard, along with a quantitative analysis of its performance.

In Part I, Chapter 1 provides the background information necessary for understanding WiMAX. We provide a brief history of broadband wireless, enumerate its applications, discuss the market drivers and competitive landscape, and present a discussion of the business and technical challenges to building broadband wireless networks. Chapter 2 provides an overview of WiMAX and serves as a summary of the rest of the book. This chapter is written as a standalone tutorial on WiMAX and should be accessible to anyone interested in the technology.

We begin Part II of the book with Chapter 3, where the immense challenge presented by a time-varying broadband wireless channel is explained. We quantify the principal effects in broadband wireless channels, present practical statistical models, and provide an overview of diversity countermeasures to overcome the challenges. Chapter 4 is a tutorial on OFDM, where the elegance of multicarrier modulation and the theory of how it works are explained. The chapter emphasizes a practical understanding of OFDM system design and discusses implementation issues for WiMAX systems such as the peak-to-average ratio. Chapter 5 presents a rigorous tutorial on multiple antenna techniques covering a broad gamut of techniques from simple receiver diversity to advanced beamforming and spatial multiplexing. The practical considerations in the application of these techniques to WiMAX are also discussed. Chapter 6 focuses on OFDMA, another key-ingredient technology responsible for the superior performance of WiMAX. The chapter explains how OFDMA can be used to enhance capacity through the exploitation of multiuser diversity and adaptive modulation, and also provides a survey of different scheduling algorithms. Chapter 7 covers end-to-end aspects of broadband wireless networking such as QoS, session management, security, and mobility management. WiMAX being an IP-based network, this chapter highlights some of the relevant IP protocols used to build an end-to-end broadband wireless service. Chapters 3 though 7 are more likely to be of interest to practicing engineers, graduate students, and others wishing to understand the science behind the WiMAX standard.

In Part III of the book, Chapters 8 and 9 describe the details of the physical and media access control layers of the WiMAX standard and can be viewed as a distilled summary of the far more lengthy IEEE 802.16e-2005 and IEEE 802.16-2004 specifications. Sufficient details of these layers of WiMAX are provided in these chapters to enable the reader to gain a solid understanding of the salient features and capabilities of WiMAX and build computer simulation models for performance analysis. Chapter 10 describes the networking aspects of WiMAX, and can be thought of as a condensed summary of the end-to-end network systems architecture developed by the WiMAX Forum. Chapters 11 and 12 provide an extensive characterization of the expected performance of WiMAX based on the research and simulation-based modeling work of the authors. Chapter 11 focuses on the link-level performance aspects, while Chapter 12 presents system-level performance results for multicellular deployment of WiMAX.

Acknowledgments

We would like to thank our publisher Bernard Goodwin, Catherine Nolan, and the rest of the staff at Prentice Hall, who encouraged us to write this book even when our better instincts told us the time and energy commitment would be overwhelming. We also thank our reviewers, Roberto Christi, Amitava Ghosh, Nihar Jindal, and Mark Reed for their valuable comments and feedback.

We thank the series editor Ted Rappaport, who strongly supported this project from the very beginning and provided us with valuable advice on how to plan and execute a co-authored book. The authors sincerely appreciate the support and encouragement received from David Wolter and David Deas at AT&T Labs, which was vital to the completion and timely publication of this book.

The authors wish to express their sincere gratitude to WiMAX Forum and their attorney, Bill Bruce Holloway, for allowing us to use some of their materials in preparing this book.

Jeffrey G. Andrews: I would like to thank my co-authors Arunabha Ghosh and Rias Muhamed for their dedication to this book; without their talents and insights, this book never would have been possible.

Several of my current and former Ph.D. students and postdocs contributed their time and in-depth knowledge to Part II of the book. In particular, I would like to thank Runhua Chen, whose excellent work with Arunabha and I has been useful to many parts of the book, including the performance predictions. He additionally contributed to parts of Chapter 3, as did Wan Choi and Aamir Hasan. Jaeweon Kim and Kitaek Bae contributed their extensive knowledge on peak-to-average ratio reduction techniques to Chapter 4. Jin Sam Kwak, Taeyoon Kim, and Kaibin Huang made very high quality contributions to Chapter 5 on beamforming, channel estimation, and linear precoding and feedback, respectively. My first Ph.D. student, Zukang Shen, whose research on OFDMA was one reason I became interested in WiMAX, contributed extensively to Chapter 6. Han Gyu Cho also provided valuable input to the OFDMA content.

As this is my first book, I would like to take this chance to thank some of the invaluable mentors and teachers who got me excited about science, mathematics, and then eventually wireless communications and networking. Starting with my public high school in Arizona, I owe two teachers particular thanks: Jeff Lockwood, my physics and astronomy teacher, and Elizabeth Callahan, a formative influence on my writing and in my interest in learning for its own sake. In college, I would like to single out John Molinder, Phil Cha, and Gary Evans. Dr. Molinder in particular taught my first classes on signal processing and communications and encouraged me to go into wireless. From my five years at Stanford, I am particularly grateful to my advisor, Teresa Meng. Much like a college graduate reflecting with amazement on his parents' effort in raising him, since graduating I have truly realized how fortunate I was to have such an optimistic, trusting, and well-rounded person as an advisor. I also owe very special thanks to my associate advisor and friend, Andrea Goldsmith, from whom I have probably learned more about wireless than anyone else. I would also like to acknowledge my University of Texas at Austin colleague, Robert Heath, who has taught me a tremendous amount about MIMO. In no particular order, I would also like to recognize my colleagues past and present, Moe Win, Steven Weber, Sanjay Shakkottai, Mike Honig, Gustavo de Veciana, Sergio Verdu, Alan Gatherer, Mihir Ravel, Sriram Vishwanath, Wei Yu, Tony Ambler, Jeff Hsieh, Keith Chugg, Avneesh Agrawal, Arne Mortensen, Tom Virgil, Brian Evans, Art Kerns, Ahmad Bahai, Mark Dzwonzcyk, Jeff Levin, Martin Haenggi, Bob Scholtz, John Cioffi, and Nihar Jindal, for sharing their knowledge and providing support and encouragement over the years.

On the personal side, I would like to thank my precious wife, Catherine, who actually was brave enough to marry me during the writing of this book. A professor herself, she is the most supportive and loving companion anyone could ever ask for. I would also like to thank my parents, Greg and Mary, who have always inspired and then supported me to the fullest in all my pursuits and have just as often encouraged me to do less rather than more. I would also like to acknowledge my grandmother, Ruth Andrews, for her love and support over the years. Finally, I would also like to thank some of my most important sources of ongoing intellectual nourishment: my close friends from Sahuaro and Harvey Mudd, and my brother, Brad.

Arunabha Ghosh: I would like to thank my co-authors Rias Muhamed and Jeff Andrews without whose expertise, hard work, and valuable feedback it would have been impossible to bring this book to completion.

I would also like to thank my collaborators, Professor Robert Heath and Mr. Runhua Chen from the University of Texas at Austin. Both Professor Heath and Mr. Chen possess an incredible degree of intuition and understanding in the area of MIMO communication systems and play a very significant role in my research activity at AT&T Labs. Their feedback and suggestions particularly to the close loop MIMO solutions that can be implemented with the IEEE 802.16e-2005 framework is a vital part of this book and one of its key distinguishing features.

I also thank several of my colleagues from AT&T Labs including Rich Kobylinski, Milap Majmundar, N. K. Shankarnarayanan, Byoung-Jo Kim, and Paul Henry. Without their support and valuable feedback it would not have been possible for me to contribute productively to a book on WiMAX. Rich, Milap, and Paul also played a key role for their contributions in Chapters 11 and 12. I would also like to especially thank Caroline Chan, Wen Tong, and Peiying Zhu from Nortel Networks' Wireless Technology Lab. Their feedback and understanding of the closed-loop MIMO techniques for WiMAX were vital for Chapters 8, 11, and 12.

Finally and most important of all I would like to thank my wife, Debolina, who has been an inspiration to me. Writing this book has been quite an undertaking for both of us as a family and it is her constant support and encouragement that really made is possible for me to accept the challenge. I would also like to thank my parents, Amitabha and Meena, my brother, Siddhartha, and my sister in-law, Mili, for their support.

Rias Muhamed: I sincerely thank my co-authors Arunabha Ghosh and Jeff Andrews for giving me the opportunity to write this book. Jeff and Arun have been outstanding collaborators, whose knowledge, expertise, and commitment to the book made working with them a very rewarding and pleasurable experience. I take this opportunity to express my appreciation for all my colleagues at AT&T Labs, past and present, from whom I have learned a great deal. A number of them, including Frank Wang, Haihao Wu, Anil Doradla, and Milap Majmundar, provided valuable reviews, advice, and suggestions for improvements. I am also thankful to Linda Black at AT&T Labs for providing the market research data used in Chapter 1. Several others have also directly or indirectly provided help with this book, and I am grateful to all of them.

Special thanks are due to Byoung-Jo "J" Kim, my colleague and active participant in the WiMAX Network Working Group (NWG) for providing a thorough and timely review of Chapter 10. I also acknowledge with gratitude Prakash Iyer, the chairman of WiMAX NWG, for his review.

Most of all, I thank my beloved wife, Shalin, for her immeasurable support, encouragement, and patience while working on this project. For more than a year, she and my precious three-year-old daughter Tanaz had to sacrifice too many evening and weekend activities as I remained preoccupied with writing this book. Without their love and understanding, this book would not have come to fruition.

I would be remiss if I fail to express my profound gratitude to my parents for the continuous love, support, and encouragement they have offered for all my pursuits. My heartfelt thanks are also due to my siblings and my in-laws for all the encouragement I have received from them.

About the Authors

Jeffrey G. Andrews, Ph.D.

Jeffrey G. Andrews is an assistant professor in the Department of Electrical and Computer Engineering at the University of Texas at Austin, where he is the associate director of the Wireless Networking and Communications Group. He received a B.S. in engineering with high distinction from Harvey Mudd College in 1995, and the M.S. and Ph.D. in electrical engineering from Stanford University in 1999 and 2002. Dr. Andrews serves as an editor for the *IEEE Transactions on Wireless Communications* and has industry experience at companies including Qualcomm, Intel, Palm, and Microsoft. He received the National Science Foundation CAREER award in 2007.

Arunabha Ghosh, Ph.D.

Arunabha Ghosh is a principal member of technical staff in the Wireless Communications Group in AT&T Labs Inc. He received his B.S. with highest distinction from Indian Institute of Technology at Kanpur in 1992 and his Ph.D. from University of Illinois at Urbana-Champaign in 1998. Dr. Ghosh has worked extensively in the area of closed loop MIMO solutions for WiMAX and has chaired several task groups within the WiMAX Forum for the development of mobile WiMAX Profiles.

Rias Muhamed

Rias Muhamed is a lead member of technical staff in the Wireless Networks Group at AT&T Labs Inc. He received his B.S. in electronics and communications engineering from Pondicherry University, India, in 1990, his M.S. in electrical engineering from Virginia Tech in 1996, and his M.B.A. from St. Edwards University at Austin in 2000. Rias has led the technology assessment activities at AT&T Labs in the area of Fixed Wireless Broadband for several years and has worked on a variety of wireless systems and networks.

PART I

Overview of WiMAX

Introduction to Broadband Wireless

B roadband wireless sits at the confluence of two of the most remarkable growth stories of the telecommunications industry in recent years. Both wireless and broadband have on their own enjoyed rapid mass-market adoption. Wireless mobile services grew from 11 million subscribers worldwide in 1990 to more than 2 billion in 2005 [1]. During the same period, the Internet grew from being a curious academic tool to having about a billion users. This staggering growth of the Internet is driving demand for higher-speed Internet-access services, leading to a parallel growth in broadband adoption. In less than a decade, broadband subscription worldwide has grown from virtually zero to over 200 million [2]. Will combining the convenience of wireless with the rich performance of broadband be the next frontier for growth in the industry? Can such a combination be technically and commercially viable? Can wireless deliver broadband applications and services that are of interest to the endusers? Many industry observers believe so.

Before we delve into broadband wireless, let us review the state of broadband access today. *Digital subscriber line* (DSL) technology, which delivers broadband over twisted-pair telephone wires, and *cable modem* technology, which delivers over coaxial cable TV plant, are the predominant mass-market broadband access technologies today. Both of these technologies typically provide up to a few megabits per second of data to each user, and continuing advances are making several tens of megabits per second possible. Since their initial deployment in the late 1990s, these services have enjoyed considerable growth. The United States has more than 50 million broadband subscribers, including more than half of home Internet users. Worldwide, this number is more than 200 million today and is projected to grow to more than 400 million by 2010 [2]. The availability of a wireless solution for broadband could potentially accelerate this growth.

What are the applications that drive this growth? Broadband users worldwide are finding that it dramatically changes how we share information, conduct business, and seek entertainment.

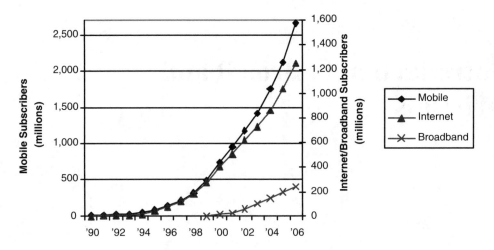

Figure 1.1 Worldwide subscriber growth 1990–2006 for mobile telephony, Internet usage, and broadband access [1, 2, 3]

Broadband access not only provides faster Web surfing and quicker file downloads but also enables several multimedia applications, such as real-time audio and video streaming, multimedia conferencing, and interactive gaming. Broadband connections are also being used for voice telephony using voice-over-Internet Protocol (VoIP) technology. More advanced broadband access systems, such as fiber-to-the-home (FTTH) and very high data rate digital subscriber loop (VDSL), enable such applications as entertainment-quality video, including high-definition TV (HDTV) and video on demand (VoD). As the broadband market continues to grow, several new applications are likely to emerge, and it is difficult to predict which ones will succeed in the future.

So what is broadband wireless? Broadband wireless is about bringing the broadband experience to a wireless context, which offers users certain unique benefits and convenience. There are two fundamentally different types of broadband wireless services. The first type attempts to provide a set of services similar to that of the traditional fixed-line broadband but using wireless as the medium of transmission. This type, called *fixed wireless broadband,* can be thought of as a competitive alternative to DSL or cable modem. The second type of broadband wireless, called *mobile broadband,* offers the additional functionality of portability, nomadicity,[1] and mobility. Mobile broadband attempts to bring broadband applications to new user experience scenarios and hence can offer the end user a very different value proposition. WiMAX (worldwide interoperability for microwave access) technology, the subject of this book, is designed to accommodate both fixed and mobile broadband applications.

---

1. *Nomadicity* implies the ability to connect to the network from different locations via different base stations; *mobility* implies the ability to keep ongoing connections active while moving at vehicular speeds.

In this chapter, we provide a brief overview of broadband wireless. The objective is to present the background and context necessary for understanding WiMAX. We review the history of broadband wireless, enumerate its applications, and discuss the business drivers and challenges. In Section 1.7, we also survey the technical challenges that need to be addressed while developing and deploying broadband wireless systems.

1.1 Evolution of Broadband Wireless

The history of broadband wireless as it relates to WiMAX can be traced back to the desire to find a competitive alternative to traditional wireline-access technologies. Spurred by the deregulation of the telecom industry and the rapid growth of the Internet, several competitive carriers were motivated to find a wireless solution to bypass incumbent service providers. During the past decade or so, a number of wireless access systems have been developed, mostly by start-up companies motivated by the disruptive potential of wireless. These systems varied widely in their performance capabilities, protocols, frequency spectrum used, applications supported, and a host of other parameters. Some systems were commercially deployed only to be decommissioned later. Successful deployments have so far been limited to a few niche applications and markets. Clearly, broadband wireless has until now had a checkered record, in part because of the fragmentation of the industry due to the lack of a common standard. The emergence of WiMAX as an industry standard is expected to change this situation.

Given the wide variety of solutions developed and deployed for broadband wireless in the past, a full historical survey of these is beyond the scope of this section. Instead, we provide a brief review of some of the broader patterns in this development. A chronological listing of some of the notable events related to broadband wireless development is given in Table 1.1.

WiMAX technology has evolved through four stages, albeit not fully distinct or clearly sequential: (1) narrowband wireless local-loop systems, (2) first-generation line-of-sight (LOS) broadband systems, (3) second-generation non-line-of-sight (NLOS) broadband systems, and (4) standards-based broadband wireless systems.

1.1.1 Narrowband Wireless Local-Loop Systems

Naturally, the first application for which a wireless alternative was developed and deployed was voice telephony. These systems, called *wireless local-loop* (WLL), were quite successful in developing countries such as China, India, Indonesia, Brazil, and Russia, whose high demand for basic telephone services could not be served using existing infrastructure. In fact, WLL systems based on the digital-enhanced cordless telephony (DECT) and code division multiple access (CDMA) standards continue to be deployed in these markets.

In markets in which a robust local-loop infrastructure already existed for voice telephony, WLL systems had to offer additional value to be competitive. Following the commercialization of the Internet in 1993, the demand for Internet-access services began to surge, and many saw providing high-speed Internet-access as a way for wireless systems to differentiate themselves. For example, in February 1997, AT&T announced that it had developed a wireless access system for the 1,900MHz PCS (personal communications services) band that could deliver two voice lines and a 128kbps data connection to subscribers. This system, developed under the code name "Project Angel," also had the distinction of being one of the first commercial wireless systems to use adaptive antenna technology. After field trials for a few years and a brief commercial offering, AT&T discontinued the service in December 2001, citing cost run-ups and poor take-rate as reasons.

During the same time, several small start-up companies focused solely on providing Internet-access services using wireless. These wireless Internet service provider (WISP) companies typically deployed systems in the license-exempt 900MHz and 2.4GHz bands. Most of these systems required antennas to be installed at the customer premises, either on rooftops or under the eaves of their buildings. Deployments were limited mostly to select neighborhoods and small towns. These early systems typically offered speeds up to a few hundred kilobits per second. Later evolutions of license-exempt systems were able to provide higher speeds.

1.1.2 First-Generation Broadband Systems

As DSL and cable modems began to be deployed, wireless systems had to evolve to support much higher speeds to be competitive. Systems began to be developed for higher frequencies, such as the 2.5GHz and 3.5GHz bands. Very high speed systems, called *local multipoint distribution systems* (LMDS), supporting up to several hundreds of megabits per second, were also developed in millimeter wave frequency bands, such as the 24GHz and 39GHz bands. LMDS-based services were targeted at business users and in the late 1990s enjoyed rapid but short-lived success. Problems obtaining access to rooftops for installing antennas, coupled with its shorter-range capabilities, squashed its growth.

In the late 1990s, one of the more important deployments of broadband wireless happened in the so-called *multichannel multipoint distribution services* (MMDS) band at 2.5GHz. The MMDS band was historically used to provide *wireless cable* broadcast video services, especially in rural areas where cable TV services were not available. The advent of satellite TV ruined the wireless cable business, and operators were looking for alternative ways to use this spectrum. A few operators began to offer one-way wireless Internet-access service, using telephone line as the return path. In September 1998, the Federal Communications Commission (FCC) relaxed the rules of the MMDS band in the United States to allow two-way communication services, sparking greater industry interest in the MMDS band. MCI WorldCom and Sprint each paid approximately $1 billion to purchase licenses to use the MMDS spectrum, and several companies started developing high-speed fixed wireless solutions for this band.

The first generation of these fixed broadband wireless solutions were deployed using the same towers that served wireless cable subscribers. These towers were typically several hundred feet tall and enabled LOS coverage to distances up to 35 miles, using high-power transmitters. First-generation MMDS systems required that subscribers install at their premises outdoor antennas high enough and pointed toward the tower for a clear LOS transmission path. Sprint and MCI launched two-way wireless broadband services using first-generation MMDS systems

Table 1.1 Important Dates in the Development of Broadband Wireless

Date	Event
February 1997	AT&T announces development of fixed wireless technology code named "Project Angel"
February 1997	FCC auctions 30MHz spectrum in 2.3GHz band for wireless communications services (WCS)
September 1997	American Telecasting (acquired later by Sprint) announces wireless Internet access services in the MMDS band offering 750kbps downstream with telephone dial-up modem upstream
September 1998	FCC relaxes rules for MMDS band to allow two-way communications
April 1999	MCI and Sprint acquire several wireless cable operators to get access to MMDS spectrum
July 1999	First working group meeting of IEEE 802.16 group
March 2000	AT&T launches first commercial high-speed fixed wireless service after years of trial
May 2000	Sprint launches first MMDS deployment in Phoenix, Arizona, using first-generation LOS technology
June 2001	WiMAX Forum established
October 2001	Sprint halts MMDS deployments
December 2001	AT&T discontinues fixed wireless services
December 2001	IEEE 802.16 standards completed for > 11GHz.
February 2002	Korea allocates spectrum in the 2.3GHz band for wireless broadband (WiBro)
January 2003	IEEE 802.16a standard completed
June 2004	IEEE 802.16-2004 standard completed and approved
September 2004	Intel begins shipping the first WiMAX chipset, called Rosedale
December 2005	IEEE 802.16e standard completed and approved
January 2006	First WiMAX Forum–certified product announced for fixed applications
June 2006	WiBro commercial services launched in Korea
August 2006	Sprint Nextel announces plans to deploy mobile WiMAX in the United States

in a few markets in early 2000. The outdoor antenna and LOS requirements proved to be significant impediments. Besides, since a fairly large area was being served by a single tower, the capacity of these systems was fairly limited. Similar first-generation LOS systems were deployed internationally in the 3.5GHz band.

1.1.3 Second-Generation Broadband Systems

Second-generation broadband wireless systems were able to overcome the LOS issue and to provide more capacity. This was done through the use of a cellular architecture and implementation of advanced-signal processing techniques to improve the link and system performance under multipath conditions. Several start-up companies developed advanced proprietary solutions that provided significant performance gains over first-generation systems. Most of these new systems could perform well under non-line-of-sight conditions, with customer-premise antennas typically mounted under the eaves or lower. Many solved the NLOS problem by using such techniques as *orthogonal frequency division multiplexing* (OFDM), *code division multiple access* (CDMA), and multiantenna processing. Some systems, such as those developed by SOMA Networks and Navini Networks, demonstrated satisfactory link performance over a few miles to desktop subscriber terminals without the need for an antenna mounted outside. A few megabits per second throughput over cell ranges of a few miles had become possible with second-generation fixed wireless broadband systems.

1.1.4 Emergence of Standards-Based Technology

In 1998, the Institute of Electrical and Electronics Engineers (IEEE) formed a group called 802.16 to develop a standard for what was called a *wireless metropolitan area network,* or wireless MAN. Originally, this group focused on developing solutions in the 10GHz to 66GHz band, with the primary application being delivering high-speed connections to businesses that could not obtain fiber. These systems, like LMDS, were conceived as being able to tap into fiber rings and to distribute that bandwidth through a point-to-multipoint configuration to businesses within line-of-sight. The IEEE 802.16 group produced a standard that was approved in December 2001. This standard, Wireless MAN-SC, specified a physical layer that used single-carrier modulation techniques and a media access control (MAC) layer with a burst time division multiplexing (TDM) structure that supported both frequency division duplexing (FDD) and time division duplexing (TDD).

After completing this standard, the group started work on extending and modifying it to work in both licensed and license-exempt frequencies in the 2GHz to 11GHz range, which would enable NLOS deployments. This amendment, IEEE 802.16a, was completed in 2003, with OFDM schemes added as part of the physical layer for supporting deployment in multipath environments. By this time, OFDM had established itself as a method of choice for dealing with multipath for broadband and was already part of the revised IEEE 802.11 standards. Besides the OFDM physical layers, 802.16a also specified additional MAC-layer options, including support for orthogonal frequency division multiple access (OFDMA).

Further revisions to 802.16a were made and completed in 2004. This revised standard, IEEE 802.16-2004, replaces 802.16, 802.16a, and 802.16c with a single standard, which has also been adopted as the basis for HIPERMAN (high-performance metropolitan area network) by ETSI (European Telecommunications Standards Institute). In 2003, the 802.16 group began work on enhancements to the specifications to allow vehicular mobility applications. That revision,

---

Sidebar 1.1 A Brief History of OFDM

Although OFDM has become widely used only recently, the concept dates back some 40 years. This brief history of OFDM cites some landmark dates.

1966: Chang shows that multicarrier modulation can solve the multipath problem without reducing data rate [4]. This is generally considered the first official publication on multicarrier modulation. Some earlier work was Holsinger's 1964 MIT dissertation [5] and some of Gallager's early work on waterfilling [6].

1971: Weinstein and Ebert show that multicarrier modulation can be accomplished using a DFT [7].

1985: Cimini at Bell Labs identifies many of the key issues in OFDM transmission and does a proof-of-concept design [8].

1993: DSL adopts OFDM, also called discrete multitone, following successful field trials/competitions at Bellcore versus equalizer-based systems.

1999: The IEEE 802.11 committee on wireless LANs releases the 802.11a standard for OFDM operation in 5GHz UNI band.

2002: The IEEE 802.16 committee releases an OFDM-based standard for wireless broadband access for metropolitan area networks under revision 802.16a.

2003: The IEEE 802.11 committee releases the 802.11g standard for operation in the 2.4GHz band.

2003: The multiband OFDM standard for ultrawideband is developed, showing OFDM's usefulness in low-SNR systems.

802.16e, was completed in December 2005 and was published formally as IEEE 802.16e-2005. It specifies scalable OFDM for the physical layer and makes further modifications to the MAC layer to accommodate high-speed mobility.

As it turns out, the IEEE 802.16 specifications are a collection of standards with a very broad scope. In order to accommodate the diverse needs of the industry, the standard incorporated a wide variety of options. In order to develop interoperable solutions using the 802.16 family of standards, the scope of the standard had to be reduced by establishing consensus on what options of the standard to implement and test for interoperability. The IEEE developed the specifications but left to the industry the task of converting them into an interoperable standard that can be certified. The WiMAX Forum was formed to solve this problem and to promote solutions based on the IEEE 802.16 standards. The WiMAX Forum was modeled along the lines of the Wi-Fi Alliance, which has had remarkable success in promoting and providing interoperability testing for products based on the IEEE 802.11 family of standards.

The WiMAX Forum enjoys broad participation from the entire cross-section of the industry, including semiconductor companies, equipment manufacturers, system integraters, and service providers. The forum has begun interoperability testing and announced its first certified product based on IEEE 802.16-2004 for fixed applications in January 2006. Products based on IEEE 802.18e-2005 are expected to be certified in early 2007. Many of the vendors that previously developed proprietary solutions have announced plans to migrate to fixed and/or mobile WiMAX. The arrival of WiMAX-certified products is a significant milestone in the history of broadband wireless.

1.2 Fixed Broadband Wireless: Market Drivers and Applications

Applications using a fixed wireless solution can be classified as *point-to-point* or *point-to-multi-point*. Point-to-point applications include interbuilding connectivity within a campus and micro-wave backhaul. Point-to-multipoint applications include (1) broadband for residential, small office/home office (SOHO), and small- to medium-enterprise (SME) markets, (2) T1 or fractional T1-like services to businesses, and (3) wireless backhaul for Wi-Fi hotspots. Figure 1.2 illustrates the various point-to-multipoint applications.

Consumer and small-business broadband: Clearly, one of the largest applications of WiMAX in the near future is likely to be broadband access for residential, SOHO, and SME markets. Broadband services provided using fixed WiMAX could include high-speed Internet access, telephony services using voice over IP, and a host of other Internet-based applications. Fixed wireless offers several advantages over traditional wired solutions. These advantages include lower entry and deployment costs; faster and easier deployment and revenue realization; ability to build out the network as needed; lower operational costs for network maintenance, management, and operation; and independence from the incumbent carriers.

From a customer premise equipment (CPE)[2] or subscriber station (SS) perspective, two types of deployment models can be used for fixed broadband services to the residential, SOHO, and SME markets. One model requires the installation of an outdoor antenna at the customer premise; the other uses an all-in-one integrated radio modem that the customer can install indoors like traditional DSL or cable modems. Using outdoor antennas improves the radio link and hence the performance of the system. This model allows for greater coverage area per base station, which reduces the density of base stations required to provide broadband coverage, thereby reducing capital expenditure. Requiring an outdoor antenna, however, means that installation will require a truck-roll with a trained professional and also implies a higher SS cost. Clearly, the two deployment scenarios show a trade-off between capital expenses and operating expense: between base station capital infrastructure costs and SS and installation costs. In developed countries, such as the United States, the high labor cost of truck-roll, coupled with consumer dislike for outdoor antennas, will likely favor an indoor SS deployment, at least for the residential application. Further, an indoor self-install SS will also allow a business model that can exploit the retail distribution channel and offer consumers a variety of SS choices. In devel-

---

2.   The CPE is referred to as a subscriber station (SS) in fixed WiMAX.

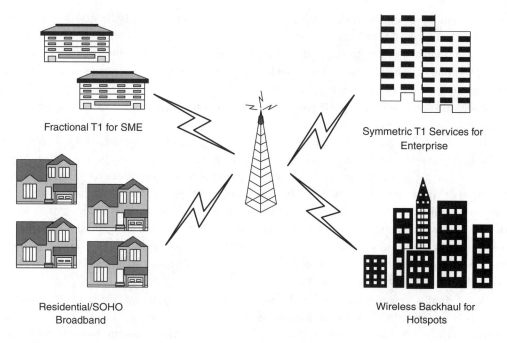

Figure 1.2 Point-to-multipoint WiMAX applications

oping countries, however, where labor is cheaper and aesthetic and zoning considerations are not so powerful, an outdoor-SS deployment model may make more economic sense.

In the United States and other developed countries with good wired infrastructure, fixed wireless broadband is more likely to be used in rural or underserved areas, where traditional means of serving them is more expensive. Services to these areas may be provided by incumbent telephone companies or by smaller players, such as WISPs, or local communities and utilities. It is also possible that competitive service providers could use WiMAX to compete directly with DSL and cable modem providers in urban and suburban markets. In the United States, the FCC's August 2005 decision to rollback cable plant sharing needs is likely to increase the appeal of fixed wireless solutions to competitive providers as they look for alternative means to reach subscribers. The competitive landscape in the United States is such that traditional cable TV companies and telephone companies are competing to offer a full bundle of telecommunications and entertainment services to customers. In this environment, satellite TV companies may be pushed to offering broadband services including voice and data in order to stay competitive with the telephone and cable companies, and may look to WiMAX as a potential solution to achieve this.

T1 emulation for business: The other major opportunity for fixed WiMAX in developed markets is as a solution for competitive T1/E1, fractional T1/E1, or higher-speed services for the business market. Given that only a small fraction of commercial buildings worldwide have access to fiber, there is a clear need for alternative high-bandwidth solutions for enterprise customers. In the business market, there is demand for symmetrical T1/E1 services that cable and DSL have so far not met the technical requirements for. Traditional telco services continue to serve this demand with relatively little competition. Fixed broadband solutions using WiMAX could potentially compete in this market and trump landline solutions in terms of time to market, pricing, and dynamic provisioning of bandwidth.

Backhaul for Wi-Fi hotspots: Another interesting opportunity for WiMAX in the developed world is the potential to serve as the backhaul connection to the burgeoning Wi-Fi hotspots market. In the United States and other developed markets, a growing number of Wi-Fi hotspots are being deployed in public areas such as convention centers, hotels, airports, and coffee shops. The Wi-Fi hotspot deployments are expected to continue to grow in the coming years. Most Wi-Fi hotspot operators currently use wired broadband connections to connect the hotspots back to a network point of presence. WiMAX could serve as a faster and cheaper alternative to wired backhaul for these hotspots. Using the point-to-multipoint transmission capabilities of WiMAX to serve as backhaul links to hotspots could substantially improve the business case for Wi-Fi hotspots and provide further momentum for hotspot deployment. Similarly, WiMAX could serve as 3G (third-generation) cellular backhaul.

A potentially larger market for fixed broadband WiMAX exists outside the United States, particularly in urban and suburban locales in developing economies—China, India, Russia, Indonesia, Brazil and several other countries in Latin America, Eastern Europe, Asia, and Africa—that lack an installed base of wireline broadband networks. National governments that are eager to quickly catch up with developed countries without massive, expensive, and slow network rollouts could use WiMAX to leapfrog ahead. A number of these countries have seen sizable deployments of legacy WLL systems for voice and narrowband data. Vendors and carriers of these networks will find it easy to promote the value of WiMAX to support broadband data and voice in a fixed environment.

1.3 Mobile Broadband Wireless: Market Drivers and Applications

Although initial WiMAX deployments are likely to be for fixed applications, the full potential of WiMAX will be realized only when used for innovative nomadic and mobile broadband applications. WiMAX technology in its IEEE 802.16e-2005 incarnation will likely be deployed by fixed operators to capture part of the wireless mobility value chain in addition to plain broadband access. As endusers get accustomed to high-speed broadband at home and work, they will demand similar services in a nomadic or mobile context, and many service providers could use WiMAX to meet this demand.

The first step toward mobility would come by simply adding nomadic capabilities to fixed broadband. Providing WiMAX services to portable devices will allow users to experience bandwidth not just at home or work but also at other locations. Users could take their broadband connection with them as they move around from one location to another. Nomadic access may not allow for seamless roaming and handover at vehicular speeds but would allow pedestrian-speed mobility and the ability to connect to the network from any location within the service area.

In many parts of the world, existing fixed-line carriers that do not own cellular, PCS, or 3G spectrum could turn to WiMAX for provisioning mobility services. As the industry moves along the path of quadruple-play service bundles—voice, data, video, and mobility—some service providers that do not have a mobility component in their portfolios—cable operators, satellite companies, and incumbent phone companies—are likely to find WiMAX attractive. For many of these companies, having a mobility plan will be not only a new revenue opportunity but also a defensive play to mitigate churn by enhancing the value of their product set.

Existing mobile operators are less likely to adopt WiMAX and more likely to continue along the path of 3G evolution for higher data rate capabilities. There may be scenarios, however, in which traditional mobile operators may deploy WiMAX as an overlay solution to provide even higher data rates in targetted urban centers or metrozones. This is indeed the case with Korea Telecom, which has begun deploying WiBro service in metropolitan areas to complement its ubiquitous CDMA2000 service by offering higher performance for multimedia messaging, video, and entertainment services. WiBro is a mobile broadband solution developed by Korea's Electronics and Telecommunications Research Institute (ETRI) for the 2.3GHz band. In Korea, WiBro systems today provide end users with data rates ranging from 512kbps to 3Mbps. The WiBro technology is now compatible with IEEE 802.16e-2005 and mobile WiMAX.

In addition to higher-speed Internet access, mobile WiMAX can be used to provide voice-over-IP services in the future. The low-latency design of mobile WiMAX makes it possible to deliver VoIP services effectively. VoIP technologies may also be leveraged to provide innovative new services, such as voice chatting, push-to-talk, and multimedia chatting.

New and existing operators may also attempt to use WiMAX to offer differentiated personal broadband services, such as mobile entertainment. The flexible channel bandwidths and multiple levels of quality-of-service (QoS) support may allow WiMAX to be used by service providers for differentiated high-bandwidth and low-latency entertainment applications. For example, WiMAX could be embedded into a portable gaming device for use in a fixed and mobile environment for interactive gaming. Other examples would be streaming audio services delivered to MP3 players and video services delivered to portable media players. As traditional telephone companies move into the entertainment area with IP-TV (Internet Protocol television), portable WiMAX could be used as a solution to extend applications and content beyond the home.

1.4 WiMAX and Other Broadband Wireless Technologies

WiMAX is not the only solution for delivering broadband wireless services. Several proprietary solutions, particularly for fixed applications, are already in the market. A few proprietary solutions, such as i-Burst technology from ArrayComm and Flash-OFDM from Flarion (acquired by QualComm) also support mobile applications. In addition to the proprietary solutions, there are standards-based alternative solutions that at least partially overlap with WiMAX, particularly for the portable and mobile applications. In the near term, the most significant of these alternatives are third-generation cellular systems and IEEE 802.11-based Wi-Fi systems. In this section, we compare and contrast the various standards-based broadband wireless technologies and high-light the differentiating aspects of WiMAX.

1.4.1 3G Cellular Systems

Around the world, mobile operators are upgrading their networks to 3G technology to deliver broadband applications to their subscribers. Mobile operators using GSM (global system for mobile communications) are deploying UMTS (universal mobile telephone system) and HSDPA (high speed downlink packet access) technologies as part of their 3G evolution. Traditional CDMA operators are deploying 1x EV-DO (1x evolution data optimized) as their 3G solution for broadband data. In China and parts of Asia, several operators look to TD-SCDMA (time division-synchronous CDMA) as their 3G solution. All these 3G solutions provide data through-put capabilities on the order of a few hundred kilobits per second to a few megabits per second. Let us briefly review the capabilities of these overlapping technologies before comparing them with WiMAX.

HSDPA is a downlink-only air interface defined in the Third-generation Partnership Project (3GPP) UMTS Release 5 specifications. HSDPA is capable of providing a peak user data rate (layer 2 throughput) of 14.4Mbps, using a 5MHz channel. Realizing this data rate, however, requires the use of all 15 codes, which is unlikely to be implemented in mobile terminals. Using 5 and 10 codes, HSDPA supports peak data rates of 3.6Mbps and 7.2Mbps, respectively. Typical average rates that users obtain are in the range of 250kbps to 750kbps. Enhancements, such as spatial processing, diversity reception in mobiles, and multiuser detection, can provide signifi-cantly higher performance over basic HSDPA systems.

It should be noted that HSDPA is a downlink-only interface; hence until an uplink comple-ment of this is implemented, the peak data rates achievable on the uplink will be less than 384kbps, in most cases averaging 40kbps to 100kbps. An uplink version, HSUPA (high-speed uplink packet access), supports peak data rates up to 5.8Mbps and is standardized as part of the 3GPP Release 6 specifications; deployments are expected in 2007. HSDPA and HSUPA together are referred to as HSPA (high-speed packet access).

1x EV-DO is a high-speed data standard defined as an evolution to second-generation IS-95 CDMA systems by the 3GPP2 standards organization. The standard supports a peak downlink data rate of 2.4Mbps in a 1.25MHz channel. Typical user-experienced data rates are in the order of 100kbps to 300kbps. Revision A of 1x EV-DO supports a peak rate of 3.1Mbps to a mobile user; Revision B will support 4.9Mbps. These versions can also support uplink data rates of up to 1.8Mbps. Revision B also has options to operate using higher channel bandwidths (up to 20MHz), offering potentially up to 73Mbps in the downlink and up to 27Mbps in the uplink.

In addition to providing high-speed data services, 3G systems are evolving to support multi-media services. For example, 1x EV-DO Rev A enables voice and video telephony over IP. To make these service possible, 1xEV-DO Rev A reduces air-link latency to almost 30ms, intro-duces intrauser QoS, and fast intersector handoffs. Multicast and broadcast services are also supported in 1x EV-DO. Similarly, development efforts are under way to support IP voice, video, and gaming, as well as multicast and broadcast services over UMTS/HSPA networks.

It should also be noted that 3GPP is developing the next major revision to the 3G standards. The objective of this long-term evolution (LTE) is to be able to support a peak data rate of 100Mbps in the downlink and 50Mbps in the uplink, with an average spectral efficiency that is three to four times that of Release 6 HSPA. In order to achieve these high data rates and spectral efficiency, the air interface will likely be based on OFDM/OFDMA and MIMO (multiple input/ multiple output), with similarities to WiMAX.

Similarly, 3GPP2 also has longer-term plans to offer higher data rates by moving to higher-bandwidth operation. The objective is to support up to 70Mbps to 200Mbps in the downlink and up to 30Mbps to 45Mbps in the uplink in EV-DO Revision C, using up to 20MHz of bandwidth. It should be noted that neither LTE nor EV-DO Rev C systems are expected to be available until about 2010.

1.4.2 Wi-Fi Systems

In addition to 3G, Wi-Fi based-systems may be used to provide broadband wireless. Wi-Fi is based on the IEEE 802.11 family of standards and is primarily a local area networking (LAN) technology designed to provide in-building broadband coverage. Current Wi-Fi systems based on IEEE 802.11a/g support a peak physical-layer data rate of 54Mbps[3] and typically provide indoor coverage over a distance of 100 feet. Wi-Fi has become the defacto standard for "last feet" broadband connectivity in homes, offices, and public hotspot locations. In the past couple of years, a number of municipalities and local communities around the world have taken the initiative to get Wi-Fi systems deployed in outdoor settings to provide broadband access to city centers and metrozones as well as to rural and underserved areas. It is this application of Wi-Fi that overlaps with the fixed and nomadic application space of WiMAX.

Metro-area Wi-Fi deployments rely on higher power transmitters that are deployed on lamp-posts or building tops and radiating at or close to the maximum allowable power limits for operating in the license-exempt band. Even with high power transmitters, Wi-Fi systems can typically provide a coverage range of only about 1,000 feet from the access point. Consequently, metro-Wi-Fi applications require dense deployment of access points, which makes it impractical for large-scale ubiquitous deployment. Nevertheless, they could be deployed to provide broadband access to hotzones within a city or community. Wi-Fi offers remarkably higher peak data rates than do 3G systems, primarily since it operates over a larger 20MHz bandwidth. The inefficient CSMA (carrier sense multiple access) protocol used by Wi-Fi, along with the interference constraints of operating in the license-exempt band, is likely to significantly reduce the capacity of outdoor Wi-Fi systems. Further, Wi-Fi systems are not designed to support high-speed mobility. One significant advantage of Wi-Fi over WiMAX and 3G is the wide availability of terminal devices. A vast majority of laptops shipped today have a built-in Wi-Fi interface. Wi-Fi interfaces

---

3. This typically translates to only around 20Mbps to 25Mbps layer 2 peak throughput owing to CSMA overhead.

are now also being built into a variety of devices, including personal data assistants (PDAs), cordless phones, cellular phones, cameras, and media players. The large embedded base of terminals makes it easy for consumers to use the services of broadband networks built using Wi-Fi. As with 3G, the capabilities of Wi-Fi are being enhanced to support even higher data rates and to provide better QoS support. In particular, using multiple-antenna spatial multiplexing technology, the emerging IEEE 802.11n standard will support a peak layer 2 throughput of at least 100Mbps. IEEE 802.11n is also expected to provide significant range improvements through the use of transmit diversity and other advanced techniques.

1.4.3 WiMAX versus 3G and Wi-Fi

How does WiMAX compare with the existing and emerging capabilities of 3G and Wi-Fi? The throughput capabilities of WiMAX depend on the channel bandwidth used. Unlike 3G systems, which have a fixed channel bandwidth, WiMAX defines a selectable channel bandwidth from 1.25MHz to 20MHz, which allows for a very flexible deployment. When deployed using the more likely 10MHz TDD (time division duplexing) channel, assuming a 3:1 downlink-to-uplink split and 2×2 MIMO, WiMAX offers 46Mbps peak downlink throughput and 7Mbps uplink. The reliance of Wi-Fi and WiMAX on OFDM modulation, as opposed to CDMA as in 3G, allows them to support very high peak rates. The need for spreading makes very high data rates more difficult in CDMA systems.

More important than peak data rate offered over an individual link is the average throughput and overall system capacity when deployed in a multicellular environment. From a capacity standpoint, the more pertinent measure of system performance is spectral efficiency. In Chapter 12, we provide a detailed analysis of WiMAX system capacity and show that WiMAX can achieve spectral efficiencies higher than what is typically achieved in 3G systems. The fact that WiMAX specifications accommodated multiple antennas right from the start gives it a boost in spectral efficiency. In 3G systems, on the other hand, multiple-antenna support is being added in the form of revisions. Further, the OFDM physical layer used by WiMAX is more amenable to MIMO implementations than are CDMA systems from the standpoint of the required complexity for comparable gain. OFDM also makes it easier to exploit frequency diversity and multiuser diversity to improve capacity. Therefore, when compared to 3G, WiMAX offers higher peak data rates, greater flexibility, and higher average throughput and system capacity.

Another advantage of WiMAX is its ability to efficiently support more symmetric links—useful for fixed applications, such as T1 replacement—and support for flexible and dynamic adjustment of the downlink-to-uplink data rate ratios. Typically, 3G systems have a fixed asymmetric data rate ratio between downlink and uplink.

What about in terms of supporting advanced IP applications, such as voice, video, and multimedia? How do the technologies compare in terms of prioritizing traffic and controlling quality? The WiMAX media access control layer is built from the ground up to support a variety of traffic mixes, including real-time and non-real-time constant bit rate and variable bit rate traffic, prioritized data, and best-effort data. Such 3G solutions as HSDPA and 1x EV-DO were also designed for a variety of QoS levels.

Perhaps the most important advantage for WiMAX may be the potential for lower cost owing to its lightweight IP architecture. Using an IP architecture simplifies the core network— 3G has a complex and separate core network for voice and data—and reduces the capital and operating expenses. IP also puts WiMAX on a performance/price curve that is more in line with general-purpose processors (Moore's Law), thereby providing greater capital and operational efficiencies. IP also allows for easier integration with third-party application developers and makes convergence with other networks and applications easier.

In terms of supporting roaming and high-speed vehicular mobility, WiMAX capabilities are somewhat unproven when compared to those of 3G. In 3G, mobility was an integral part of the design; WiMAX was designed as a fixed system, with mobility capabilities developed as an add-on feature.

In summary, WiMAX occupies a somewhat middle ground between Wi-Fi and 3G technologies when compared in the key dimensions of data rate, coverage, QoS, mobility, and price. Table 1.2 provides a summary comparison of WiMAX with 3G and Wi-Fi technologies.

1.4.4 Other Comparable Systems

So far, we have limited our comparison of WiMAX to 3G and Wi-Fi technologies. Two other standards based-technology solutions could emerge in the future with some overlap with WiMAX: the IEEE 802.20 and IEEE 802.22 standards under development. The IEEE 802.20 standard is aimed at broadband solutions specifically for vehicular mobility up to 250 kmph. This standard is likely to be defined for operation below 3.5GHz to deliver peak user data rates in excess of 4Mbps and 1.2Mbps in the downlink and uplink, respectively. The development effort for this standard began a few years ago but it has not made much progress, owing to lack of consensus on technology and issues with the standardization process. The IEEE 802.22 standard is aimed specifically at bringing broadband access to rural and remote areas through wireless regional area networks (WRAN). The basic goal of 802.22 is to define a cognitive radio that can take advantage of unused TV channels that exist in these sparsely populated areas. Operating in the VHF and low UHF bands provides favorable propagation conditions that can lead to greater range. This development effort is motivated by the fact that the FCC plans to allow the use of this spectrum without licenses as long as a cognitive radio solution that identifies and operates in unused portions of the spectrum is used. IEEE 802.22 is in early stages of development and is expected to provide fixed broadband applications over larger coverage areas with low user densities.

1.5 Spectrum Options for Broadband Wireless

The availability of frequency spectrum is key to providing broadband wireless services. Several frequency bands can be used for deploying WiMAX. Each band has unique characteristics that have a significant impact on system performance. The operating frequency band often dictates

Table 1.2 Comparison of WiMAX with Other Broadband Wireless Technologies

Parameter	Fixed WiMAX	Mobile WiMAX	HSPA	1x EV-DO Rev A	Wi-Fi
Standards	IEEE 802.16-2004	IEEE 802.16e-2005	3GPP Release 6	3GPP2	IEEE 802.11a/g/n
Peak down link data rate	9.4Mbps in 3.5MHz with 3:1 DL-to-UL ratio TDD; 6.1Mbps with 1:1	46Mbps[a] with 3:1 DL- to-UL ratio TDD; 32Mbps with 1:1	14.4Mbps using all 15 codes; 7.2Mbps with 10 codes	3.1Mbps; Rev. B will support 4.9Mbps	54 Mbps[b] shared using 802.11a/g; more than 100Mbps peak layer 2 through-put using 802.11n
Peak uplink data rate	3.3Mbps in 3.5MHz using 3:1 DL-to-UL ratio; 6.5Mbps with 1:1	7Mbps in 10MHz using 3:1 DL-to-UL ratio; 4Mbps using 1:1	1.4Mbps initially; 5.8Mbps later	1.8Mbps	
Bandwidth	3.5MHz and 7MHz in 3.5GHz band; 10MHz in 5.8GHz band	3.5MHz, 7MHz, 5MHz, 10MHz, and 8.75MHz initially	5MHz	1.25MHz	20MHz for 802.11a/g; 20/40MHz for 802.11n
Modulation	QPSK, 16 QAM, 64 QAM	QPSK, 16 QAM, 64 QAM	QPSK, 16 QAM	QPSK, 8 PSK, 16 QAM	BPSK, QPSK, 16 QAM, 64 QAM
Multiplexing	TDM	TDM/OFDMA	TDM/CDMA	TDM/CDMA	CSMA
Duplexing	TDD, FDD	TDD initially	FDD	FDD	TDD
Frequency	3.5GHz and 5.8GHz initially	2.3GHz, 2.5GHz, and 3.5GHz initially	800/900/1,800/ 1,900/ 2,100MHz	800/900/ 1,800/ 1,900MHz	2.4GHz, 5GHz
Coverage (typical)	3–5 miles	< 2 miles	1–3 miles	1–3 miles	< 100 ft indoors; < 1000 ft outdoors
Mobility	Not applicable	Mid	High	High	Low

a. Assumes 2×2 MIMO and a 10MHz channel.

b. Due to inefficient CSMA MAC, this typically translates to only ~20Mbps to 25Mbps layer 2 throughput.

fundamental bounds on achievable data rates and coverage range. Table 1.3 summarizes the various frequency bands that could be used for broadband wireless deployment.

From a global perspective, the 2.3GHz, 2.5GHz, 3.5GHz, and 5.7GHz bands are most likely to see WiMAX deployments. The WiMAX Forum has identified these bands for initial interoperability certifications. A brief description of these bands follows.

Licensed 2.5GHz: The bands between 2.5GHz and 2.7GHz have been allocated in the United States, Canada, Mexico, Brazil, and some southeast Asian countries. In many countries, this band

Table 1.3 Summary of Potential Spectrum Options for Broadband Wireless

Designation	Frequency Allocation	Amount of Spectrum	Notes
Fixed wireless access (FWA): 3.5GHz	3.4GHz – 3.6GHz mostly; 3.3GHz – 3.4GHz and 3.6GHz – 3.8GHz also available in some countries	Total 200MHz mostly; varies from 2 × 5MHz to 2 × 56MHz paired across nations	Not generally available in the United States. A 50MHz chunk from 3.65GHz – 3.70GHz being allocated for unlicensed operation in United States.
Broadband radio services (BRS): 2.5GHz	2.495GHz – 2.690GHz	194MHz total; 22.5MHz licenses, where a 16.5MHz is paired with 6MHz	Allocation shown is for United States after the recent change in band plan. Available in a few other countries as well.
Wireless Communications Services (WCS) 2.3GHz	2.305GHz–2.320GHz; 2.345GHz – 2.360GHz	Two 2 × 5MHz paired; two unpaired 5MHz	Allocation shown for United States. Also available in Korea, Australia, New Zealand.
License exempt: 2.4GHz	2.405GHz – 2.4835GHz	One 80MHz block	Allocation shown for United States but available worldwide. Heavily crowded band; used by Wi-Fi.
License exempt: 5GHz	5.250GHz–5.350GHz; 5.725GHz – 5.825GHz	200MHz available in United States; additional 255MHz to be allocated	Called U-NII in United States. Generally available worldwide; lower bands have severe power restrictions.
UHF band: 700MHz	698MHz – 746MHz (lower); 747MHz – 792MHz (upper)	30MHz upper band; 48MHz lower band	Allocations shown for United States, only 18MHz of lower band auctioned so far. Other nations may follow.
Advanced wireless services (AWS)	1.710GHz – 1.755GHz 2.110GHz – 2.155GHz	2 × 45MHz paired	Auctioned in the United States. In other parts of the world, this is used for 3G.

is restricted to fixed applications; in some countries, two-way communication is not permitted. Among all the available bands, this one offers the most promise for broadband wireless, particularly within the United States. The FCC allowed two-way transmissions in this band in 1998 and in mid-2004 realigned the channel plan. This band, now called the *broadband radio services* (BRS) band, was previously called the MMDS band. The BRS band now has 195MHz, including guard bands and MDS (multi-point distribution services) channels, available in the United States between 2.495GHz and 2.690GHz. Regulations allow a variety of services, including fixed, portable, and mobile services. Both FDD and TDD operations are allowed. Licenses were issued for eight 22.5MHz slices of this band, where a 16.5MHz block is paired with a 6MHz block, with the separation between the two blocks varying from 10MHz to 55MHz. The rules of this band also allow for license aggregation. A majority of this spectrum in the United States is controlled by Sprint,

Nextel, and Clearwire. Regulatory changes may be required in many countries to make this band more available and attractive, particularly for mobile WiMAX.

Licensed 2.3GHz: This band, called the WCS band in the United States, is also available in many other countries such as Australia, South Korea, and New Zealand. In fact, the WiBro services being deployed in South Korea uses this band. In the United States, this band includes two paired 5MHz bands and two unpaired 5MHz bands in the 2.305GHz to 2.320GHz and 2.345GHz to 2.360GHz range. A major constraint in this spectrum is the tight out-of-band emission requirements enforced by the FCC to protect the adjacent DARS (digital audio radio services) band (2.320GHz to 2.345GHz). This makes broadband services, particularly mobile services, difficult in the sections of this band closest to the DARS band.

Licensed 3.5GHz: This is the primary band allocated for fixed wireless broadband access in several countries across the globe, with the notable exception of the United States. In the United States, the FCC has recently allocated 50MHz of spectrum in the 3.65GHz to 3.70GHz band for high-power unlicensed use with restrictions on transmission protocols that precludes WiMAX. Internationally, the allocated band is in the general vicinity of 3.4GHz to 3.6GHz, with some newer allocation in 3.3GHz to 3.4GHz and 3.6GHz to 3.8GHz as well. The available bandwidth varies from country to country, but it is generally around 200MHz. The available band is usually split into many individual licenses, varying from 2 × 5MHz to 2 × 56MHz. Spectrum aggregation rules also vary from country to country. While some countries only allow FDD operations, others allow either FDD or TDD. In most countries, the current rules in this band do not allow for nomadic and mobile broadband applications. It is hoped that the regulations in this band will, over time, become more flexible, and the WiMAX Forum has committed to working with regulatory authorities around the world to achieve this flexibility. The heavier radio propagation losses at 3.5GHz, however, is likely to make it more difficult to provide nomadic and mobile services in this band.

License-exempt 5GHz: The license-exempt frequency band 5.25GHz to 5.85GHz is of interest to WiMAX. This band is generally available worldwide. In the United States, it is part of the unlicensed national information infrastructure (U-NII) band and has 200MHz of spectrum for outdoor use. An additional 255MHz of spectrum in this band has been identified by the FCC for future unlicensed use. Being free for anyone to use, this band could enable grassroots deployments of WiMAX, particularly in underserved, low-population-density rural and remote markets. The large bandwidth available may enable operators to coordinate frequencies and mitigate the interference concerns surrounding the use of license-exempt bands, particularly in underserved markets. The relatively high frequency, coupled with the power restrictions in this band, will, however, make it extremely difficult to provide nomadic or mobile services. Even fixed applications will, in most cases, require installing external antennas at the subscriber premise. Within the 5GHz band, it is the upper 5.725GHz–5.850GHz band that is most attractive to WiMAX. Many countries allow higher power output—4 W EIRP (effective isotropic radiated power)—in this band compared to an EIRP of 1W or less in the lower 5GHz bands. In the United States, the FCC is considering proposals to further increase power output—perhaps to the tune of 25 W—in license-exempt bands in rural areas to facilitate less costly deployments in underserved areas. It should be noted that there is another 80MHz of license-exempt spectrum, in the 2.4GHz band, which could also be used for WiMAX. Given the already high usage in this band, particularly from Wi-Fi, it is not very likely that WiMAX will be deployed in the 2.4GHz band, particularly for point-to-multipoint applications.

Although the 2.3GHz, 2.5GHz, 3.5GHz, and 5.7GHz bands are the most attractive for WiMAX in the near term, other bands could see future WiMAX deployments. Examples of these are the UHF (ultra high frequency) and AWS bands.

UHF bands: Around the world, as television stations transition from analog to digital broadcasting, a large amount of spectrum below 800MHz could become available. For example, in the United States, the FCC has identified frequency bands 698MHz–746MHz to be vacated by broadcasters as they transition to digital TV. Of these bands, 18MHz of spectrum has already been auctioned, and the remaining 60MHz is expected to be auctioned in a couple of years. The slow pace of digital TV adoption has delayed these auctions, and it is not likely that this spectrum will be usable for broadband wireless until at least 2009–2010. The FCC has also begun looking into the possibility of allocating more spectrum in the sub-700MHz bands, perhaps for unlicensed use as well. UHF band spectrum has excellent propagation characteristics compared to the other microwave bands and hence is valuable, particularly for portable and mobile services. The larger coverage range possible in this band makes the economics of deployment particularly attractive for suburban and rural applications.

AWS band: In August 2006, the FCC auctioned 1.710GHz–1.755GHz paired with 2.110GHz–2.155GHz as spectrum for advanced wireless services (AWS) in the United States. This band offers 90MHz of attractive spectrum that could be viable for WiMAX in the longer term.

Beyond these, it is possible that WiMAX could be deployed in bands designated for 3G. Particularly in Europe, greenfield 3G operators could choose to deploy WiMAX if regulatory relief to do so is obtained. Another interesting possibility is the 1.5GHz L-band used by mobile satellite today. Clearly, WiMAX systems could be deployed in a number of spectrum bands. The challenge is get the allocations and regulations across the globe harmonized in order to gain the advantage of economies of scale. In the next section, we discuss this and other business challenges to broadband wireless in general and WiMAX in particular.

1.6 Business Challenges for Broadband Wireless and WiMAX

Despite the marketing hype and the broad industry support for the development of WiMAX, its success is not a forgone conclusion. In fact, broadband wireless in general and WiMAX in particular face a number of challenges that could impede their adoption in the marketplace.

The rising bar of traditional broadband: In the fixed broadband application space, WiMAX will have to compete effectively with traditional wired alternatives, such as DSL and cable, to achieve widespread adoption in mature markets, such as the United States. DSL and cable modem technologies continue to evolve at a rapid pace, providing increasing data rate capabilities. For example, DSL services in the United States already offer 3Mbps–6Mbps of downstream throughput to the end user, and solutions based on the newer VDSL2 standard will soon deliver up to 50Mbps–100Mbps, depending on the loop length. With incumbent carriers pushing fiber deeper into the networks, the copper loop lengths are getting shorter, allowing for significantly improved data rates. Cable modem technologies offer even higher speeds than DSL. Even on the upstream, where bandwidth had been traditionally limited, data rates on the order of several megabits per second per user are becoming a reality in both DSL and cable. The extremely high data rates supported by these wired broadband solutions allow providers to offer not only data, voice, and multimedia applications but also entertainment TV, including HDTV.

It will be extremely difficult for broadband wireless systems to match the rising throughput performance of traditional broadband. WiMAX will have to rely on portability and mobility as differentiators as opposed to data rate. WiMAX may have an advantage in terms of network infrastructure cost, but DSL and cable benefit from the declining cost curves on their CPE, due to their mature-market state. Given these impediments, fixed WiMAX is more likely to be deployed in rural or underserved areas in countries with a mature broadband access market. In developing countries, where existing broadband infrastructure is weak, the business challenges for fixed WiMAX are less daunting, and hence it is much more likely to succeed.

Differences in global spectrum availability: As discussed earlier, there are considerable differences in the allocation and regulations of broadband spectrum worldwide. Although 2.5GHz, 3.5GHz, and 5.8GHz bands are allotted in many regions of the world, many growth markets require new allocations. Given the diverse requirements and regulatory philosophy of various national governments, it will be a challenge for the industry to achieve global harmonization. For WiMAX to be a global success like Wi-Fi, regulatory bodies need to allow full flexibility in terms of the services that can be offered in the various spectrum bands.

Competition from 3G: For mobile WiMAX, the most significant challenge comes from 3G technologies that are being deployed worldwide by mobile operators. Incumbent mobile operators are more likely to seek performance improvements through 3G evolution than to adopt WiMAX. New entrants and innovative challengers entering the mobile broadband market using WiMAX will have to face stiff competition from 3G operators and will have to find a way to differentiate themselves from 3G in a manner that is attractive to the users. They may have to develop innovative applications and business models to effectively compete against 3G.

Device development: For mobile WiMAX to be successful, it is important to have a wide variety of terminal devices. Embedding WiMAX chips into computers could be a good first step but may not be sufficient. Perhaps WiMAX can differentiate from 3G by approaching the market with innovative devices. Some examples could include WiMAX embedded into MP3 players, video players, or handheld PCs. Device-development efforts should also include multimode devices. A variety of broadband systems will likely be deployed, and it is critical that diverse networks interoperate to make ubiquitous personal broadband services a reality. Ensuring that device development happens concomitant with network deployment will be a challenge.

1.7 Technical Challenges for Broadband Wireless

So far, we have discussed the history, applications, and business challenges of broadband wireless. We now address the technical challenges of developing and deploying a successful broadband wireless system. The discussion presented in this section sets the stage for the rest of the book, especially Part II, where the technical foundations of WiMAX are discussed in detail.

To gain widespread success, broadband wireless systems must deliver multimegabit per second throughput to end users, with robust QoS to support a variety of services, such as voice, data, and multimedia. Given the remarkable success of the Internet and the large variety of emerging IP-based applications, it is critical that broadband wireless systems be built to support these IP-based applications and services efficiently. Fixed broadband systems must, ideally, deliver these services to indoor locations, using subscriber stations that can be easily self-installed by the enduser. Mobile broadband systems must deliver broadband applications to laptops and handheld devices while moving at high speeds. Customers now demand that all this be done without sacrificing quality, reliability, or security. For WiMAX to be successful, it must deliver significantly better performance than current alternatives, such as 3G and Wi-Fi. This is indeed a high bar.

Meeting these stringent service requirements while being saddled with a number of constraints imposed by wireless make the system design of broadband wireless a formidable technical challenge. Some of the key technical design challenges are

- Developing reliable transmission and reception schemes to push broadband data through a hostile wireless channel
- Achieving high spectral efficiency and coverage in order to deliver broadband services to a large number of users, using limited available spectrum
- Supporting and efficiently multiplexing services with a variety of QoS (throughput, delay, etc.) requirements
- Supporting mobility through seamless handover and roaming
- Achieving low power consumption to support handheld battery-operated devices
- Providing robust security
- Adapting IP-based protocols and architecture for the wireless environment to achieve lower cost and convergence with wired networks

As is often the case in engineering, solutions that effectively overcome one challenge may aggravate another. Design trade-offs have to be made to find the right balance among competing requirements—for example, coverage and capacity. Advances in computing power, hardware miniaturization, and signal-processing algorithms, however, enable increasingly favorable trade-offs, albeit within the fundamental bounds imposed by laws of physics and information theory. Despite these advances, researchers continue to be challenged as wireless consumers demand even greater performance.

We briefly explain each of the technical challenges, and touch on approaches that have been explored to overcome them. We begin with the challenges imposed by the wireless radio channel.

1.7.1 Wireless Radio Channel

The first and most fundamental challenge for broadband wireless comes from the transmission medium itself. In wired communications channels, a physical connection, such as a copper wire or fiber-optic cable, guides the signal from the transmitter to the receiver, but wireless communication systems rely on complex radio wave propagation mechanisms for traversing the intervening space. The requirements of most broadband wireless services are such that signals have to travel under challenging NLOS conditions. Several large and small obstructions, terrain undulations, relative motion between the transmitter and the receiver, interference from other signals, noise, and various other complicating factors together weaken, delay, and distort the transmitted signal in an unpredictable and time-varying fashion. It is a challenge to design a digital communication system that performs well under these conditions, especially when the service requirements call for very high data rates and high-speed mobility. The wireless channel for broadband communication introduces several major impairments.

Distance-dependent decay of signal power: In NLOS environments, the received signal power typically decays with distance at a rate much faster than in LOS conditions. This distance-dependent power loss, called *pathloss*, depends on a number of variables, such as terrain, foliage, obstructions, and antenna height. Pathloss also has an inverse-square relationship with carrier frequency. Given that many broadband wireless systems will be deployed in bands above 2GHz under NLOS conditions, systems will have to overcome significant pathloss.

Blockage due to large obstructions: Large obstructions, such as buildings, cause localized blockage of signals. Radio waves propagate around such blockages via diffraction but incur severe loss of power in the process. This loss, referred to as *shadowing*, is in addition to the distance-dependent decay and is a further challenge to overcome.

Large variations in received signal envelope: The presence of several reflecting and scattering objects in the channel causes the transmitted signal to propagate to the receiver via multiple paths. This leads to the phenomenon of *multipath fading*, which is characterized by large (tens of dBs) variations in the amplitude of the received radio signal over very small distances or small durations. Broadband wireless systems need to be designed to cope with these large and rapid variations in received signal strength. This is usually done through the use of one or more diversity techniques, some of which are covered in more detail in Chapters 4, 5, and 6.

Intersymbol interference due to time dispersion: In a multipath environment, when the time delay between the various signal paths is a significant fraction of the transmitted signal's symbol period, a transmitted symbol may arrive at the receiver during the next symbol period and cause *intersymbol interference* (ISI). At higher data rates, the symbol time is shorter; hence, it takes only a smaller delay to cause ISI. This makes ISI a bigger concern for broadband wireless and mitigating it more challenging. Equalization is the conventional method for dealing with ISI but at high data rates requires too much processing power. OFDM has become the solution of choice for mitigating ISI in broadband systems, including WiMAX, and is covered in Chapter 4 in detail.

Frequency dispersion due to motion: The relative motion between the transmitter and the receiver causes carrier frequency dispersion called *Doppler spread*. Doppler spread is directly related to vehicle speed and carrier frequency. For broadband systems, Doppler spread typically leads to loss of signal-to-noise ratio (SNR) and can make carrier recovery and synchronization more difficult. Doppler spread is of particular concern for OFDM systems, since it can corrupt the orthogonality of the OFDM subcarriers.

Noise: Additive white Gaussian noise (AWGN) is the most basic impairment present in any communication channel. Since the amount of thermal noise picked up by a receiver is proportional to the bandwidth, the noise floor seen by broadband receivers is much higher than those seen by traditional narrowband systems. The higher noise floor, along with the larger pathloss, reduces the coverage range of broadband systems.

Interference: Limitations in the amount of available spectrum dictate that users share the available bandwidth. This sharing can cause signals from different users to interfere with one another. In capacity-driven networks, interference typically poses a larger impairment than noise and hence needs to be addressed.

Each of these impairments should be well understood and taken into consideration while designing broadband wireless systems. In Chapter 3, we present a more rigorous characterization of the radio channel, which is essential to the development of effective solutions for broadband wireless.

1.7.2 Spectrum Scarcity

The second challenge to broadband wireless comes from the scarcity of radio-spectrum resources. As discussed in Section 1.5, regulatory bodies around the world have allocated only a limited amount of spectrum for commercial use. The need to accommodate an ever-increasing number of users and offering bandwidth-rich applications using a limited spectrum challenges the system designer to continuously search for solutions that use the spectrum more efficiently. Spectral-efficiency considerations impact many aspects of broadband wireless system design.

The most fundamental tool used to achieve higher system-wide spectral efficiency is the concept of a *cellular architecture*, whereby instead of using a single high-powered transmitter to cover a large geographic area, several lower-power transmitters that each cover a smaller area, called a *cell,* are used. The cells themselves are often subdivided into a few sectors through the use of directional antennas. Typically, a small group of cells or sectors form a cluster, and the available frequency spectrum is divided among the cells or sectors in a cluster and allocated intelligently to minimize interference to one another. The pattern of frequency allocation within a cluster is then repeated throughout the desired service area and is termed *frequency reuse*.

For higher capacity and spectral efficiency, frequency reuse must be maximized. Increasing reuse, however, leads to a larger potential for interference. Therefore, to facilitate tighter reuse, the challenge is to design transmission and reception schemes that can operate under lower signal-to-interference-plus-noise ratio (SINR) conditions or implement effective methods to deal with interference. One effective way to deal with interference is to use multiple-antenna processing.

Beyond using the cellular architecture and maximizing frequency reuse, several other signal-processing techniques can be used to maximize the spectral efficiency and hence capacity of the system. Many of these techniques exploit channel information to maximize capacity. Examples of these are included below.

Adaptive modulation and coding: The idea is to vary the modulation and coding rate on a per user and/or per packet basis based on the prevailing SINR conditions. By using the highest level modulation and coding rate that can be supported by the SINR, the user data rates—and hence capacity—can be maximized. Adaptive modulation and coding is part of the WiMAX standard and are discussed in detail in Chapter 6.

Spatial multiplexing: The idea behind spatial multiplexing is that multiple independent streams can be transmitted in parallel over multiple antennas and can be separated at the receiver using multiple receive chains through appropriate signal processing. This can be done as long as the multipath channels as seen by the various antennas are sufficiently decorrelated, as would be the case in a scattering-rich environment. Spatial multiplexing provides data rate and capacity gains proportional to the number of antennas used. This and other multiantenna techniques are covered in Chapter 5.

Efficient multiaccess techniques: Besides ensuring that each user uses the spectrum as efficiently as possible, effective methods must be devised to share the resources among the multiple users efficiently. This is the challenge addressed at the MAC layer of the system. Greater efficiencies in spectrum use can be achieved by coupling channel-quality information in the resource-allocation process. MAC-layer techniques are discussed in more detail in Chapter 6.

It should be emphasized that capacity and spectral efficiency cannot be divorced from the need to provide adequate coverage. If one were concerned purely with high spectral efficiency or capacity, an obvious way to achieve that would be to decrease the cell radius or to pack more base stations per unit area. Obviously, this is an expensive way to improve capacity. Therefore, it is important to look at spectral efficiency more broadly to include the notion of coverage area. The big challenge for broadband wireless system design is to come up with the right balance between capacity and coverage that offers good quality and reliability at a reasonable cost.

1.7.3 Quality of Service

QoS is a broad and loose term that refers to the "collective effect of service," as perceived by the user. For the purposes of this discussion, QoS more narrowly refers to meeting certain requirements—typically, throughput, packet error rate, delay, and jitter—associated with a given application. Broadband wireless networks must support a variety of applications, such as voice, data, video, and multimedia, and each of these has different traffic patterns and QoS requirements, as shown in Table 1.4. In addition to the application-specific QoS requirements, networks often need to also enforce policy-based QoS, such as giving differentiated services to users based on

Table 1.4 Sample Traffic Parameters for Broadband Wireless Applications

Parameter	Interactive Gaming	Voice	Streaming Media	Data	Video
Data rate	50Kbps–85Kbps	4Kbps–64Kbps	5Kbps–384Kbps	0.01Mbps–100Mbps	> 1Mbps
Example applications	Interactive gaming	VoIP	Music, speech, video clips	Web browsing, e-mail, instant messaging (IM), telnet, file downloads	IPTV, movie download, peer-to-peer video sharing
Traffic flow	Real time	Real-time continuous	Continuous, bursty	Non–real time, bursty	Continuous
Packet loss	Zero	< 1%	< 1% for audio; < 2% for video	Zero	$< 10^{-8}$
Delay variation	Not applicable	< 20 ms	< 2 sec	Not applicable	< 2 sec
Delay	< 50 ms–150 ms	< 100 ms	< 250 ms	Flexible	< 100 ms

their subscribed service plans. The variability in the QoS requirements across applications, services, and users makes it a challenge to accommodate all these on a single-access network, particularly wireless networks, where bandwidth is at a premium.

The problem of providing QoS in broadband wireless systems is one of managing radio resources effectively. Effective scheduling algorithms that balance the QoS requirements of each application and user with the available radio resources need to be developed. In other words, capacity needs to be allocated in the right proportions among users and applications at the right time. This is the challenge that the MAC-layer protocol must meet: simultaneously handling multiple types of traffic flows—bursty and continuous—of varying throughputs and latency requirements. Also needed are an effective signaling mechanism for users and applications to indicate their QoS requirements and for the network to differentiate among various flows.

Delivering QoS is more challenging for mobile broadband than for fixed. The time variability and unpredictability of the channel become more acute, and complication arises from the need to hand over sessions from one cell to another as the user moves across their coverage boundaries. Handovers cause packets to be lost and introduce additional latency. Reducing handover latency and packet loss is also an important aspect of delivering QoS. Handover also necessitates coordination of radio resources across multiple cells.

So far, our discussion of QoS has been limited to delivering it across the wireless link. From a user perspective, however, the perceived quality is based on the end-to-end performance of the network. To be effective, therefore, QoS has to be delivered end-to-end across the network, which may include, besides the wireless link, a variety of aggregation, switching, and routing elements between the communication end points. IP-based networks are expected to form the bulk of the core network; hence, IP-layer QoS is critical to providing end-to-end service quality. A more detailed discussion of end-to-end QoS is provided in Chapter 7.

1.7.4 Mobility

For the end user, mobility is one of the truly distinctive values that wireless offers. The fact that the subscriber station moves over a large area brings several networking challenges. Two of the main challenges are (1) providing a means to reach inactive users for session initiation and packet delivery, regardless of their location within the network, and (2) maintaining an ongoing session without interruption while on the move, even at vehicular speeds. The first challenge is referred to as *roaming*; the second, *handoff*. Together, the two are referred to as *mobility management,* and performing them well is critical to providing a good user experience.

Roaming: The task of locating roaming subscriber stations is typically accomplished through the use of centralized databases that store up-to-date information about their location. These databases are kept current though location-update messages that subscriber stations send to the network as it moves from one location area to another. To reach a subscriber station for session setup, the network typically pages for it over the base stations in and around the location area. The number of base stations over which the page is sent depends on the updating rate and movement of the subscriber stations. The radio-resource management challenge here is the trade-off between spending radio resources on transmitting location-update messages from non-active subscriber stations more frequently versus paging terminals over a larger set of base stations at session setup.

Handoff: To meet the second challenge of mobility, the system should provide a method for seamlessly handing over an ongoing session from one base station to another as the user moves across them. A handoff process typically involves detecting and deciding when to do a handoff, allocating radio resources for it, and executing it. It is required that all handoffs be performed successfully and that they happen as infrequently and imperceptibly as possible. The challenge for handoff-decision algorithms is the need to carefully balance the dropping probability and handoff rate. Being too cautious in making handoff decisions can lead to dropped sessions; excessive handoff can lead to an unnecessary signaling load. The other challenge is to ensure that sufficient radio resources are set aside so that ongoing sessons are not dropped midsession during handoff. Some system designs reserve bandwidth resources for accepting handoff or at least prioritize handoff requests over session-initiation requests.

Another aspect of mobility management that will become increasingly important in the future is layer 3, IP mobility. Traditionally, in mobile networks, mobility is handled by the layer 2 protocol, and the fact that the terminal is moving is hidden from the IP network. The terminal continues to have a fixed IP address, regardless of its changing its point of attachment to the network. Although this is not an issue for most IP applications, it poses a challenge for certain IP applications, such as Web-caching and multicasting. IP-based mobility-management solutions can solve this problem, but it is tricky to make them work in a wireless environment. IP-based mobility management is also required to support roaming and handover across heterogeneous networks, such as between a WiMAX network and a Wi-Fi network.

A more detailed discussion of the challenges of mobility is presented in Chapter 7.

1.7.5 Portability

Like mobility, portability is another unique value provided by wireless. Portability is desired for not only full-mobility applications but also nomadic applications. Portability dictates that the subscriber device be battery powered and lightweight and therefore consume as little power as possible. Unfortunately, advances in battery technology have been fairly limited, especially when compared to processor technology. The problem is compounded by the fact that mobile terminals are required to pack greater processing power and functionality within a decreasing real estate. Given the limitations in battery power, it is important that it be used most efficiently. The need for reducing power consumption challenges designers to look for power-efficient transmission schemes, power-saving protocols, computationally less intensive signal-processing algorithms, low-power circuit-design and fabrication, and battery technologies with longer life.

The requirement of low-power consumption drives physical-layer design toward the direction of using power-efficient modulation schemes: signal sets that can be detected and decoded at lower signal levels. Unfortunately, power-efficient modulation and coding schemes tend to be less spectrally efficient. Since spectral efficiency is also a very important requirement for broadband wireless, it is a challenge to make the appropriate trade-off between them. This often results in portable wireless systems offering asymmetric data rates on the downlink and the uplink. The power-constrained uplink often supports lower bits per second per Hertz than the downlink.

It is not only the transmitter power that drains the battery. Digital signal processors used in terminal devices are also notorious for their power consumption. This motivates the designer to come up with computationally more efficient signal-processing algorithms for implementation in the portable device. Protocol design efforts at power conservation focus on incorporating low-power sleep and idle modes with methods to wake up the device as and when required. Fast-switching technologies to ensure that the transmitter circuitry is turned on only when required and on an instantaneous demand basis can also be used to reduce overall power consumption.

1.7.6 Security

Security is an important consideration in any communications system design but is particularly so in wireless communication systems. The fact that connections can be established in a untethered fashion makes it easier to intrude in an inconspicuous and undetectable manner than is the case for wired access. Further, the shared wireless medium is often perceived by the general public to be somewhat less secure than its wired counterpart. Therefore, a robust level of security must be built into the design of broadband wireless systems.

From the perspective of an end user, the primary security concerns are privacy and data integrity. Users need assurance that no one can eavesdrop on their sessions and that the data sent across the communication link is not tampered with. This is usually achieved through the use of encryption.

From the service provider's perspective, an important security consideration is preventing unauthorized use of the network services. This is usually done using strong authentication and access control methods. Authentication and access control can be implemented at various levels of the network: the physical layer, the network layer, and the service layer. The service provider's need to prevent fraud should be balanced against the inconvenience that it may impose on the user.

Besides privacy and fraud, other security concerns include denial-of-service attacks in which malignant users attempt to degrade network performance, session hijacking, and virus insertion. Chapter 7 presents a more detailed discussion of the various security issues and solutions.

1.7.7 Supporting IP in Wireless

The Internet Protocol (IP) has become the networking protocol of choice for modern communication systems. Internet-based protocols are now beginning to be used to support not only data but also voice, video, and multimedia. Voice over IP is quickly emerging as a formidable competitor to traditional circuit-switched voice and appears likely to displace it over time. Video over IP and IPTV are also emerging as potential rivals to traditional cable TV. Because more and more applications will migrate to IP, IP-based protocols and architecture must be considered for broadband wireless systems.

A number of arguments favor the use of IP-based protocols and architecture for broadband wireless. First, IP-based systems tend to be cheaper because of the economies of scale they enjoy from widespread adoption in wired communication systems. Adopting an IP architecture can make it easier to develop new services and applications rapidly. The large IP application development community can be leveraged. An IP-based architecture for broadband wireless will enable easier support for such applications as IP multicast and anycast. An IP-based architecture makes it easy to integrate broadband wireless systems with other access technologies and thereby enable converged services.

IP-based protocols are simple and flexible but not very efficient or robust. These deficiencies were not such a huge concern as IP evolved largely in the wired communications space, where transmission media, such as fiber-optic channels, offered abundant bandwidth and very high reliability. In wireless systems, however, introducing IP poses several challenges: (1) making IP-based protocols more bandwidth efficient, (2) adapting them to deliver the required QoS (delay, jitter, throughput, etc.) when operating in bandwidth-limited and unreliable media, and (3) adapting them to handle terminals that move and change their point of attachment to the network. Some of these issues and solutions are also presented in Chapter 7.

1.7.8 Summary of Technical Challenges

Table 1.5 summarizes the various technical challenges associated with meeting the service requirements for broadband wireless, along with potential solutions. Many of the solutions listed are described in more detail in Part II of the book.

Table 1.5 Summary of Technical Design Challenges to Broadband Wireless

Service Requirements	Technical Challenge	Potential Solution
Non-line-of-sight coverage	Mitigation of multipath fading and interference	Diversity, channel coding, etc.
High data rate and capacity	Achieving high spectral efficiency	Cellular architecture, adaptive modulation and coding, spatial multiplexing, etc.
	Overcoming intersymbol interference	OFDM, equalization, etc.
	Interference mitigation	Adaptive antennas, sectorization, dynamic channel allocation, CDMA, etc.
Quality of service	Supporting voice, data, video, etc. on a single access network	Complex MAC layer
	Radio resource management	Efficient scheduling algorithms
	End-to-end quality of service	IP QoS: DiffServ, IntServ, MPLS, etc.
Mobility	Ability to be reached regardless of location	Roaming database, location update, paging
	Session continuity while moving from the coverage area of one base station to another	Seamless handover
	Session continuity across diverse networks	IP-based mobility: mobile IP
Portability	Reduce battery power consumption on portable subscriber terminals	Power-efficient modulation; sleep, idle modes and fast switching between modes; low-power circuit; efficient signal-processing algorithms
Security	Protect privacy and integrity of user data	Encryption
	Prevent unauthorized access to network	Authentication and access control
Low cost	Provide efficient and reliable communication using IP architecture and protocols	Adaptation of IP-based protocols for wireless; adapt layer 2 protocols for IP

1.8 Summary and Conclusions

In this chapter, we outlined a high-level overview of broadband wireless by presenting its history, applications, business challenges, and technical design issues.

- Broadband wireless could be a significant growth market for the telecom industry.
- Broadband wireless has had a checkered history, and the emergence of the WiMAX standard offers a significant new opportunity for success.
- Broadband wireless systems can be used to deliver a variety of applications and services to both fixed and mobile users.
- WiMAX could potentially be deployed in a variety of spectrum bands: 2.3GHz, 2.5GHz, 3.5GHz, and 5.8GHz.
- WiMAX faces a number of competitive challenges from both fixed-line and third-generation mobile broadband alternatives.
- The service requirements and special constraints of wireless broadband make the technical design of broadband wireless quite challenging.

1.9 Bibliography

[1] ITU. Telecommunications indicators update—2004. www.itu.int/ITU-D/ict/statistics/.

[2] In-stat Report. Paxton. The broadband boom continues: Worldwide subscribers pass 200 million, No. IN0603199MBS, March 2006.

[3] Schroth. The evolution of WiMAX service providers and applications. *Yankee Group Report*. September 2005.

[4] R. W. Chang. Synthesis of band-limited orthogonal signals for multichannel data transmission. *Bell Systems Technical Journal*, 45:1775–1796, December 1966.

[5] J. L. Holsinger. Digital communication over fixed time-continuous channels with memory, with special application to telephone channels. PhD thesis, Massachusetts Institute of Technology, 1964.

[6] R. G. Gallager. *Information Theory and Reliable Communications*. Wiley, 1968. 33.

[7] S. Weinstein and P. Ebert. Data transmission by frequency-division multiplexing using the discrete Fourier transform. *IEEE Transactions on Communications*, 19(5):628–634, October 1971.

[8] L. J. Cimini. Analysis and simulation of a digital mobile channel using orthogonal frequency division multiplexing. *IEEE Transactions on Communications*, 33(7):665–675, July 1985.

Overview of WiMAX

After years of development and uncertainty, a standards-based interoperable solution is emerging for wireless broadband. A broad industry consortium, the Worldwide Interoperability for Microwave Access (WiMAX) Forum has begun certifying broadband wireless products for interoperability and compliance with a standard. WiMAX is based on wireless metropolitan area networking (WMAN) standards developed by the IEEE 802.16 group and adopted by both IEEE and the ETSI HIPERMAN group. In this chapter, we present a concise technical overview of the emerging WiMAX solution for broadband wireless. The purpose here is to provide an executive summary before offering a more detailed exposition of WiMAX in later chapters.

We begin the chapter by summarizing the activities of the IEEE 802.16 group and its relation to WiMAX. Next, we discuss the salient features of WiMAX and briefly describe the physical- and MAC-layer characteristics of WiMAX. Service aspects, such as quality of service, security, and mobility, are discussed, and a reference network architecture is presented. The chapter ends with a brief discussion of expected WiMAX performance.

2.1 Background on IEEE 802.16 and WiMAX

The IEEE 802.16 group was formed in 1998 to develop an air-interface standard for wireless broadband. The group's initial focus was the development of a line-of-sight (LOS) based point-to-multipoint wireless broadband system for operation in the 10GHz–66GHz millimeter wave band. The resulting standard—the original 802.16 standard, completed in December 2001—was based on a single-carrier physical (PHY) layer with a burst time division multiplexed (TDM) media access control (MAC) layer. Many of the concepts related to the MAC layer were adapted for wireless from the popular cable modem DOCSIS (data over cable service interface specification) standard.

The IEEE 802.16 group subsequently produced 802.16a, an amendment to the standard, to include non-line-of-sight (NLOS) applications in the 2GHz–11GHz band, using an *orthogonal frequency division multiplexing* (OFDM)-based physical layer. Additions to the MAC layer, such as support for *orthogonal frequency division multiple access* (OFDMA), were also included. Further revisions resulted in a new standard in 2004, called IEEE 802.16-2004, which replaced all prior versions and formed the basis for the first WiMAX solution. These early WiMAX solutions based on IEEE 802.16-2004 targeted fixed applications, and we will refer to these as fixed WiMAX [1]. In December 2005, the IEEE group completed and approved IFEEE 802.16e-2005, an amendment to the IEEE 802.16-2004 standard that added mobility support. The IEEE 802.16e-2005 forms the basis for the WiMAX solution for nomadic and mobile applications and is often referred to as mobile WiMAX [2].

The basic characteristics of the various IEEE 802.16 standards are summarized in Table 2.1. Note that these standards offer a variety of fundamentally different design options. For example, there are multiple physical-layer choices: a single-carrier-based physical layer called Wireless-MAN-SCa, an OFDM-based physical layer called WirelessMAN-OFDM, and an OFDMA-based physical layer called Wireless-OFDMA. Similarly, there are multiple choices for MAC architecture, duplexing, frequency band of operation, etc. These standards were developed to suit a variety of applications and deployment scenarios, and hence offer a plethora of design choices for system developers. In fact, one could say that IEEE 802.16 is a collection of standards, not one single interoperable standard.

For practical reasons of interoperability, the scope of the standard needs to be reduced, and a smaller set of design choices for implementation need to be defined. The WiMAX Forum does this by defining a limited number of system profiles and certification profiles. A *system profile* defines the subset of mandatory and optional physical- and MAC-layer features selected by the WiMAX Forum from the IEEE 802.16-2004 or IEEE 802.16e-2005 standard. It should be noted that the mandatory and optional status of a particular feature within a WiMAX system profile may be different from what it is in the original IEEE standard. Currently, the WiMAX Forum has two different system profiles: one based on IEEE 802.16-2004, OFDM PHY, called the fixed system profile; the other one based on IEEE 802.16e-2005 scalable OFDMA PHY, called the mobility system profile. A *certification profile* is defined as a particular instantiation of a system profile where the operating frequency, channel bandwidth, and duplexing mode are also specified. WiMAX equipment are certified for interoperability against a particular certification profile.

The WiMAX Forum has thus far defined five fixed certification profiles and fourteen mobility certification profiles (see Table 2.2). To date, there are two fixed WiMAX profiles against which equipment have been certified. These are 3.5GHz systems operating over a 3.5MHz channel, using the fixed system profile based on the IEEE 802.16-2004 OFDM physical layer with a point-to-multipoint MAC. One of the profiles uses frequency division duplexing (FDD), and the other uses time division duplexing (TDD).

Table 2.1 Basic Data on IEEE 802.16 Standards

	802.16	802.16-2004	802.16e-2005
Status	Completed December 2001	Completed June 2004	Completed December 2005
Frequency band	10GHz–66GHz	2GHz–11GHz	2GHz–11GHz for fixed; 2GHz–6GHz for mobile applications
Application	Fixed LOS	Fixed NLOS	Fixed and mobile NLOS
MAC architecture	Point-to-multipoint, mesh	Point-to-multipoint, mesh	Point-to-multipoint, mesh
Transmission scheme	Single carrier only	Single carrier, 256 OFDM or 2,048 OFDM	Single carrier, 256 OFDM or scalable OFDM with 128, 512, 1,024, or 2,048 subcarriers
Modulation	QPSK, 16 QAM, 64 QAM	QPSK, 16 QAM, 64 QAM	QPSK, 16 QAM, 64 QAM
Gross data rate	32Mbps–134.4Mbps	1Mbps–75Mbps	1Mbps–75Mbps
Multiplexing	Burst TDM/TDMA	Burst TDM/TDMA/ OFDMA	Burst TDM/TDMA/ OFDMA
Duplexing	TDD and FDD	TDD and FDD	TDD and FDD
Channel bandwidths	20MHz, 25MHz, 28MHz	1.75MHz, 3.5MHz, 7MHz, 14MHz, 1.25MHz, 5MHz, 10MHz, 15MHz, 8.75MHz	1.75MHz, 3.5MHz, 7MHz, 14MHz, 1.25MHz, 5MHz, 10MHz, 15MHz, 8.75MHz
Air-interface designation	WirelessMAN-SC	WirelessMAN-SCa WirelessMAN-OFDM WirelessMAN-OFDMA WirelessHUMAN[a]	WirelessMAN-SCa WirelessMAN-OFDM WirelessMAN-OFDMA WirelessHUMAN[a]
WiMAX implementation	None	256 - OFDM as Fixed WiMAX	Scalable OFDMA as Mobile WiMAX

a. WirelessHUMAN (wireless high-speed unlicensed MAN) is similar to OFDM-PHY (physical layer) but mandates dynamic frequency selection for license-exempt bands.

Table 2.2 Fixed and Mobile WiMAX Initial Certification Profiles

Band Index	Frequency Band	Channel Bandwidth	OFDM FFT Size	Duplexing	Notes
**Fixed WiMAX Profiles**
1	3.5 GHz	3.5MHz	256	FDD	Products already certified
		3.5MHz	256	TDD	
		7MHz	256	FDD	
		7MHz	256	TDD	
2	5.8GHz	10MHz	256	TDD	
**Mobile WiMAX Profiles**
1	2.3GHz–2.4GHz	5MHz	512	TDD	Both bandwidths must be supported by mobile station (MS)
		10MHz	1,024	TDD	
		8.75MHz	1,024	TDD	
2	2.305GHz–2.320GHz, 2.345GHz–2.360GHz	3.5MHz	512	TDD	
		5MHz	512	TDD	
		10MHz	1,024	TDD	
3	2.496GHz–2.69GHz	5MHz	512	TDD	Both bandwidths must be supported by mobile station (MS)
		10MHz	1,024	TDD	
4	3.3GHz–3.4GHz	5MHz	512	TDD	
		7MHz	1,024	TDD	
		10MHz	1,024	TDD	
5	3.4GHz–3.8GHz, 3.4GHz–3.6GHz, 3.6GHz–3.8GHz	5MHz	512	TDD	
		7MHz	1,024	TDD	
		10MHz	1,024	TDD	

With the completion of the IEEE 802.16e-2005 standard, interest within the WiMAX group has shifted sharply toward developing and certifying mobile WiMAX[1] system profiles based on this newer standard. All mobile WiMAX profiles use scalable OFDMA as the physical layer. At least initially, all mobility profiles will use a point-to-multipoint MAC. It should also be noted that all the current candidate mobility certification profiles are TDD based. Although TDD is often preferred, FDD profiles may be needed in the future to comply with regulatory pairing and interoperator coexistence requirements in certain bands.

---

1. Although designated as mobile WiMAX, it is designed for fixed, nomadic, and mobile usage scenarios.

For the reminder of this chapter, we focus solely on WiMAX and therefore discuss only aspects of IEEE 802.16 family of standards that may be relevant to current and future WiMAX certification. It should be noted that the IEEE 802.16e-2004 and IEEE 802.16-2005 standards specifications are limited to the control and data plane aspects of the air-interface. Some aspects of network management are defined in IEEE 802.16g. For a complete end-to-end system, particularly in the context of mobility, several additional end-to-end service management aspects need to be specified. This task is being performed by the WiMAX Forums Network Working Group (NWG). The WiMAX NWG is developing an end-to-end network architecture and filling in some of the missing pieces. We cover the end-to-end architecture in Section 2.6.

2.2 Salient Features of WiMAX

WiMAX is a wireless broadband solution that offers a rich set of features with a lot of flexibility in terms of deployment options and potential service offerings. Some of the more salient features that deserve highlighting are as follows:

OFDM-based physical layer: The WiMAX physical layer (PHY) is based on orthogonal frequency division multiplexing, a scheme that offers good resistance to multipath, and allows WiMAX to operate in NLOS conditions. OFDM is now widely recognized as the method of choice for mitigating multipath for broadband wireless. Chapter 4 provides a detailed overview of OFDM.

Very high peak data rates: WiMAX is capable of supporting very high peak data rates. In fact, the peak PHY data rate can be as high as 74Mbps when operating using a 20MHz[2] wide spectrum. More typically, using a 10MHz spectrum operating using TDD scheme with a 3:1 downlink-to-uplink ratio, the peak PHY data rate is about 25Mbps and 6.7Mbps for the downlink and the uplink, respectively. These peak PHY data rates are achieved when using 64 QAM modulation with rate 5/6 error-correction coding. Under very good signal conditions, even higher peak rates may be achieved using multiple antennas and spatial multiplexing.

Scalable bandwidth and data rate: WiMAX has a scalable physical-layer architecture that allows for the data rate to scale easily with available channel bandwidth. This scalability is supported in the OFDMA mode, where the FFT (fast fourier transform) size may be scaled based on the available channel bandwidth. For example, a WiMAX system may use 128-, 512-, or 1,048-bit FFTs based on whether the channel bandwidth is 1.25MHz, 5MHz, or 10MHz, respectively. This scaling may be done dynamically to support user roaming across different networks that may have different bandwidth allocations.

Adaptive modulation and coding (AMC): WiMAX supports a number of modulation and forward error correction (FEC) coding schemes and allows the scheme to be changed on a per user and per frame basis, based on channel conditions. AMC is an effective mechanism to maximize throughput in a time-varying channel. The adaptation algorithm typically calls for the use

---

2. Initial WiMAX profiles do not include 20MHz support; 74Mbps is combined uplink/downlink PHY throughput.

of the highest modulation and coding scheme that can be supported by the signal-to-noise and interference ratio at the receiver such that each user is provided with the highest possible data rate that can be supported in their respective links. AMC is discussed in Chapter 6.

Link-layer retransmissions: For connections that require enhanced reliability, WiMAX supports automatic retransmission requests (ARQ) at the link layer. ARQ-enabled connections require each transmitted packet to be acknowledged by the receiver; unacknowledged packets are assumed to be lost and are retransmitted. WiMAX also optionally supports hybrid-ARQ, which is an effective hybrid between FEC and ARQ.

Support for TDD and FDD: IEEE 802.16-2004 and IEEE 802.16e-2005 supports both time division duplexing and frequency division duplexing, as well as a half-duplex FDD, which allows for a low-cost system implementation. TDD is favored by a majority of implementations because of its advantages: (1) flexibility in choosing uplink-to-downlink data rate ratios, (2) ability to exploit channel reciprocity, (3) ability to implement in nonpaired spectrum, and (4) less complex transceiver design. All the initial WiMAX profiles are based on TDD, except for two fixed WiMAX profiles in 3.5GHz.

Orthogonal frequency division multiple access (OFDMA): Mobile WiMAX uses OFDM as a multiple-access technique, whereby different users can be allocated different subsets of the OFDM tones. As discussed in detail in Chapter 6, OFDMA facilitates the exploitation of frequency diversity and multiuser diversity to significantly improve the system capacity.

Flexible and dynamic per user resource allocation: Both uplink and downlink resource allocation are controlled by a scheduler in the base station. Capacity is shared among multiple users on a demand basis, using a burst TDM scheme. When using the OFDMA-PHY mode, multiplexing is additionally done in the frequency dimension, by allocating different subsets of OFDM subcarriers to different users. Resources may be allocated in the spatial domain as well when using the optional advanced antenna systems (AAS). The standard allows for bandwidth resources to be allocated in time, frequency, and space and has a flexible mechanism to convey the resource allocation information on a frame-by-frame basis.

Support for advanced antenna techniques: The WiMAX solution has a number of hooks built into the physical-layer design, which allows for the use of multiple-antenna techniques, such as beamforming, space-time coding, and spatial multiplexing. These schemes can be used to improve the overall system capacity and spectral efficiency by deploying multiple antennas at the transmitter and/or the receiver. Chapter 5 presents detailed overview of the various multiple-antenna techniques.

Quality-of-service: The WiMAX MAC layer has a connection-oriented architecture that is designed to support a variety of applications, including voice and multimedia services. The system offers support for constant bit rate, variable bit rate, real-time, and non-real-time traffic flows, in addition to best-effort data traffic. WiMAX MAC is designed to support a large number of users, with multiple connections per terminal, each with its own QoS requirement.

Robust security: WiMAX supports strong encryption, using Advanced Encryption Standard (AES), and has a robust privacy and key-management protocol. The system also offers a very flexible authentication architecture based on Extensible Authentication Protocol (EAP), which allows for a variety of user credentials, including username/password, digital certificates, and smart cards.

Support for mobility: The mobile WiMAX variant of the system has mechanisms to support secure seamless handovers for delay-tolerant full-mobility applications, such as VoIP. The system also has built-in support for power-saving mechanisms that extend the battery life of handheld subscriber devices. Physical-layer enhancements, such as more frequent channel estimation, uplink subchannelization, and power control, are also specified in support of mobile applications.

IP-based architecture: The WiMAX Forum has defined a reference network architecture that is based on an all-IP platform. All end-to-end services are delivered over an IP architecture relying on IP-based protocols for end-to-end transport, QoS, session management, security, and mobility. Reliance on IP allows WiMAX to ride the declining costcurves of IP processing, facilitate easy convergence with other networks, and exploit the rich ecosystem for application development that exists for IP.

2.3 WiMAX Physical Layer

The WiMAX physical layer is based on orthogonal frequency division multiplexing. OFDM is the transmission scheme of choice to enable high-speed data, video, and multimedia communications and is used by a variety of commercial broadband systems, including DSL, Wi-Fi, Digital Video Broadcast-Handheld (DVB-H), and MediaFLO, besides WiMAX. OFDM is an elegant and efficient scheme for high data rate transmission in a non-line-of-sight or multipath radio environment. In this section, we cover the basics of OFDM and provide an overview of the WiMAX physical layer. Chapter 8 provides a more detailed discussion of the WiMAX PHY.

2.3.1 OFDM Basics

OFDM belongs to a family of transmission schemes called *multicarrier modulation*, which is based on the idea of dividing a given high-bit-rate data stream into several parallel lower bit-rate streams and modulating each stream on separate carriers—often called subcarriers, or tones. Multicarrier modulation schemes eliminate or minimize intersymbol interference (ISI) by making the symbol time large enough so that the channel-induced delays—delay spread being a good measure of this in wireless channels[3]—are an insignificant (typically, <10 percent) fraction of the symbol duration. Therefore, in high-data-rate systems in which the symbol duration is small—being inversely proportional to the data rate—splitting the data stream into many parallel streams increases the symbol duration of each stream such that the delay spread is only a small fraction of the symbol duration.

OFDM is a spectrally efficient version of multicarrier modulation, where the subcarriers are selected such that they are all orthogonal to one another over the symbol duration, thereby

---

3. Delay spread is discussed in Chapter 3.

avoiding the need to have nonoverlapping subcarrier channels to eliminate intercarrier interference. Choosing the first subcarrier to have a frequency such that it has an integer number of cycles in a symbol period, and setting the spacing between adjacent subcarriers (subcarrier bandwidth) to be $B_{SC} = B/L$, where B is the nominal bandwidth (equal to data rate), and L is the number of subcarriers, ensures that all tones are orthogonal to one another over the symbol period. It can be shown that the OFDM signal is equivalent to the inverse discrete Fourier transform (IDFT) of the data sequence block taken L at a time. This makes it extremely easy to implement OFDM transmitters and receivers in discrete time using IFFT (inverse fast Fourier) and FFT, respectively.[4]

In order to completely eliminate ISI, guard intervals are used between OFDM symbols. By making the guard interval larger than the expected multipath delay spread, ISI can be completely eliminated. Adding a guard interval, however, implies power wastage and a decrease in bandwidth efficiency. The amount of power wasted depends on how large a fraction of the OFDM symbol duration the guard time is. Therefore, the larger the symbol period—for a given data rate, this means more subcarriers—the smaller the loss of power and bandwidth efficiency.

The size of the FFT in an OFDM design should be chosen carefully as a balance between protection against multipath, Doppler shift, and design cost/complexity. For a given bandwidth, selecting a large FFT size would reduce the subcarrier spacing and increase the symbol time. This makes it easier to protect against multipath delay spread. A reduced subcarrier spacing, however, also makes the system more vulnerable to intercarrier interference owing to Doppler spread in mobile applications. The competing influences of delay and Doppler spread in an OFDM design require careful balancing. Chapter 4 provides a more detailed and rigorous treatment of OFDM.

2.3.2 OFDM Pros and Cons

OFDM enjoys several advantages over other solutions for high-speed transmission.

- **Reduced computational complexity:** OFDM can be easily implemented using FFT/IFFT, and the processing requirements grow only slightly faster than linearly with data rate or bandwidth. The computational complexity of OFDM can be shown to be $O(B \log B T_m)$, where B is the bandwidth and T_m is the delay spread. This complexity is much lower than that of a standard equalizer-based system, which has a complexity $O(B^2 T_m)$.

- **Graceful degradation of performance under excess delay:** The performance of an OFDM system degrades gracefully as the delay spread exceeds the value designed for. Greater coding and low constellation sizes can be used to provide fallback rates that are significantly more robust against delay spread. In other words, OFDM is well suited for

---

4.   FFT (fast Fourier transform) is a computationally efficient way of computing DFT (discrete Fourier transform).

adaptive modulation and coding, which allows the system to make the best of the available channel conditions. This contrasts with the abrupt degradation owing to error propagation that single-carrier systems experience as the delay spread exceeds the value for which the equalizer is designed.

- **Exploitation of frequency diversity:** OFDM facilitates coding and interleaving across subcarriers in the frequency domain, which can provide robustness against burst errors caused by portions of the transmitted spectrum undergoing deep fades. In fact, WiMAX defines subcarrier permutations that allow systems to exploit this.
- **Use as a multiaccess scheme:** OFDM can be used as a multiaccess scheme, where different tones are partitioned among multiple users. This scheme is referred to as OFDMA and is exploited in mobile WiMAX. This scheme also offers the ability to provide fine granularity in channel allocation. In relatively slow time-varying channels, it is possible to significantly enhance the capacity by adapting the data rate per subscriber according to the signal-to-noise ratio of that particular subcarrier.
- **Robust against narrowband interference:** OFDM is relatively robust against narrowband interference, since such interference affects only a fraction of the subcarriers.
- **Suitable for coherent demodulation:** It is relatively easy to do pilot-based channel estimation in OFDM systems, which renders them suitable for coherent demodulation schemes that are more power efficient.

Despite these advantages, OFDM techniques also face several challenges. First, there is the problem associated with OFDM signals having a high peak-to-average ratio that causes nonlinearities and clipping distortion. This can lead to power inefficiencies that need to be countered. Second, OFDM signals are very susceptible to phase noise and frequency dispersion, and the design must mitigate these imperfections. This also makes it critical to have accurate frequency synchronization. Chapter 4 provides a good overview of available solutions to overcome these OFDM challenges.

2.3.3 OFDM Parameters in WiMAX

As mentioned previously, the fixed and mobile versions of WiMAX have slightly different implementations of the OFDM physical layer. Fixed WiMAX, which is based on IEEE 802.16-2004, uses a 256 FFT-based OFDM physical layer. Mobile WiMAX, which is based on the IEEE 802.16e-2005[5] standard, uses a scalable OFDMA-based physical layer. In the case of mobile WiMAX, the FFT sizes can vary from 128 bits to 2,048 bits.

Table 2.3 shows the OFDM-related parameters for both the OFDM-PHY and the OFDMA-PHY. The parameters are shown here for only a limited set of profiles that are likely to be deployed and do not constitute an exhaustive set of possible values.

---

5. Although the scalable OFDMA scheme is referred to as mobile WiMAX, it can be used in fixed, nomadic, and mobile applications.

Fixed WiMAX OFDM-PHY: For this version the FFT size is fixed at 256, of which 192 sub-carriers are used for carrying data, 8 are used as pilot subcarriers for channel estimation and synchronization purposes, and the rest are used as guard band subcarriers.[6] Since the FFT size is fixed, the subcarrier spacing varies with channel bandwidth. When larger bandwidths are used, the subcarrier spacing increases, and the symbol time decreases. Decreasing symbol time implies that a larger fraction needs to be allocated as guard time to overcome delay spread. As Table 2.3 shows, WiMAX allows a wide range of guard times that allow system designers to make appropriate trade-offs between spectral efficiency and delay spread robustness. For maximum delay spread robustness, a 25 percent guard time can be used, which can accommodate delay spreads up to 16 μs when operating in a 3.5MHz channel and up to 8 μs when operating in a 7MHz channel. In relatively benign multipath channels, the guard time overhead may be reduced to as little as 3 percent.

Table 2.3 OFDM Parameters Used in WiMAX

Parameter	Fixed WiMAX OFDM-PHY	Mobile WiMAX Scalable OFDMA-PHY[a]			
FFT size	256	128	**512**	1,024	2,048
Number of used data subcarriers[b]	192	72	**360**	720	1,440
Number of pilot subcarriers	8	12	**60**	120	240
Number of null/guardband subcarriers	56	44	**92**	184	368
Cyclic prefix or guard time (Tg/Tb)	1/32, 1/16, **1/8**, 1/4				
Oversampling rate (Fs/BW)	Depends on bandwidth: 7/6 for 256 OFDM, 8/7 for multiples of 1.75MHz, and 28/25 for multiples of 1.25MHz, 1.5MHz, 2MHz, or 2.75MHz.				
Channel bandwidth (MHz)	3.5	1.25	**5**	10	20
Subcarrier frequency spacing (kHz)	15.625	**10.94**			
Useful symbol time (μs)	64	**91.4**			
Guard time assuming 12.5% (μs)	8	**11.4**			
OFDM symbol duration (μs)	72	**102.9**			
Number of OFDM symbols in 5 ms frame	69	**48.0**			

a. Boldfaced values correspond to those of the initial mobile WiMAX system profiles.

b. The mobile WiMAX subcarrier distribution listed is for downlink PUSC (partial usage of subcarrier).

---

6. Since FFT size can take only values equal to 2^n, dummy subcarriers are padded to the left and right of the useful subcarriers.

Mobile WiMAX OFDMA-PHY: In Mobile WiMAX, the FFT size is scalable from 128 to 2,048. Here, when the available bandwidth increases, the FFT size is also increased such that the subcarrier spacing is always 10.94kHz. This keeps the OFDM symbol duration, which is the basic resource unit, fixed and therefore makes scaling have minimal impact on higher layers. A scalable design also keeps the costs low. The subcarrier spacing of 10.94kHz was chosen as a good balance between satisfying the delay spread and Doppler spread requirements for operating in mixed fixed and mobile environments. This subcarrier spacing can support delay-spread values up to 20 μs and vehicular mobility up to 125 kmph when operating in 3.5GHz. A subcarrier spacing of 10.94kHz implies that 128, 512, 1,024, and 2,048 FFT are used when the channel bandwidth is 1.25MHz, 5MHz, 10MHz, and 20MHz, respectively. It should, however, be noted that mobile WiMAX may also include additional bandwidth profiles. For example, a profile compatible with WiBro will use an 8.75MHz channel bandwidth and 1,024 FFT. This obviously will require a different subcarrier spacing and hence will not have the same scalability properties.

2.3.4 Subchannelization: OFDMA

The available subcarriers may be divided into several groups of subcarriers called subchannels. Fixed WiMAX based on OFDM-PHY allows a limited form of subchannelization in the uplink only. The standard defines 16 subchannels, where 1, 2, 4, 8, or all sets can be assigned to a subscriber station (SS) in the uplink. Uplink subchannelization in fixed WiMAX allows subscriber stations to transmit using only a fraction (as low as 1/16) of the bandwidth allocated to it by the base station, which provides link budget improvements that can be used to enhance range performance and/or improve battery life of subscriber stations. A 1/16 subchannelization factor provides a 12 dB link budget enhancement.

Mobile WiMAX based on OFDMA-PHY, however, allows subchannelization in both the uplink and the downlink, and here, subchannels form the minimum frequency resource-unit allocated by the base station. Therefore, different subchannels may be allocated to different users as a multiple-access mechanism. This type of multiaccess scheme is called orthogonal frequency division multiple access (OFDMA), which gives the mobile WiMAX PHY its name.

Subchannels may be constituted using either contiguous subcarriers or subcarriers pseudo-randomly distributed across the frequency spectrum. Subchannels formed using distributed subcarriers provide more frequency diversity, which is particularly useful for mobile applications. WiMAX defines several subchannelization schemes based on distributed carriers for both the uplink and the downlink. One, called *partial usage of subcarriers* (PUSC), is mandatory for all mobile WiMAX implementations. The initial WiMAX profiles define 15 and 17 subchannels for the downlink and the uplink, respectively, for PUSC operation in 5MHz bandwidth. For 10MHz operation, it is 30 and 35 channels, respectively.

The subchannelization scheme based on contiguous subcarriers in WiMAX is called band *adaptive modulation and coding* (AMC). Although frequency diversity is lost, band AMC allows system designers to exploit multiuser diversity, allocating subchannels to users based on their frequency response. Multiuser diversity can provide significant gains in overall system capacity, if the system strives to provide each user with a subchannel that maximizes its received SINR. In general, contiguous subchannels are more suited for fixed and low-mobility applications.

2.3.5 Slot and Frame Structure

The WiMAX PHY layer is also responsible for slot allocation and framing over the air. The minimum time-frequency resource that can be allocated by a WiMAX system to a given link is called a *slot*. Each slot consists of one subchannel over one, two, or three OFDM symbols, depending on the particular subchannelization scheme used. A contiguous series of slots assigned to a given user is called that user's *data region*; scheduling algorithms could allocate data regions to different users, based on demand, QoS requirements, and channel conditions.

Figure 2.1 shows an OFDMA and OFDM frame when operating in TDD mode. The frame is divided into two subframes: a downlink frame followed by an uplink frame after a small guard interval. The downlink-to-uplink-subframe ratio may be varied from 3:1 to 1:1 to support different traffic profiles. WiMAX also supports frequency division duplexing, in which case the frame structure is the same except that both downlink and uplink are transmitted simultaneously over different carriers. Some of the current fixed WiMAX systems use FDD. Most WiMAX deployments, however, are likely to be in TDD mode because of its advantages. TDD allows for a more flexible sharing of bandwidth between uplink and downlink, does not require paired spectrum, has a reciprocal channel that can be exploited for spatial processing, and has a simpler transceiver design. The downside of TDD is the need for synchronization across multiple base stations to ensure interference-free coexistence. Paired band regulations and interoperator coexistence issues, however, may force some operators to deploy WiMAX in FDD mode.

As shown in Figure 2.1, the downlink subframe begins with a downlink preamble that is used for physical-layer procedures, such as time and frequency synchronization and initial channel estimation. The downlink preamble is followed by a frame control header (FCH), which provides frame configuration information, such as the MAP message length, the modulation and coding scheme, and the usable subcarriers. Multiple users are allocated data regions within the frame, and these allocations are specified in the uplink and downlink MAP messages (DL-MAP and UL-MAP) that are broadcast following the FCH in the downlink subframe. MAP messages include the burst profile for each user, which defines the modulation and coding scheme used in that link. Since MAP contains critical information that needs to reach all users, it is often sent over a very reliable link, such as BPSK with rate 1/2 coding and repetition coding. Although the MAP messages are an elegant way for the base station to inform the various users of its allocations and burst profiles on a per-frame basis, it could form a significant overhead, particularly when there are a large number of users with small packets (e.g., VoIP) for which allocations need to be specified. To mitigate the overhead concern, mobile WiMAX systems can optionally use multiple sub-MAP messages where the dedicated control messages to different users are transmitted at higher rates, based on their individual SINR conditions. The broadcast MAP messages may also optionally be compressed for additional efficiency.

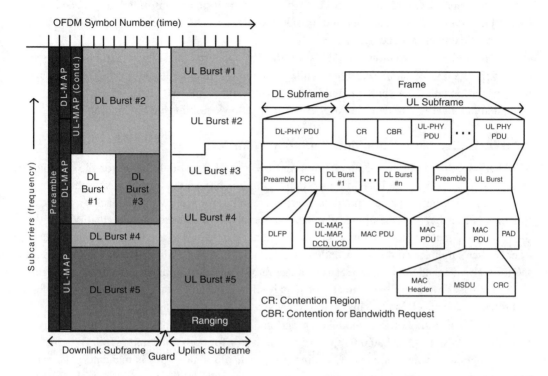

Figure 2.1 A sample TDD frame structure for mobile WiMAX

WiMAX is quite flexible in terms of how multiple users and packets are multiplexed on a single frame. A single downlink frame may contain multiple bursts of varying size and type carrying data for several users. The frame size is also variable on a frame-by-frame basis from 2 ms to 20 ms, and each burst can contain multiple concatenated fixed-size or variable-size packets or fragments of packets received from the higher layers. At least initially, however, all WiMAX equipment will support only 5 ms frames.

The uplink subframe is made up of several uplink bursts from different users. A portion of the uplink subframe is set aside for contention-based access that is used for a variety of purposes. This subframe is used mainly as a ranging channel to perform closed-loop frequency, time, and power adjustments during network entry as well as periodically afterward. The ranging channel may also be used by subscriber stations or mobile stations (SS/MS)[7] to make uplink bandwidth requests. In addition, best-effort data may be sent on this contention-based channel, particularly when the amount of data to send is too small to justify requesting a dedicated channel. Besides the ranging channel and traffic bursts, the uplink subframe has a channel-quality

---

7. The subscriber terminal mobile station (MS) is mobile WiMAX, and subscriber station (SS) is fixed WiMAX. Henceforth, for simplicity, we use MS to denote both.

indicator channel (CQICH) for the MS to feed back channel-quality information that can be used by the base station (BS) scheduler and an acknowledgment (ACK) channel for the MS to feed back downlink acknowledgements.

To handle time variations, WiMAX optionally supports repeating preambles more frequently. In the uplink, short preambles, called midambles, may be used after 8, 16, or 32 symbols; in the downlink, a short preamble can be inserted at the beginning of each burst. It is estimated that having a midamble every 10 symbols allows mobility up to 150 kmph.

2.3.6 Adaptive Modulation and Coding in WiMAX

WiMAX supports a variety of modulation and coding schemes and allows for the scheme to change on a burst-by-burst basis per link, depending on channel conditions. Using the channel-quality feedback indicator, the mobile can provide the base station with feedback on the downlink channel quality. For the uplink, the base station can estimate the channel quality, based on the received signal quality. The base station scheduler can take into account the channel quality of each user's uplink and downlink and assign a modulation and coding scheme that maximizes the throughput for the available signal-to-noise ratio. Adaptive modulation and coding significantly increases the overall system capacity, as it allows real-time trade-off between throughput and robustness on each link. This topic is discussed in more detail in Chapter 6.

Table 2.4 lists the various modulation and coding schemes supported by WiMAX. In the downlink, QPSK, 16 QAM, and 64 QAM are mandatory for both fixed and mobile WiMAX; 64 QAM is optional in the uplink. FEC coding using convolutional codes is mandatory. Convolutional codes are combined with an outer Reed-Solomon code in the downlink for OFDM-PHY. The standard optionally supports turbo codes and low-density parity check (LDPC) codes at a variety of code rates as well. A total of 52 combinations of modulation and coding schemes are defined in WiMAX as burst profiles. More details on burst profiles are provided in Chapter 8.

2.3.7 PHY-Layer Data Rates

Because the physical layer of WiMAX is quite flexible, data rate performance varies based on the operating parameters. Parameters that have a significant impact on the physical-layer data rate are channel bandwidth and the modulation and coding scheme used. Other parameters, such as number of subchannels, OFDM guard time, and oversampling rate, also have an impact.

Table 2.5 lists the PHY-layer data rate at various channel bandwidths, as well as modulation and coding schemes. The rates shown are the aggregate physical-layer data rate that is shared among all users in the sector for the TDD case, assuming a 3:1 downlink-to-uplink bandwidth ratio. The calculations here assume a frame size of 5 ms, a 12.5 percent OFDM guard interval overhead, and a PUSC subcarrier permutation scheme. It is also assumed that all usable OFDM data symbols are available for user traffic except one symbol used for downlink frame overhead. The numbers shown here do not assume spatial multiplexing using multiple antennas at the transmitter or the receiver, the use of which can further increase the peak rates in rich multipath channels.

Table 2.4 Modulation and Coding Supported in WiMAX

	Downlink	Uplink
Modulation	BPSK, QPSK, 16 QAM, 64 QAM; BPSK optional for OFDMA-PHY	BPSK, QPSK, 16 QAM; 64 QAM optional
Coding	Mandatory: convolutional codes at rate 1/2, 2/3, 3/4, 5/6 Optional: convolutional turbo codes at rate 1/2, 2/3, 3/4, 5/6; repetition codes at rate 1/2, 1/3, 1/6, LDPC, RS-Codes for OFDM-PHY	Mandatory: convolutional codes at rate 1/2, 2/3, 3/4, 5/6 Optional: convolutional turbo codes at rate 1/2, 2/3, 3/4, 5/6; repetition codes at rate 1/2, 1/3, 1/6, LDPC

Table 2.5 PHY-Layer Data Rate at Various Channel Bandwidths

Channel bandwidth	3.5MHz		1.25MHz		5MHz		10MHz		8.75MHz[a]	
PHY mode	256 OFDM		128 OFDMA		512 OFDMA		1,024 OFDMA		1,024 OFDMA	
Oversampling	8/7		28/25		28/25		28/25		28/25	
Modulation and Code Rate	**PHY-Layer Data Rate (kbps)**									
	DL	UL	DL	UL	DL	UL	DL	UL	DL	UL
BPSK, 1/2	946	326	Not applicable							
QPSK, 1/2	1,882	653	504	154	2,520	653	5,040	1,344	4,464	1,120
QPSK, 3/4	2,822	979	756	230	3,780	979	7,560	2,016	6,696	1,680
16 QAM, 1/2	3,763	1,306	1,008	307	5,040	1,306	10,080	2,688	8,928	2,240
16 QAM, 3/4	5,645	1,958	1,512	461	7,560	1,958	15,120	4,032	13,392	3,360
64 QAM, 1/2	5,645	1,958	1,512	461	7,560	1,958	15,120	4,032	13,392	3,360
64 QAM, 2/3	7,526	2,611	2,016	614	10,080	2,611	20,160	5,376	17,856	4,480
64 QAM, 3/4	8,467	2,938	2,268	691	11,340	2,938	22,680	6,048	20,088	5,040
64 QAM, 5/6	9,408	3,264	2,520	768	12,600	3,264	25,200	6,720	22,320	5,600

a. The version deployed as WiBro in South Korea.

2.4 MAC-Layer Overview

The primary task of the WiMAX MAC layer is to provide an interface between the higher trans-port layers and the physical layer. The MAC layer takes packets from the upper layer—these packets are called *MAC service data units* (MSDUs)—and organizes them into *MAC protocol data units* (MPDUs) for transmission over the air. For received transmissions, the MAC layer does the reverse. The IEEE 802.16-2004 and IEEE 802.16e-2005 MAC design includes a *convergence sublayer* that can interface with a variety of higher-layer protocols, such as ATM,

TDM Voice, Ethernet, IP, and any unknown future protocol. Given the predominance of IP and Ethernet in the industry, the WiMAX Forum has decided to support only IP and Ethernet at this time. Besides providing a mapping to and from the higher layers, the convergence sublayer supports MSDU header suppression to reduce the higher layer overheads on each packet.

The WiMAX MAC is designed from the ground up to support very high peak bit rates while delivering quality of service similar to that of ATM and DOCSIS. The WiMAX MAC uses a variable-length MPDU and offers a lot of flexibility to allow for their efficient transmission. For example, multiple MPDUs of same or different lengths may be aggregated into a single burst to save PHY overhead. Similarly, multiple MSDUs from the same higher-layer service may be concatenated into a single MPDU to save MAC header overhead. Conversely, large MSDUs may be fragmented into smaller MPDUs and sent across multiple frames.

Figure 2.2 shows examples of various MAC PDU (packet data unit) frames. Each MAC frame is prefixed with a generic MAC header (GMH) that contains a connection identifier[8] (CID), the length of frame, and bits to qualify the presence of CRC, subheaders, and whether the payload is encrypted and if so, with which key. The MAC payload is either a transport or a management message. Besides MSDUs, the transport payload may contain bandwidth requests or retransmission requests. The type of transport payload is identified by the subheader that immediately precedes it. Examples of subheaders are packing subheaders and fragmentation subheaders. WiMAX MAC also supports ARQ, which can be used to request the retransmission of unfragmented MSDUs and fragments of MSDUs. The maximum frame length is 2,047 bytes, which is represented by 11 bits in the GMH.

2.4.1 Channel-Access Mechanisms

In WiMAX, the MAC layer at the base station is fully responsible for allocating bandwidth to all users, in both the uplink and the downlink. The only time the MS has some control over bandwidth allocation is when it has multiple sessions or connections with the BS. In that case, the BS allocates bandwidth to the MS in the aggregate, and it is up to the MS to apportion it among the multiple connections. All other scheduling on the downlink *and* uplink is done by the BS. For the downlink, the BS can allocate bandwidth to each MS, based on the needs of the incoming traffic, without involving the MS. For the uplink, allocations have to be based on requests from the MS.

The WiMAX standard supports several mechanisms by which an MS can request and obtain uplink bandwidth. Depending on the particular QoS and traffic parameters associated with a service, one or more of these mechanisms may be used by the MS. The BS allocates dedicated or shared resources periodically to each MS, which it can use to request bandwidth. This process is called *polling*. Polling may be done either individually (unicast) or in groups (multicast). Multicast polling is done when there is insufficient bandwidth to poll each MS individually. When polling is done in multicast, the allocated slot for making bandwidth requests is a shared slot,

---

8. See Section 2.4.2 for the definition of a connection identifier.

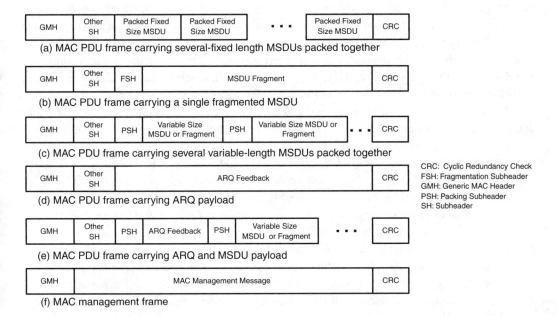

Figure 2.2 Examples of various MAC PDU frames

which every polled MS attempts to use. WiMAX defines a contention access and resolution mechanism for the case when more than one MS attempts to use the shared slot. If it already has an allocation for sending traffic, the MS is not polled. Instead, it is allowed to request more bandwidth by (1) transmitting a stand-alone bandwidth request MPDU, (2) sending a bandwidth request using the ranging channel, or (3) piggybacking a bandwidth request on generic MAC packets.

2.4.2 Quality of Service

Support for QoS is a fundamental part of the WiMAX MAC-layer design. WiMAX borrows some of the basic ideas behind its QoS design from the DOCSIS cable modem standard. Strong QoS control is achieved by using a connection-oriented MAC architecture, where all downlink and uplink connections are controlled by the serving BS. Before any data transmission happens, the BS and the MS establish a unidirectional logical link, called a *connection,* between the two MAC-layer peers. Each connection is identified by a *connection identifier* (CID), which serves as a temporary address for data transmissions over the particular link. In addition to connections for transferring user data, the WiMAX MAC defines three management connections—the basic, primary, and secondary connections—that are used for such functions as ranging.

WiMAX also defines a concept of a service flow. A *service flow* is a unidirectional flow of packets with a particular set of QoS parameters and is identified by a *service flow identifier* (SFID). The QoS parameters could include traffic priority, maximum sustained traffic rate, maximum burst rate, minimum tolerable rate, scheduling type, ARQ type, maximum delay, tolerated jitter, service data unit type and size, bandwidth request mechanism to be used, transmission PDU formation rules, and so on. Service flows may be provisioned through a network management system or created dynamically through defined signaling mechanisms in the standard. The base station is responsible for issuing the SFID and mapping it to unique CIDs. Service flows can also be mapped to DiffServ code points or MPLS flow labels to enable end-to-end IP-based QoS.

To support a wide variety of applications, WiMAX defines five scheduling services (Table 2.6) that should be supported by the base station MAC scheduler for data transport over a connection:

1. **Unsolicited grant services (UGS):** This is designed to support fixed-size data packets at a constant bit rate (CBR). Examples of applications that may use this service are T1/E1 emulation and VoIP without silence suppression. The mandatory service flow parameters that define this service are maximum sustained traffic rate, maximum latency, tolerated jitter, and request/transmission policy.[9]

2. **Real-time polling services (rtPS):** This service is designed to support real-time service flows, such as MPEG video, that generate variable-size data packets on a periodic basis. The mandatory service flow parameters that define this service are minimum reserved traffic rate, maximum sustained traffic rate, maximum latency, and request/transmission policy.

3. **Non-real-time polling service (nrtPS):** This service is designed to support delay-tolerant data streams, such as an FTP, that require variable-size data grants at a minimum guaranteed rate. The mandatory service flow parameters to define this service are minimum reserved traffic rate, maximum sustained traffic rate, traffic priority, and request/transmission policy.

4. **Best-effort (BE) service:** This service is designed to support data streams, such as Web browsing, that do not require a minimum service-level guarantee. The mandatory service flow parameters to define this service are maximum sustained traffic rate, traffic priority, and request/transmission policy.

5. **Extended real-time variable rate (ERT-VR) service:** This service is designed to support real-time applications, such as VoIP with silence suppression, that have variable data rates but require guaranteed data rate and delay. This service is defined only in IEEE 802.16e-2005, not in IEEE 802.16-2004. This is also referred to as extended real-time polling service (ErtPS).

Although it does not define the scheduler per se, WiMAX does define several parameters and features that facilitate the implementation of an effective scheduler:

• Support for a detailed parametric definition of QoS requirements and a variety of mechanisms to effectively signal traffic conditions and detailed QoS requirements in the uplink.

---

9. This policy includes how to request for bandwidth and the rules around PDU formation, such as whether fragmentation is allowed.

Table 2.6 Service Flows Supported in WiMAX

Service Flow Designation	Defining QoS Parameters	Application Examples
Unsolicited grant services (UGS)	Maximum sustained rate Maximum latency tolerance Jitter tolerance	Voice over IP (VoIP) without silence suppression
Real-time Polling service (rtPS)	Minimum reserved rate Maximum sustained rate Maximum latency tolerance Traffic priority	Streaming audio and video, MPEG (Motion Picture Experts Group) encoded
Non-real-time Polling service (nrtPS)	Minimum reserved rate Maximum sustained rate Traffic priority	File Transfer Protocol (FTP)
Best-effort service (BE)	Maximum sustained rate Traffic priority	Web browsing, data transfer
Extended real-time Polling service (ErtPS)	Minimum reserved rate Maximum sustained rate Maximum latency tolerance Jitter tolerance Traffic priority	VoIP with silence suppression

- Support for three-dimensional dynamic resource allocation in the MAC layer. Resources can be allocated in time (time slots), frequency (subcarriers), and space (multiple antennas) on a frame-by-frame basis.
- Support for fast channel-quality information feedback to enable the scheduler to select the appropriate coding and modulation (burst profile) for each allocation.
- Support for contiguous subcarrier permutations, such as AMC, that allow the scheduler to exploit multiuser diversity by allocating each subscriber to its corresponding strongest subchannel.

It should be noted that the implementation of an effective scheduler is critical to the overall capacity and performance of a WiMAX system.

2.4.3 Power-Saving Features

To support battery-operated portable devices, mobile WiMAX has power-saving features that allow portable subscriber stations to operate for longer durations without having to recharge. Power saving is achieved by turning off parts of the MS in a controlled manner when it is not actively transmitting or receiving data. Mobile WiMAX defines signaling methods that allow the MS to retreat into a sleep mode or idle mode when inactive. *Sleep mode* is a state in which the MS effectively turns itself off and becomes unavailable for predetermined periods. The periods of absence are negotiated with the serving BS. WiMAX defines three power-saving classes, based on the manner in which sleep mode is executed. When in Power Save Class 1 mode, the sleep window is exponentially increased from a minimum value to a maximum value. This is typically done when the MS is doing best-effort and non-real-time traffic. Power Save Class 2 has a fixed-length sleep window and is used for UGS service. Power Save Class 3 allows for a one-time sleep window and is typically used for multicast traffic or management traffic when the MS knows when the next traffic is expected. In addition to minimizing MS power consumption, sleep mode conserves BS radio resources. To facilitate handoff while in sleep mode, the MS is allowed to scan other base stations to collect handoff-related information.

Idle mode allows even greater power savings, and support for it is optional in WiMAX. Idle mode allows the MS to completely turn off and to not be registered with any BS and yet receive downlink broadcast traffic. When downlink traffic arrives for the idle-mode MS, the MS is paged by a collection of base stations that form a paging group. The MS is assigned to a paging group by the BS before going into idle mode, and the MS periodically wakes up to update its paging group. Idle mode saves more power than sleep mode, since the MS does not even have to register or do handoffs. Idle mode also benefits the network and BS by eliminating handover traffic from inactive MSs.

2.4.4 Mobility Support

In addition to fixed broadband access, WiMAX envisions four mobility-related usage scenarios:

1. **Nomadic**. The user is allowed to take a fixed subscriber station and reconnect from a different point of attachment.
2. **Portable**. Nomadic access is provided to a portable device, such as a PC card, with expectation of a best-effort handover.
3. **Simple mobility**. The subscriber may move at speeds up to 60 kmph with brief interruptions (less than 1 sec) during handoff.
4. **Full mobility**: Up to 120 kmph mobility and seamless handoff (less than 50 ms latency and <1% packet loss) is supported.

It is likely that WiMAX networks will initially be deployed for fixed and nomadic applications and then evolve to support portability to full mobility over time.

The IEEE 802.16e-2005 standard defines a framework for supporting mobility management. In particular, the standard defines signaling mechanisms for tracking subscriber stations as they move from the coverage range of one base station to another when active or as they move from one paging group to another when idle. The standard also has protocols to enable a seamless handover of ongoing connections from one base station to another. The WiMAX Forum has used the framework defined in IEEE 802.16e-2005 to further develop mobility management within an end-to-end network architecture framework. The architecture also supports IP-layer mobility using mobile IP.

Three handoff methods are supported in IEEE 802.16e-2005; one is mandatory and other two are optional. The mandatory handoff method is called the *hard handover* (HHO) and is the only type required to be implemented by mobile WiMAX initially. HHO implies an abrupt transfer of connection from one BS to another. The handoff decisions are made by the BS, MS, or another entity, based on measurement results reported by the MS. The MS periodically does a radio frequency (RF) scan and measures the signal quality of neighboring base stations. Scanning is performed during *scanning intervals* allocated by the BS. During these intervals, the MS is also allowed to optionally perform initial ranging and to associate with one or more neighboring base stations. Once a handover decision is made, the MS begins synchronization with the downlink transmission of the target BS, performs ranging if it was not done while scanning, and then terminates the connection with the previous BS. Any undelivered MPDUs at the BS are retained until a timer expires.

The two optional handoff methods supported in IEEE 802.16e-2005 are *fast base station switching* (FBSS) and *macro diversity handover* (MDHO). In these two methods, the MS maintains a valid connection simultaneously with more than one BS. In the FBSS case, the MS maintains a list of the BSs involved, called the *active set*. The MS continuously monitors the active set, does ranging, and maintains a valid connection ID with each of them. The MS, however, communicates with only one BS, called the *anchor BS*. When a change of anchor BS is required, the connection is switched from one base station to another without having to explicitly perform handoff signaling. The MS simply reports the selected anchor BS on the CQICH.

Macro diversity handover is similar to FBSS, except that the MS communicates on the downlink and the uplink with all the base stations in the active set—called a *diversity set* here— simultaneously. In the downlink, multiple copies received at the MS are combined using any of the well-known diversity-combining techniques (see Chapter 5). In the uplink, where the MS sends data to multiple base stations, selection diversity is performed to pick the best uplink.

Both FBSS and MDHO offer superior performance to HHO, but they require that the base stations in the active or diversity set be synchronized, use the same carrier frequency, and share network entry–related information. Support for FBHH and MDHO in WiMAX networks is not fully developed yet and is not part of WiMAX Forum Release 1 network specifications.

2.4.5 Security Functions

Unlike Wi-Fi, WiMAX systems were designed at the outset with robust security in mind. The standard includes state-of-the-art methods for ensuring user data privacy and preventing unauthorized access, with additional protocol optimization for mobility. Security is handled by a privacy sublayer within the WiMAX MAC. The key aspects of WiMAX security are as follow.

Support for privacy: User data is encrypted using cryptographic schemes of proven robustness to provide privacy. Both AES (Advanced Encryption Standard) and 3DES (Triple Data Encryption Standard) are supported. Most system implementations will likely use AES, as it is the new encryption standard approved as compliant with Federal Information Processing Standard

(FIPS) and is easier to implement.[10] The 128-bit or 256-bit key used for deriving the cipher is generated during the authentication phase and is periodically refreshed for additional protection.

Device/user authentication: WiMAX provides a flexible means for authenticating subscriber stations and users to prevent unauthorized use. The authentication framework is based on the Internet Engineering Task Force (IETF) EAP, which supports a variety of credentials, such as username/password, digital certificates, and smart cards. WiMAX terminal devices come with built-in X.509 digital certificates that contain their public key and MAC address. WiMAX operators can use the certificates for device authentication and use a username/password or smart card authentication on top of it for user authentication.

Flexible key-management protocol: The Privacy and Key Management Protocol Version 2 (PKMv2) is used for securely transferring keying material from the base station to the mobile station, periodically reauthorizing and refreshing the keys. PKM is a client-server protocol: The MS acts as the client; the BS, the server. PKM uses X.509 digital certificates and RSA (Rivest-Shamer-Adleman) public-key encryption algorithms to securely perform key exchanges between the BS and the MS.

Protection of control messages: The integrity of over-the-air control messages is protected by using message digest schemes, such as AES-based CMAC or MD5-based HMAC.[11]

Support for fast handover: To support fast handovers, WiMAX allows the MS to use preauthentication with a particular target BS to facilitate accelerated reentry. A three-way handshake scheme is supported to optimize the reauthentication mechanisms for supporting fast handovers, while simultaneously preventing any man-in-the-middle attacks.

2.4.6 Multicast and Broadcast Services

The mobile WiMAX MAC layer has support for multicast and broadcast services (MBS). MBS-related functions and features supported in the standard include

- Signaling mechanisms for MS to request and establish MBS
- Subscriber station access to MBS over a single or multiple BS, depending on its capability and desire
- MBS associated QoS and encryption using a globally defined traffic encryption key
- A separate zone within the MAC frame with its own MAP information for MBS traffic
- Methods for delivering MBS traffic to idle-mode subscriber stations
- Support for macro diversity to enhance the delivery performance of MBS traffic

---

10. See Chapter 7 for more details on encryption.
11. CMAC (cipher-based message authentication code); HMAC (hash-based message authentication codes); MD5 (Message-Digest 5 Algorithm). All protocols are standardized within the IETF.

2.5 Advanced Features for Performance Enhancements

WiMAX defines a number of optional advanced features for improving the performance. Among the more important of these advanced features are support for multiple-antenna techniques, hybrid-ARQ, and enhanced frequency reuse.

2.5.1 Advanced Antenna Systems

The WiMAX standard provides extensive support for implementing advanced multiantenna solutions to improve system performance. Significant gains in overall system *capacity and spectral* efficiency can be achieved by deploying the optional *advanced antenna systems* (AAS) defined in WiMAX. AAS includes support for a variety of multiantenna solutions, including transmit diversity, beamforming, and spatial multiplexing.

Transmit diversity: WiMAX defines a number of space-time block coding schemes that can be used to provide transmit diversity in the downlink. For transmit diversity, there could be two or more transmit antennas and one or more receive antennas. The space-time block code (STBC) used for the 2×1 antenna case is the Alamouti codes, which are orthogonal and amenable to maximum likelihood detection. The Alamouti STBC is quite easy to implement and offers the same diversity gain as a 1×2 receiver diversity with maximum ratio combining, albeit with a 3 dB penalty owing to redundant transmissions. But transmit diversity offers the advantage that the complexity is shifted to the base station, which helps to keep the MS cost low. In addition to the 2×1 case, WiMAX also defines STBCs for the three- and four-antenna cases.

Beamforming: Multiple antennas in WiMAX may also be used to transmit the same signal appropriately weighted for each antenna element such that the effect is to focus the transmitted beam in the direction of the receiver and away from interference, thereby improving the received SINR. Beamforming can provide significant improvement in the coverage range, capacity, and reliability. To perform transmit beamforming, the transmitter needs to have accurate knowledge of the channel, which in the case of TDD is easily available owing to channel reciprocity but for FDD requires a feedback channel to learn the channel characteristics. WiMAX supports beamforming in both the uplink and the downlink. For the uplink, this often takes the form of receive beamforming.

Spatial multiplexing: WiMAX also supports spatial multiplexing, where multiple independent streams are transmitted across multiple antennas. If the receiver also has multiple antennas, the streams can be separated out using space-time processing. Instead of increasing diversity, multiple antennas in this case are used to increase the data rate or capacity of the system. Assuming a rich multipath environment, the capacity of the system can be increased linearly with the number of antennas when performing spatial multiplexing. A 2×2 MIMO system therefore doubles the peak throughput capability of WiMAX. If the mobile station has only one antenna, WiMAX can still support spatial multiplexing by coding across multiple users in the uplink. This is called multiuser collaborative spatial multiplexing. Unlike transmit diversity and beamforming, spatial multiplexing works only under good SINR conditions.

2.5.2 Hybrid-ARQ

Hybrid-ARQ is an ARQ system that is implemented at the physical layer together with FEC, providing improved link performance over traditional ARQ at the cost of increased implementation complexity. The simplest version of H-ARQ is a simple combination of FEC and ARQ, where blocks of data, along with a CRC code, are encoded using an FEC coder before transmission; retransmission is requested if the decoder is unable to correctly decode the received block. When a retransmitted coded block is received, it is combined with the previously detected coded block and fed to the input of the FEC decoder. Combining the two received versions of the code block improves the chances of correctly decoding. This type of H-ARQ is often called type I *chase combining*.

The WiMAX standard supports this by combining an *N*-channel *stop and wait ARQ* along with a variety of supported FEC codes. Doing multiple parallel channels of H-ARQ at a time can improve the throughput, since when one H-ARQ process is waiting for an acknowledgment, another process can use the channel to send some more data. WiMAX supports signaling mechanisms to allow asynchronous operation of H-ARQ and supports a dedicated acknowledgment channel in the uplink for ACK/NACK signaling. Asynchronous operations allow variable delay between retransmissions, which provides greater flexibility for the scheduler.

To further improve the reliability of retransmission, WiMAX also optionally supports type II H-ARQ, which is also called *incremental redundancy*. Here, unlike in type I H-ARQ, each (re)transmission is coded differently to gain improved performance. Typically, the code rate is effectively decreased every retransmission. That is, additional parity bits are sent every iteration, equivalent to coding across retransmissions.

2.5.3 Improved Frequency Reuse

Although it is possible to operate WiMAX systems with a universal frequency reuse plan,[12] doing so can cause severe outage owing to interference, particularly along the intercell and intersector edges. To mitigate this, WiMAX allows for coordination of subchannel allocation to users at the cell edges such that there is minimal overlap. This allows for a more dynamic frequency allocation across sectors, based on loading and interference conditions, as opposed to traditional fixed frequency planning. Those users under good SINR conditions will have access to the full channel bandwidth and operate under a frequency reuse of 1. Those in poor SINR conditions will be allocated nonoverlapping subchannels such that they operate under a frequency reuse of 2, 3, or 4, depending on the number of nonoverlapping subchannel groups that are allocated to be shared among these users. This type of subchannel allocation leads to the effective reuse factor taking fractional values greater than 1. The variety of subchannelization schemes supported by WiMAX makes it possible to do this in a very flexible manner. Obviously, the downside is that cell edge users cannot have access to the full bandwidth of the channel, and hence their peak rates will be reduced.

---

12. This corresponds to all sectors and cells using the same frequency. Reuse factor is equal to 1.

2.6 Reference Network Architecture

The IEEE 802.16e-2005 standard provides the air interface for WiMAX but does not define the full end-to-end WiMAX network. The WiMAX Forum's Network Working Group, is responsible for developing the end-to-end network requirements, architecture, and protocols for WiMAX, using IEEE 802.16e-2005 as the air interface.

The WiMAX NWG has developed a network reference model to serve as an architecture framework for WiMAX deployments and to ensure interoperability among various WiMAX equipment and operators. The network reference model envisions a unified network architecture for supporting fixed, nomadic, and mobile deployments and is based on an IP service model. Figure 2.3 shows a simplified illustration of an IP-based WiMAX network architecture. The overall network may be logically divided into three parts: (1) mobile stations used by the end user to access the network, (2) the access service network (ASN), which comprises one or more base stations and one or more ASN gateways that form the radio access network at the edge, and (3) the connectivity service network (CSN), which provides IP connectivity and all the IP core network functions.

The architecture framework is defined such that the multiple players can be part of the WiMAX service value chain. More specifically, the architecture allows for three separate business entities: (1) network access provider (NAP), which owns and operates the ASN; (2) network services provider (NSP), which provides IP connectivity and WiMAX services to subscribers using the ASN infrastructure provided by one or more NAPs; and (3) application service provider (ASP), which can provide value-added services such as multimedia applications using IMS (IP multimedia subsystem) and corporate VPN (virtual private networks) that run on top of IP. This separation between NAP, NSP, and ASP is designed to enable a richer ecosystem for WiMAX service business, leading to more competition and hence better services.

The network reference model developed by the WiMAX Forum NWG defines a number of functional entities and interfaces between those entities. (The interfaces are referred to as reference points.) Figure 2.3 shows some of the more important functional entities.

Base station (BS): The BS is responsible for providing the air interface to the MS. Additional functions that may be part of the BS are micromobility management functions, such as handoff triggering and tunnel establishment, radio resource management, QoS policy enforcement, traffic classification, DHCP (Dynamic Host Control Protocol) proxy, key management, session management, and multicast group management.

Access service network gateway (ASN-GW): The ASN gateway typically acts as a layer 2 traffic aggregation point within an ASN. Additional functions that may be part of the ASN gateway include intra-ASN location management and paging, radio resource management and admission control, caching of subscriber profiles and encryption keys, AAA client functionality, establishment and management of mobility tunnel with base stations, QoS and policy enforcement, foreign agent functionality for mobile IP, and routing to the selected CSN.

Connectivity service network (CSN): The CSN provides connectivity to the Internet, ASP, other public networks, and corporate networks. The CSN is owned by the NSP and includes

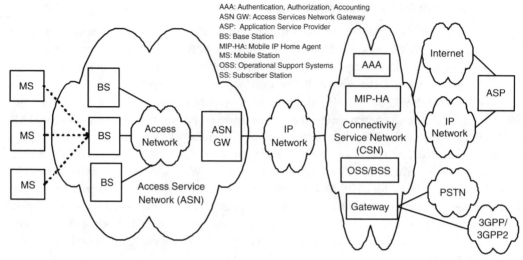

Figure 2.3 IP-Based WiMAX Network Architecture

AAA servers that support authentication for the devices, users, and specific services. The CSN also provides per user policy management of QoS and security. The CSN is also responsible for IP address management, support for roaming between different NSPs, location management between ASNs, and mobility and roaming between ASNs. Further, CSN can also provide gateways and interworking with other networks, such as PSTN (public switched telephone network), 3GPP, and 3GPP2.

The WiMAX architecture framework allows for the flexible decomposition and/or combination of functional entities when building the physical entities. For example, the ASN may be decomposed into base station transceivers (BST), base station controllers (BSC), and an ASN-GW analogous to the GSM model of BTS, BSC, and Serving GPRS Support Node (SGSN). It is also possible to collapse the BS and ASN-GW into a single unit, which could be thought of as a WiMAX router. Such a design is often referred to as a distributed, or flat, architecture. By not mandating a single physical ASN or CSN topology, the reference architecture allows for vendor/operator differentiation.

In addition to functional entities, the reference architecture defines interfaces, called *reference points*, between function entities. The interfaces carry control and management protocols—mostly IETF-developed network and transport-layer protocols—in support of several functions, such as mobility, security, and QoS, in addition to bearer data. Figure 2.4 shows an example.

The WiMAX network reference model defines reference points between: (1) MS and the ASN, called R1, which in addition to the air interface includes protocols in the management plane, (2) MS and CSN, called R2, which provides authentication, service authorization, IP configuration, and mobility management, (3) ASN and CSN, called R3, to support policy enforcement and mobility management, (4) ASN and ASN, called R4, to support inter-ASN mobility, (5) CSN and CSN, called R5, to support roaming across multiple NSPs, (6) BS and ASN-GW,

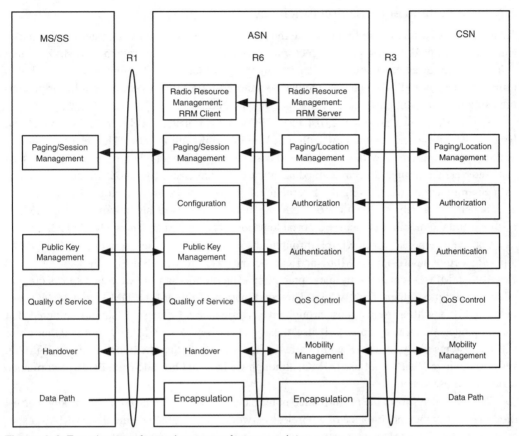

Figure 2.4 Functions performed across reference points

called R6, which consists of intra-ASN bearer paths and IP tunnels for mobility events, and (7) BS to BS, called R7, to facilitate fast, seamless handover.

A more detailed description of the WiMAX network architecture is provided in Chapter 10.

2.7 Performance Characterization

So far in this chapter, we have provided an overview description of the WiMAX broadband wireless standard, focusing on the various features, functions, and protocols. We now briefly turn to the system performance of WiMAX networks. As discussed in Chapter 1, a number of trade-offs are involved in designing a wireless system, and WiMAX offers a broad and flexible set of design choices that can be used to optimize the system for the desired service requirements. In this section, we present only a brief summary of the throughput performance and coverage range of WiMAX for a few specific deployment scenarios. Chapters 11 and 12 explore the link-and system-level performance of WiMAX is greater detail.

2.7.1 Throughput and Spectral Efficiency

Table 2.7 shows a small sampling of some the results of a simulation-based system performance study we performed. It shows the per sector average throughput achievable in a WiMAX system using a variety of antenna configurations: from an open-loop MIMO antenna system with two transmit antennas and two receiver antennas to a closed-loop MIMO system with linear precoding using four transmit antennas and two receive antennas.

The results shown are for a 1,024 FFT OFDMA-PHY using a 10MHz TDD channel and band AMC subcarrier permutation with a 1:3 uplink-to-downlink ratio. The results assume a multicellular deployment with three sectored base stations using a $(1,1)$[13] frequency reuse. This is an interference-limited design, with adjacent base stations assumed to be 2 km apart. A multipath environment modeled using the International Telecommunications Union (ITU) pedestrian B channel[14] is assumed. Results for both the fixed case where an indoor desktop CPE is assumed and the mobile case where a portable handset is assumed are shown in Table 2.7.

The average per sector downlink throughput for the baseline case—assuming a fixed desktop CPE deployment—is 16.3Mbps and can be increased to over 35Mbps by using a 4×2 closed-loop MIMO scheme with linear precoding. The mobile-handset case also shows comparable performance, albeit slightly less. The combination of OFDM, OFDMA, and MIMO provides WiMAX with a tremendous throughput performance advantage. It should be noted that early mobile WiMAX systems will use mostly open-loop 2×2 MIMO, with higher-order MIMO systems likely to follow within a few years. Also note that there may be fixed WiMAX systems deployed that do not use MIMO, although we have not provided simulated performance results for those systems.

Table 2.7 also shows the performance in terms of spectral efficiency, one of the key metrics used to quantify the performance of a wireless network. The results indicate that WiMAX, especially with MIMO implementations, can achieve significantly higher spectral efficiencies than what is offered by current 3G systems, such as HSDPA and 1xEV-DO.

It should be noted, however, that the high spectral efficiency obtained through the use of $(1,1,)$ frequency reuse does entail an increased outage probability. As discussed in Chapter 12, the outage can be higher than 10 percent in many cases unless a 4×2 closed-loop MIMO scheme is used.

2.7.2 Sample Link Budgets and Coverage Range

Table 2.8 shows a sample link budget for a WiMAX system for two deployment scenarios. In the first scenario, the mobile WiMAX case, service is provided to a portable mobile handset located outdoors; in the second case, service is provided to a fixed desktop subscriber station placed indoors. The fixed desktop subscriber is assumed to have a switched directional antenna that provides 6 dBi gain. For both cases, MIMO spatial multiplexing is not assumed; only diversity

---

13. This implies that frequencies are reused in every sector.
14. See Section 12.1 for more details.

Table 2.7 Throughput and Spectral Efficiency of WiMAX

Parameter			Antenna Configuration			
			2 × 2 Open-Loop MIMO	2 × 4 Open-Loop MIMO	4 × 2 Open-Loop MIMO	4 × 2 Closed-Loop MIMO
Per sector average throughput (Mbps) in a 10MHz channel	Fixed indoor desktop CPE	DL	16.31	27.25	23.25	35.11
		UL	2.62	2.50	3.74	5.64
	Mobile handset	DL	14.61	26.31	22.25	34.11
		UL	2.34	2.34	3.58	5.48
Spectral efficiency (bps/Hertz)	Fixed indoor desktop CPE	DL	2.17	3.63	3.10	4.68
		UL	1.05	1.00	1.50	2.26
	Mobile handset	DL	1.95	3.51	2.97	4.55
		UL	0.94	0.94	1.43	2.19

reception and transmission are assumed at the base station. The numbers shown are therefore for a basic WiMAX system.

The link budget assumes a QPSK rate 1/2 modulation and coding operating at a 10 percent block error rate (BLER) for subscribers at the edge of the cell. This corresponds to a cell edge physical-layer throughput of about 150kbps in the downlink and 35kbps on the uplink, assuming a 3:1 downlink-to-uplink ratio. Table 2.8 shows that the system offers a link margin in excess of 140 dB at this data rate. Assuming 2,300MHz carrier frequency, a base station antenna height of 30 m, and a mobile station height of 1 m, this translates to a coverage range of about 1 km using the COST-231 Hata model discussed in Chapter 12. Table 2.8 shows results for both the urban and suburban models. The pathloss for the urban model is 3 dB higher than for the suburban model.

2.8 Summary and Conclusions

This chapter presented an overview of WiMAX and set the stage for more detailed exploration in subsequent chapters.

- WiMAX is based on a very flexible and robust air interface defined by the IEEE 802.16 group.
- The WiMAX physical layer is based on OFDM, which is an elegant and effective technique for overcoming multipath distortion.
- The physical layer supports several advanced techniques for increasing the reliability of the link layer. These techniques include powerful error correction coding, including turbo coding and LDPC, hybrid-ARQ, and antenna arrays.

Table 2.8 Sample Link Budgets for a WiMAX System

Parameter	Mobile Handheld in Outdoor Scenario		Fixed Desktop in Indoor Scenario		Notes
	Downlink	Uplink	Downlink	Uplink	
Power amplifier output power	43.0 dB	27.0 dB	43.0 dB	27.0 dB	A1
Number of tx antennas	2.0	1.0	2.0	1.0	A2
Power amplifier backoff	0 dB	0 dB	0 dB	0 dB	A3; assumes that amplifier has sufficient linearity for QPSK operation without backoff
Transmit antenna gain	18 dBi	0 dBi	18 dBi	6 dBi	A4; assumes 6 dBi antenna for desktop SS
Transmitter losses	3.0 dB	0 dB	3.0 dB	0 dB	A5
Effective isotropic radiated power	61 dBm	27 dBm	61 dBm	33 dBm	$A6 = A1 + 10\log_{10}(A2) - A3 + A4 - A5$
Channel bandwidth	10MHz	10MHz	10MHz	10MHz	A7
Number of subchannels	16	16	16	16	A8
Receiver noise level	−104 dBm	−104 dBm	−104 dBm	−104 dBm	$A9 = -174 + 10\log_{10}(A7*1e6)$
Receiver noise figure	8 dB	4 dB	8 dB	4 dB	A10
Required SNR	0.8 dB	1.8 dB	0.8 dB	1.8 dB	A11; for QPSK, R1/2 at 10% BLER in ITU Ped. B channel
Macro diversity gain	0 dB	0 dB	0 dB	0 dB	A12; No macro diversity assumed
Subchannelization gain	0 dB	12 dB	0 dB	12 dB	$A13 = 10\log_{10}(A8)$
Data rate per subchannel (kbps)	151.2	34.6	151.2	34.6	A14; using QPSK, R1/2 at 10% BLER
Receiver sensitivity (dBm)	−95.2	−110.2	−95.2	−110.2	$A15 = A9 + A10 + A11 + A12 - A13$
Receiver antenna gain	0 dBi	18 dBi	6 dBi	18 dBi	A16
System gain	156.2 dB	155.2 dB	162.2 dB	161.2 dB	$A17 = A6 - A15 + A16$
Shadow-fade margin	10 dB	10 dB	10 dB	10 dB	A18
Building penetration loss	0 dB	0 dB	10 dB	10 dB	A19; assumes single wall
Link margin	146.2 dB	145.2 dB	142.2 dB	141.2 dB	$A20 = A17 - A18 - A19$
Coverage range	1.06 km (0.66 miles)		0.81 km (0.51 miles)		Assuming COST-231 Hata urban model
Coverage range	1.29 km (0.80 miles)		0.99 km (0.62 miles)		Assuming the suburban model

- WiMAX supports a number of advanced signal-processing techniques to improve overall system capacity. These techniques include adaptive modulation and coding, spatial multiplexing, and multiuser diversity.
- WiMAX has a very flexible MAC layer that can accommodate a variety of traffic types, including voice, video, and multimedia, and provide strong QoS.
- Robust security functions, such as strong encryption and mutual authentication, are built into the WiMAX standard.
- WiMAX has several features to enhance mobility-related functions such as seamless handover and low power consumption for portable devices.
- WiMAX defines a flexible all-IP-based network architecture that allows for the exploitation of all the benefits of IP. The reference network model calls for the use of IP-based protocols to deliver end-to-end functions, such as QoS, security, and mobility management.
- WiMAX offers very high spectral efficiency, particularly when using higher-order MIMO solutions.

2.9 Bibliography

[1] IEEE. Standard 802.16-2004. Part16: Air interface for fixed broadband wireless access systems. October 2004.

[2] IEEE. Standard 802.16e-2005. Part16: Air interface for fixed and mobile broadband wireless access systems—Amendment for physical and medium access control layers for combined fixed and mobile operation in licensed band. December 2005.

[3] WiMAX Forum. Mobile WiMAX—Part I: A technical overview and performance evaluation. White Paper. March 2006. www.wimaxforum.org.

[4] WiMAX Forum. Mobile WiMAX—Part II: A comparative analysis. White Paper. April 2006. www.wimaxforum.org.

[5] WiMAX Forum. WiMAX Forum Mobile System Profile. 2006–07.

PART II

Technical Foundations of WiMAX

The Challenge of Broadband Wireless Channels

A chieving high data rates in terrestrial wireless communication is difficult. High data rates for wireless local area networks, namely the IEEE 802.11 family of standards, became commercially successful only around 2000. Wide area wireless networks, namely cellular systems, are still designed and used primarily for low-rate voice services. Despite many promising technologies, the reality of a wide area network that services many users at high data rates with reasonable bandwidth and power consumption, while maintaining high coverage and quality of service, has not yet been achieved.

The goal of the IEEE 802.16 committee was to design a wireless communication system that incorporates the most promising new technologies in communications and digital signal processing to achieve a broadband Internet experience for nomadic or mobile users over a wide or metropolitan area. It is important to realize that WiMAX systems have to confront similar challenges as existing cellular systems, and their eventual performance will be bounded by the same laws of physics and information theory.

In this chapter, we explain the immense challenge presented by a time-varying broadband wireless channel. We quantify the principle effects in broadband wireless channels and present practical statistical models. We conclude with an overview of diversity countermeasures that can be used to maintain robust communication in these challenging conditions. With these diversity techniques, it is even possible in many cases to take advantage of what were originally viewed as impediments. The rest of Part II of the book focuses on the technologies that have been developed by many sources—in some cases, very recently—and adopted in WiMAX to achieve robust high data rates in such channels.

3.1 Communication System Building Blocks

All wireless digital communication systems must possess a few key building blocks, as shown in Figure 3.1. Even in a reasonably complicated wireless network, the entire system can be broken down into a collection of *links*, each consisting of a transmitter, a channel, and a receiver.

The transmitter receives packets of bits from a higher protocol layer and sends those bits as electromagnetic waves toward the receiver. The key steps in the digital domain are encoding and modulation. The encoder generally adds redundancy that will allow error correction at the receiver. The modulator prepares the digital signal for the wireless channel and may comprise a number of operations. The modulated digital signal is converted into a representative analog waveform by a digital-to-analog convertor (DAC) and then upconverted to one of the desired WiMAX radio frequency (RF) bands. This RF signal is then radiated as electromagnetic waves by a suitable antenna.

The receiver performs essentially the reverse of these operations. After downconverting the received RF signal and filtering out signals at other frequencies, the resulting baseband signal is converted to a digital signal by an analog-to-digital convertor (ADC). This digital signal can then be demodulated and decoded with energy and space-efficient integrated circuits to, ideally, reproduce the original bit stream.

Naturally, the devil is in the details. As we will see, the designer of a digital communication system has an endless number of choices. It is important to note that the IEEE 802.16 standard and WiMAX focus almost exclusively on the *digital* aspects of wireless communication, in particular at the transmitter side. The receiver implementation is unspecified; each equipment manufacturer is welcome to develop efficient proprietary receiver algorithms. Aside from agreeing on a carrier frequency and transmit spectrum mask, few requirements are placed on the RF units. The standard is interested primarily in the digital transmitter because the receiver must understand what the transmitter did in order to make sense of the received signal—but not vice versa.

Next, we describe the large-scale characteristics of broadband wireless channels and see why they present such a large design challenge.

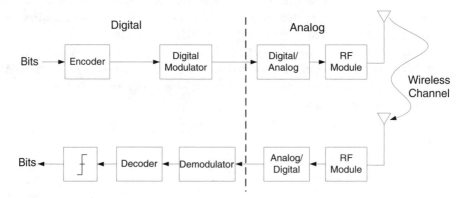

Figure 3.1 Wireless digital communication system

3.2 The Broadband Wireless Channel: Pathloss and Shadowing

The main goal of this chapter is to explain the fundamental factors affecting the received signal in a wireless system and how they can be modeled using a handful of parameters. The relative values of these parameters, which are summarized in Table 3.1 and described throughout this section, make all the difference when designing a wireless communication system. In this section, we introduce the overall channel model and discuss the *large-scale* trends that affect this model.

The overall model we use for describing the channel in discrete time is a simple tap-delay line (TDL):

$$h[k,t] = h_0\delta[k,t] + h_1\delta[k-1,t] + \ldots + h_v\delta[k-v,t]. \tag{3.1}$$

Here, the discrete-time channel is time varying—so it changes with respect to t—and has non-negligible values over a span of $v+1$ channel taps. Generally, we assume that the channel is sampled at a frequency $f_s = 1/T$, where T is the symbol period,[1] and that hence, the duration of the channel in this case is about vT. The $v+1$ sampled values are in general complex numbers.

Assuming that the channel is static over a period of $(v+1)T$ seconds, we can then describe the output of the channel as

$$y[k,t] \quad = \quad \sum_{j=-\infty}^{\infty} h[j,t]x[k-j] \tag{3.2}$$

$$\triangleq \quad h[k,t] * x[k], \tag{3.3}$$

where $x[k]$ is an input sequence of data symbols with rate $1/T$, and $*$ denotes convolution. In simpler notation, the channel can be represented as a time-varying $(v+1) \times 1$ column vector:[2]

$$\mathbf{h}(t) = [\, h_0(t) \;\; h_1(t) \ldots \;\; h_v(t)]^T. \tag{3.4}$$

Although this tapped-delay-line model is general and accurate, it is difficult to design a communication system for the channel without knowing some of the key attributes about $\mathbf{h}(t)$. Some likely questions one might have follow.

- What is the value for the total received power? In other words, what are the relative values of the h_i terms?

  Answer: As we will see, a number of effects cause the received power to vary over long (path loss), medium (shadowing), and short (fading) distances.

---

1. The symbol period T is the amount of time over which a single data symbol is transmitted. Hence, the data rate in a digital transmission system is directly proportional to $1/T$.

2. $(\cdot)^T$ denotes the standard transpose operation.

- How quickly does the channel change with the parameter t?

   Answer: The *channel-coherence time* specifies the period of time over which the channel's value is correlated. The coherence time depends on how quickly the transmitter and the receiver are moving relative to each other.

- What is the approximate value of the channel duration v?

   Answer: This value is known as the *delay spread* and is measured or approximated based on the propagation distance and environment.

The rest of the chapter explores these questions more deeply in an effort to characterize and explain these key wireless channel parameters, which are given in Table 3.1.

3.2.1 Pathloss

The first obvious difference between wired and wireless channels is the amount of transmitted power that reaches the receiver. Assuming that an isotropic antenna is used, as shown in Figure 3.2, the propagated signal energy expands over a spherical wavefront, so the energy received at an antenna distance d away is inversely proportional to the sphere surface area, $4\pi d^2$. The *free-space pathloss formula*, or Friis formula, is given more precisely as

$$P_r = P_t \frac{\lambda^2 G_t G_r}{(4\pi d)^2},$$ (3.5)

where P_r and P_t are the received and transmitted powers, and λ is the wavelength. In the context of the TDL model of Equation (3.1), P_r/P_t is the average value of the channel gain, that is, $P_r/P_t = E \parallel \mathbf{h} \parallel^2$, where $E[\cdot]$ denotes the expected value, or mathematical mean. If directional antennas are used at the transmitter or the receiver, a gain of G_t and/or G_r is achieved, and the received power is simply increased by the gain of these antennae.[3] An important observation from Equation (3.5) is that since $c = f_c \lambda \Rightarrow \lambda = c/f_c$, the received power fall offs quadratically with the carrier frequency. In other words, for a given transmit power, the range is decreased when higher-frequency waves are used. This has important implications for high-data-rate systems, since most large bandwidths are available at higher frequencies (see Sidebar 3.1).

   The terrestrial propagation environment is not free space. Intuitively, it seems that reflections from the earth or other objects would increase the received power since more energy would reach the receiver. However, because a reflected wave often experiences a 180° phase shift, the reflection at relatively large distances (usually over a kilometer) serves to create destructive interference, and the common *two-ray approximation* for pathloss is

$$P_r = P_t \frac{G_t G_r h_t^2 h_r^2}{d^4},$$ (3.6)

---

3. For an ideal isotropic radiator, $G_t = G_r = 1$.

Table 3.1 Key Wireless Channel Parameters

Symbol	Parameter
α	Pathloss exponent
σ_s	Lognormal shadowing standard deviation
f_D	Doppler spread (maximum Doppler frequency), $f_D = \dfrac{v f_c}{c}$
T_c	Channel coherence time, $T_c \approx f_D^{-1}$
τ_{max}	Channel delay spread (maximum)
τ_{RMS}	Channel delay spread (RMS)[a]
B_c	Channel coherence bandwidth, $B_c \approx \tau^{-1}$
θ_{RMS}	Angular spread (RMS)

a. Root mean square.

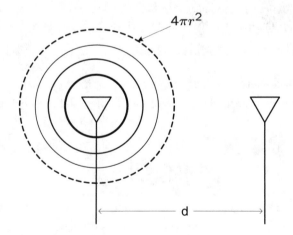

Figure 3.2 Free-space propagation

which is significantly different from free-space path loss in several respects. First, the antenna heights now assume a very important role in the propagation, as is anecdotally familiar: Radio transmitters are usually placed on the highest available object. Second, the wavelength and hence carrier frequency dependence has disappeared from the formula, which is not typically observed in practice, however. Third, and crucially, the distance dependence has changed to d^{-4}, implying that energy loss is more severe with distance in a terrestrial system than in free space.

Sidebar 3.1 Range versus Bandwidth

As noted in Chapter 1, much of the globally available bandwidth is at carrier frequencies of several GHz. Lower carrier frequencies are generally considered more desirable, and frequencies below 1GHz are often referred to as "beachfront" spectrum. The reasons for this historically have been twofold. First, high-frequency RF electronics have traditionally been more difficult to design and manufacture and hence more expensive. However, this issue is not as prominent presently, owing to advances in RF integrated circuit design. Second, as easily seen in Equation (3.5), the pathloss increases as f_c^2. A signal at 3.5GHz—one of WiMAX's candidate frequencies—will be received with about 20 times less power than at 800MHz, a popular cellular frequency. In fact, measurement campaigns have consistently shown that the effective pathloss exponent α also increases at higher frequencies, owing increased absorption and attenuation of high-frequency signals [17, 20, 21, 34].

This means that there is a direct conflict between range and bandwidth. The bandwidth at higher carrier frequencies is more plentiful and less expensive but, as we have noted, does not support large transmission ranges. Since it is crucial for WiMAX systems to have large bandwidths compared to cellular systems, at a much smaller cost per unit of bandwidth, there does not appear to be a credible alternative to accepting fairly short transmission ranges. In summary, it appears that WiMAX systems can have only two of the following three generally desirable characteristics: high data rate, high range, low cost.

In order to more accurately describe various propagation environments, empirical models are often developed using experimental data. One of the simplest and most common is the *empirical path loss formula*:

$$P_r = P_t P_o \left(\frac{d_o}{d} \right)^{\alpha},$$ (3.7)

which groups all the various effects into two parameters: the pathloss exponent α and the measured pathloss P_o at a reference distance of d_o, which is often chosen as 1 meter. Although P_o should be determined from measurements, it is often well approximated, within several dB, as simply $(4\pi/\lambda)^2$ when $d_o = 1$. This simple empirical pathloss formula is capable of reasonably representing most of the important pathloss trends with only these two parameters, at least over some range of interest (see Sidebar 3.2).

More accurate pathloss models have also been developed, including the well-known Okamura models [19], which also have a frequency-driven trend. Pathloss models that are especially relevant to WiMAX are discussed in more detail in Chapter 12.

> **Sidebar 3.2 Large PathLoss and Increased Capacity**
>
> Before continuing, it should be noted that, somewhat counterintuitively, severe pathloss environments often are desirable in a multiuser wireless network, such as WiMAX. Why? Since many users are attempting to simultaneously access the network, both the uplink and the downlink generally become *interference limited*, which means that increasing the transmit power of all users at once will not increase the overall network throughput. Instead, a lower interference level is preferable. In a cellular system with base stations, most of the interfering transmitters are farther away than the desired transmitter. Thus, their interference power will be attenuated more severely by a large path loss exponent than the desired signal. As noted in the Sidebar 3.1, a large pathloss exponent can be caused in part by a higher carrier frequency. Example 3.1 will be instructive.

Example 3.1

Consider a user in the downlink of a cellular system, where the desired base station is at a distance of 500 meters, and numerous nearby interfering base stations are transmitting at the same power level. If three interfering base stations are at a distance of 1 km, three at a distance of 2 km, and ten at a distance of 4 km, use the empirical pathloss formula to find the signal-to-interference ratio (SIR)—the noise is neglected—when $\alpha = 3$ and when $\alpha = 5$.

Solution

For $\alpha = 3$, the desired received power is

$$P_{r,d} = P_t P_o d_o^3 (0.5)^{-3}, \tag{3.8}$$

and the interference power is

$$P_{r,I} = P_t P_o d_o^3 \left[3(1)^{-3} + 3(2)^{-3} + 10(4)^{-3} \right]. \tag{3.9}$$

The SIR expressions compute to

$$SIR(\alpha = 3) = \frac{P_{r,d}}{P_{r,I}} = 28.25 = 14.5 dB,$$

$$SIR(\alpha = 5) = 99.3 = 20 dB,$$

demonstrating that the overall system performance can be substantially improved when the pathloss is in fact large. These calculations can be viewed as an upper bound, where the SINR $\gamma < SIR$, owing to the addition of noise. This means that as the pathloss worsens, microcells grow increasingly attractive, since the required signal power can be decreased down to the noise floor, and the overall performance will be better than in a system with lower pathloss at the same transmit-power level.

3.2.2 Shadowing

As we have seen, pathloss models attempt to account for the distance-dependent relationship between transmitted and received power. However, many factors other than distance can have a large effect on the total received power. For example, trees and buildings may be located between the transmitter and the receiver and cause temporary degradation in received signal strength; on the other hand, a temporary line-of-sight transmission path would result in abnormally high received power as shown in Figure 3.3. Since modeling the locations of all objects in every possible communication environment is generally impossible, the standard method of accounting for these variations in signal strength is to introduce a random effect called *shadowing*. With shadowing, the empirical pathloss formula becomes

$$P_r = P_t P_o \chi \left(\frac{d_o}{d} \right)^\alpha, \tag{3.10}$$

where χ is a sample of the *shadowing* random process. Hence, the received power is now also modeled as a random process. In effect, the distance trend in the pathloss can be thought of as the mean, or expected, received power, whereas the χ shadowing value causes a perturbation from that expected value. It should be emphasized that since shadowing is caused by macroscopic objects, it typically has a correlation distance on the order of meters or tens of meters. Hence, shadowing is often alternatively called large-scale fading.

The shadowing value χ is typically modeled as a lognormal random variable, that is,

$$\chi = 10^{x/10}, \text{where } x \sim N(0, \sigma_s^2), \tag{3.11}$$

where $N(0, \sigma_s^2)$ is a Gaussian (normal) distribution with mean 0 and variance σ_s^2. With this formulation, the standard deviation σ_s is expressed in dB. Typical values for σ_s are in the 6–12 dB range. Figure 3.4 shows the very important effect of shadowing, where $\sigma_s = 11.8$ dB and $\sigma_s = 8.9$ dB, respectively.

Shadowing is an important effect in wireless networks because it causes the received SINR to vary dramatically over long time scales. In some locations in a given cell, reliable high-rate communication may be nearly impossible. The system design and base station deployment must account for lognormal shadowing through macrodiversity, variable transmit power, and/or simply accepting that some users will experience poor performance at a certain percentage of locations (see Sidebar 3.3). Although shadowing can sometimes be beneficial—for example, if an object is blocking interference—it is generally detrimental to system performance because it requires a several-dB margin to be built into the system. Let's do a realistic numerical example to see how shadowing affects wireless system design.

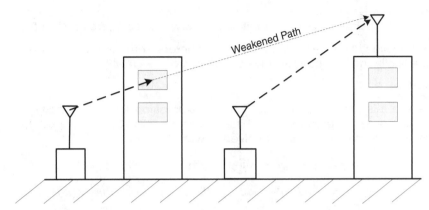

Figure 3.3 Shadowing can cause longer deviations from pathloss predictions.

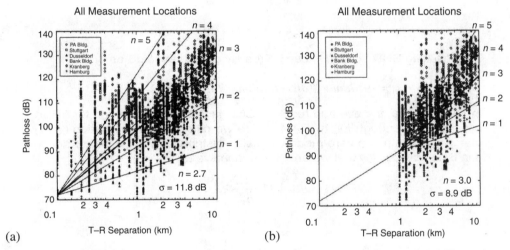

Figure 3.4 Shadowing causes large random fluctuations about the pathloss model. Figure from [28], courtesy of IEEE.

Example 3.2

Consider a WiMAX base station communicating to a subscriber; the channel parameters are $\alpha = 3$, $P_o = -40$ dB, and $d_0 = 1$ m, and $\sigma_s = 6$ dB. We assume a transmit power of $P_t = 1$ watt (30 dBm) and a bandwidth of $B = 10$ MHz. Owing to rate $1/2$ convolutional codes, a received SNR of 14.7 dB is required for 16 QAM, but just 3 dB is required for BPSK.[4] Finally, we

---

4. These values are both 3 dB from the Shannon limit.

consider only ambient noise, with a typical power-spectral density of $N_o = -173$ dBm/Hz, with an additional receiver-noise figure of $N_f = 5$ dB.[5]

The question is this: At a distance of 500 meters from the base station, what is the likelihood that can reliably send BPSK or 16 QAM?

Solution

To solve this problem, we must find an expression for received SNR, and then compute the probability that it is above the BPSK and 16 QAM thresholds. First, let's compute the received power, P_r in dB:

$$P_r(dB) \quad = \quad 10\log_{10}P_t + 10\log_{10}P_o - 10\log_{10}d^\alpha + 10\log_{10}\chi \tag{3.12}$$

$$= \quad 30dBm - 40dB - 81dB + \chi(dB) = -91dBm + \chi(dB) \tag{3.13}$$

Next, we can compute the total noise/interference power I_{tot} in dB similarly:

$$I_{tot}(dB) \quad = \quad N_o + N_f + 10\log_{10}B \tag{3.14}$$

$$= \quad -173 + 5dB + 70 = -98dBm \tag{3.15}$$

The resulting SNR $\gamma = P_r/I_{tot}$ can be readily computed in dB as

$$\gamma = -91dBm + \chi(dB) + 98dBm = 7dB + \chi(dB). \tag{3.16}$$

In this scenario, the average received SNR is 7 dB, good enough for BPSK but not good enough for 16 QAM. Since we can see from Equation (3.11) that $\chi(dB) = x$ has a zero mean Gaussian distribution with standard deviation 6, the probability that we are able to achieve BPSK is

$$P[\gamma \geq 3dB] \quad = P[\frac{\chi + 7}{\sigma_s} \geq \frac{3}{\sigma}] \tag{3.17}$$

$$= P[\frac{\chi}{6} \geq -\frac{4}{6}] \tag{3.18}$$

$$= Q(-\frac{4}{6}) = 0.75. \tag{3.19}$$

And similarly for QPSK:

$$P[\gamma \geq 14.7dB] = P[\frac{\chi + 7}{\sigma_s} \geq \frac{14.7}{\sigma_s}] \tag{3.20}$$

$$= Q(\frac{7.7}{6}) = .007. \tag{3.21}$$

---

5. The total additional noise from all sources can be considered to be 5 dB

To summarize the example: Although 75 percent of users can use BPSK modulation and hence get a PHY data rate of 10 MHz • 1 bit/symbol • 1/2 = 5 Mbps, less than 1 percent of users can reliably use 16 QAM (4 bits/symbol) for a more desirable data rate of 20Mbps. Additionally, whereas without shadowing, *all* the users could at least get low-rate BPSK through, with shadowing, 25 percent of the users appear unable to communicate at all. Interestingly, though, without shadowing, 16 QAM could never be sent; with shadowing, it can be sent a small fraction of the time. Subsequent chapters describe adaptive modulation and coding, alluded to here, in more detail and also show how other advanced techniques may be used to further increase the possible data rates in WiMAX.

Sidebar 3.3 Why is the shadowing lognormal?

Although the primary rationale for the lognormal distribution for the shadowing value χ is accumulated evidence from channel-measurement campaigns, one plausible explanation is as follows. Neglecting the pathloss for a moment, if a transmission experiences N random attenuations β_i, $i = 1, 2,..., N$ between the transmitter and receiver, the received power can be modeled as

$$P_r = P_t \prod_{i=1}^{N} \beta_i \qquad (3.22)$$

which can be expressed in dB as

$$P_r(dB) = P_t(dB) + 10 \sum_{i=1}^{N} \log_{10} \beta_i \qquad (3.23)$$

Then, using the Central Limit Theorem, it can be argued that the sum term will become Gaussian as N becomes large—and often the CLT is accurate for fairly small N—and since the expression is in dB, the shadowing is hence lognormal.

3.3 Cellular Systems

As explained in Section 3.2, owing to pathloss and, to a lesser extent, shadowing, given a maximum allowable transmit power, it is possible to reliably communicate only over some limited distance. However, we saw in Sidebar 3.2 that pathloss allows for spatial isolation of different transmitters operating on the same frequency at the same time. As a result, pathloss and short-range transmissions in fact *increase* the overall capacity of the system by allowing more simultaneous transmissions to occur. This straightforward observation is the theoretical basis for the ubiquity of modern cellular communication systems.

In this section, we briefly explore the key aspects of cellular systems and the closely related topics of sectoring and frequency reuse. Since WiMAX systems are expected to be deployed primarily in a cellular architecture, the concepts presented here are fundamental to understanding WiMAX system design and performance.

3.3.1 The Cellular Concept

In cellular systems, the service area is subdivided into smaller geographic areas called *cells*, each served by its own base station. In order to minimize interference between cells, the transmit-power level of each base station is regulated to be just enough to provide the required signal strength at the cell boundaries. Then, as we have seen, propagation pathloss allows for spatial isolation of different cells operating on the same frequency channels at the same time. Therefore, the same frequency channels can be reassigned to different cells, as long as those cells are spatially isolated.

Although perfect spatial isolation of different cells cannot be achieved, the rate at which frequencies can be reused should be determined such that the interference between base stations is kept to an acceptable level. In this context, *frequency planning* is required to determine a proper frequency-reuse factor and a geographic-reuse pattern. The frequency-reuse factor f is defined as $f \leq 1$, where $f = 1$ means that all cells reuse all the frequencies. Accordingly, $f = 1/3$ implies that a given frequency band is used by only one of every three cells.

The reuse of the same frequency channels should be intelligently planned in order to maximize the geographic distance between the cochannel base stations. Figure 3.5 shows a hexagonal cellular system model with frequency-reuse factor $f = 1/7$, where cells labeled with the same letter use the same frequency channels. In this model, a cluster is outlined in boldface and consists of seven cells with different frequency channels. Even though the hexagonal cell shape is conceptual, it has been widely used in the analysis of a cellular system, owing to its simplicity and analytical convenience.

Cellular systems allow the overall system capacity to increase by simply making the cells smaller and turning down the power. In this manner, cellular systems have a very desirable scaling property: More capacity can be supplied by installing more base stations. As the cell size decreases, the transmit power of each base station decreases correspondingly. For example, if the radius of a cell is reduced by half when the propagation pathloss exponent is 4, the transmit-power level of a base station is reduced by 12 dB (= 10 log 16 dB).

Since cellular systems support user mobility, seamless call transfer from one cell to another should be provided. The handoff process provides a means of the seamless transfer of a connection from one base station to another. Achieving smooth handoffs is a challenging aspect of cellular system design.

Although small cells give a large capacity advantage and reduce power consumption, their primary drawbacks are the need for more base stations—and their associated hardware costs—and the need for frequent handoffs. The offered traffic in each cell also becomes more variable as the cell shrinks, resulting in inefficiency. As in most aspects of wireless systems, an appropri-

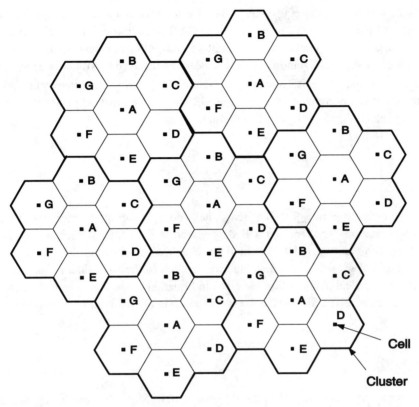

Figure 3.5 Standard figure of a hexagonal cellular system with $f = 1/7$

ate trade-off between these competing factors needs to be determined, depending on the system requirements.

3.3.2 Analysis of Cellular Systems

The performance of wireless cellular systems is significantly limited by cochannel interference (CCI), which comes from other users in the same cell or from other cells. In cellular systems, other-cell interference (OCI) is a function of the radius of the cell (R) and the distances to the center of the neighboring cochannel cell but, interestingly, is independent of the transmitted power if the size of each cell is the same. The spatial isolation between cochannel cells can be measured by defining the parameter Q, called *cochannel-reuse ratio*, as the ratio of the distance to the center of the nearest cochannel cell (D) to the radius of the cell. In a hexagonal cell structure, the cochannel-reuse ratio is given by

$$Q = \frac{D}{R} = \sqrt{3N} \,,$$

(3.24)

where N is the size of a cluster equivalent to the inverse of the frequency-reuse factor. Obviously, a higher value of Q reduces cochannel interference so that it improves the quality of the communication link and capacity. However, the overall spectral efficiency decreases with the size of a cluster N; hence, N should be minimized only to keep the received SINR above acceptable levels.

Since the background-noise power is negligible compared to the interference power in an interference-limited environment, the received SIR can be used instead of SINR. If the number of interfering cells is N_I, the SIR for a mobile station can be given by

$$\frac{S}{I} = \frac{S}{\sum_{i=1}^{N_I} I_i},$$

(3.25)

where S is the received power of the desired signal, and I_i is the interference power from the ith cochannel base station. The received SIR depends on the location of each mobile station and should be kept above an appropriate threshold for reliable communication. The received SIR at the cell boundaries is of great interest, since this corresponds to the worst-interference scenario. For example, if the empirical pathloss formula given in Equation (3.10) and universal frequency reuse are considered, the received SIR for the worst case given in Figure 3.6 is expressed as

$$\frac{S}{I} = \frac{\chi_0}{\chi_0 + \sum_{i=1}^{2} \chi_i + 2^{-\alpha} \sum_{i=3}^{5} \chi_i + (2.633)^{-\alpha} \sum_{i=6}^{11} \chi_i},$$

(3.26)

where χ_i denotes the shadowing from the ith base station. Since the sum of lognormal random variables is well approximated by a lognormal random variable [10, 27], the denominator can be approximated as a lognormal random variable, and then the received SIR follows a lognormal distribution [5]. Therefore, the outage probability that the received SIR falls below a threshold can be derived from the distribution. If the mean and the standard deviation of the lognormal distribution are μ and σ_s in dB, the outage probability is derived in the form of Q function as

$$P_o = Q\left(\frac{\gamma - \mu}{\sigma_s}\right),$$

(3.27)

where γ is the threshold SIR level in dB. Usually, the SINR at the cell boundaries is too low to achieve the outage-probability design target if universal frequency reuse is adopted. Therefore, a lower frequency-reuse factor is typically adopted in the system design to satisfy the target outage probability at the sacrifice of spectral efficiency.

Figure 3.7 highlights the OCI problem in a cellular system if universal frequency reuse is adopted. The figure shows the regions of a cell in various SIR bins of the systems with universal frequency reuse and $f = 1/3$ frequency reuse. The figure is based on a two-tier cellular structure and the simple empirical pathloss model given in Equation (3.7) with $\alpha = 3.5$. The SIR in most parts of the cell is very low if universal frequency reuse is adopted. The OCI problem can be mit-

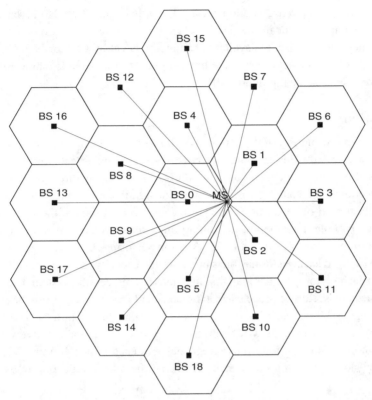

Figure 3.6 Forward-link interference in a hexagonal cellular system (worst case)

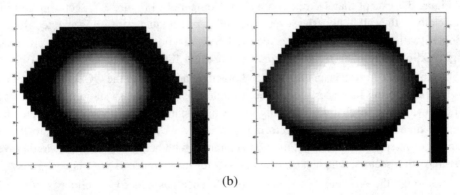

(a)   (b)

Figure 3.7 The received SIR in a cell with pathloss exponent $\alpha = 3.5$. The scale on the right indicates the SINR bins: Darker indicates lower SIR. (a) Universal frequency reuse, $f = 1$. (b) Frequency reuse, $f = 1/3$.

igated if higher frequency reuse is adopted, as shown in Figure 3.7b. However, as previously emphasized, this improvement in the quality of communication is achieved at the sacrifice of spectral efficiency: In this case, the available bandwidth is cut by a factor of 3. Frequency planning is a delicate balancing act of using the highest reuse factor possible while still having most of the cell have at least some minimum SIR.

3.3.3 Sectoring

Since the SIR is so low in most of the cell, it is desirable to find techniques to improve it without sacrificing so much bandwidth, as frequency reuse does. A popular technique is to sectorize the cells, which is effective if frequencies are reused in each cell. Using directional antennas instead of an omnidirectional antenna at the base station can significantly reduced the cochannel interference. An illustration of sectoring is shown in Figure 3.8. Although the absolute amount of bandwidth used is three times before (assuming three sector cells), the capacity increase is in fact more than three times. No capacity is lost from sectoring, because each sector can reuse time and code slots, so each sector has the same nominal capacity as an entire cell. Furthermore, the capacity in each sector is higher than that in a nonsectored cellular system, because the interference is reduced by sectoring, since users experience only interference from the sectors at their frequency. In Figure 3.8a, if each sector 1 points in the same direction in each cell, the interference caused by neighboring cells will be dramatically reduced. An alternative way to use sectors, not shown in Figure 3.8, is to reuse frequencies in each sector. In this case, all the time/code/frequency slots can be reused in each sector, but there is no reduction in the experienced interference.

Figure 3.9 shows the regions of a three-sector cell in various SIR bins of the systems with universal frequency reuse and 1/3 frequency reuse. All the configurations are the same as those of Figure 3.7 except that sectoring is added. Compared to Figure 3.7, sectoring improves SIR, especially at the cell boundaries, even when universal frequency reuse is adopted. If sectoring is adopted with frequency reuse, the received SIR can be significantly improved, as shown in Figure 3.9b, where both $f = 1/3$ frequency reuse and 120° sectoring are used.

Although sectoring is an effective and practical approach to the OCI problem, it is not without cost. Sectoring increases the number of antennas at each base station and reduces trunking efficiency, owing to channel sectoring at the base station. Even though intersector handoff is simpler than intercell handoff, sectoring also increases the overhead, owing to the increased number of intersector handoffs. Finally, in channels with heavy scattering, desired power can be lost into other sectors, which can cause inter-sector interference as well as power loss.

Although the problem of cochannel interference has existed in cellular systems for many years, its effect on future cellular systems, such as WiMAX, is likely to be far more severe, owing to the requirements for high data rate, high spectral efficiency, and the likely use of multiple antennas. This is a very tough combination [2, 6]. Recent research approaches to this difficult problem have focused on advanced signal-processing techniques at the receiver [1, 6] and the transmitter [15, 29, 35] as a means of reducing or canceling the perceived interference.

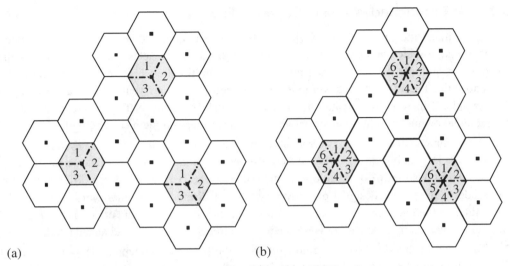

Figure 3.8 (a) Three-sector (120°) cells and (b) six-sector (60°) cells

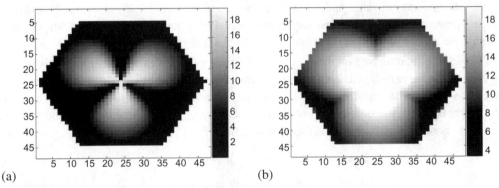

Figure 3.9 Received SINR in a sectorized cell (three sectors) with pathloss exponent = 3.5: (a) universal frequency reuse (1:1); (b) frequency reuse (1:3)

Although those techniques have important merits and are being actively researched and considered, they have some important shortcomings when viewed in a practical context of near-future cellular systems, such as WiMAX. As an alternative, network-level approaches, such as cooperative transmission [3, 37, 38, 39] and distributed antennas [4, 14, 23] can be considered. Those network-level approaches require relatively little channel knowledge and effectively reduce other-cell interference through macrodiversity, even though the gain may be smaller than that of advanced signal-processing techniques.

3.4   The Broadband Wireless Channel: Fading

One of the more intriguing aspects of wireless channels is fading. Unlike pathloss or shadowing, which are large-scale attenuation effects owing to distance or obstacles, fading is caused by the reception of multiple versions of the same signal. The multiple received versions are caused by reflections that are referred to as *multipath*. The reflections may arrive nearly simultaneously— for example, if there is local scattering around the receiver—or at relatively longer intervals— for example, owing to multiple paths between the transmitter and the receiver (Figure 3.10).

When some of the reflections arrive at nearly the same time, their combined effect is as in Figure 3.11. Depending on the phase difference between the arriving signals, the interference can be either constructive or destructive, which causes a very large observed difference in the amplitude of the received signal even over very short distances. In other words, moving the transmitter or the receiver even a very short distance can have a dramatic effect on the received amplitude, even though the pathloss and shadowing effects may not have changed at all.

To formalize this discussion, we now return to the time-varying tapped-delay-line channel model of Equation (3.1). As either the transmitter or the receiver moves relative to the other, the channel response $\mathbf{h}(t)$ will change. This channel response can be thought of as having two dimensions: a delay dimension τ and a time-dimension t, as shown in Figure 3.12. Since the channel changes over distance and hence time, the values of $h_0, h_1, \ldots, h_v$ may be totally different at time t versus time $t + \Delta t$. Because the channel is highly variant in both the τ and t dimensions, we must use statistical methods to discuss what the channel response is.[6]

The most important and fundamental function used to statistically describe broadband fading channels is the two-dimensional autocorrelation function, $A(\Delta \tau, \Delta t)$. Although it is over two dimensions and hence requires a three-dimensional plot, this autocorrelation function can usefully be thought of as two simpler functions, $A_t(\Delta t)$ and $A_\tau(\Delta \tau)$, where both $\Delta \tau$ and Δt have been set to zero. The autocorrelation function is defined as

$$
\begin{aligned}
A(\Delta \tau, \Delta t) &= E[h(\tau_1, t_1) h^*(\tau_2, t_2)] \\
&= E[h(\tau_1, t) h^*(\tau_2, t + \Delta t) \\
&= E[h(\tau, t) h^*(\tau + \Delta \tau, t +
\end{aligned}
\tag{3.28}
$$

where in the first step, we have assumed that the channel response is wide-sense stationary (WSS); (hence, the autocorrelation function depends only on $\Delta t = t_2 - t_1$). In the second step, we have assumed that the channel response of paths arriving at different times, τ_1 and τ_2, is uncorrelated. This allows the dependence on specific times τ_1 and τ_2 to be replaced simply by $\tau = \tau_1 - \tau_2$. Channels that can be described by the autocorrelation in Equation (3.28) are thus referred to as wide-sense stationary uncorrelated scattering (WSSUS), which is the most popular model for wideband fading channels and relatively accurate in many practical scenarios, largely because the scale of interest for τ (usually μ sec) and t (usually msec) generally differs by a few orders of magnitude.

---

6. Movement in the propagation environment will also cause the channel response to change over time.

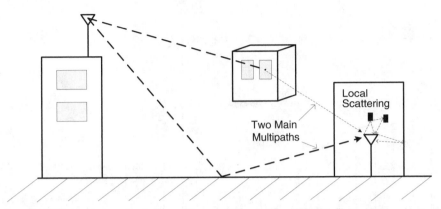

Figure 3.10 A channel with a few major paths of different lengths, with the receiver seeing a number of locally scattered versions of those paths

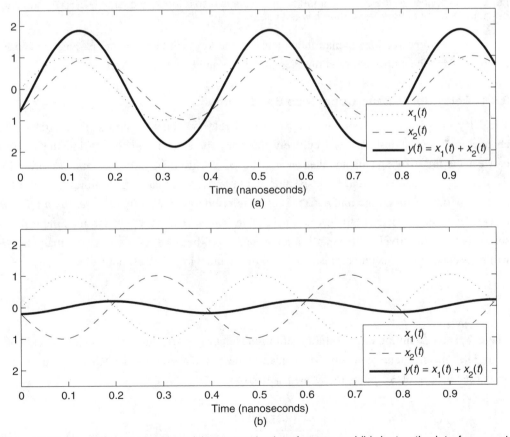

Figure 3.11 The difference between (a) constructive interference and (b) destructive interference at $f_c = 2.5$GHz is less than 0.1 nanoseconds in phase, which corresponds to about 3 cm.

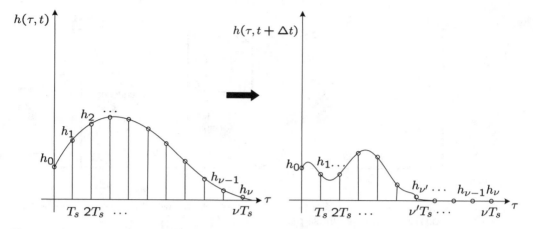

Figure 3.12 The delay τ corresponds to how *long* the channel impulse response lasts. The channel is time varying, so the channel impulse response is also a function of time— $h(\tau, t)$—and can be quite different at time $t + \Delta t$ than it was at time t.

The next three sections explain how many of the key wireless channel parameters can be estimated from the autocorrelation function $A(\Delta \tau, \Delta t)$ and how they are related.

3.4.1 Delay Spread and Coherence Bandwidth

The delay spread is a very important property of a wireless channel, specifing the duration of the channel impulse response $h(\tau, t)$. Intuitively, the delay spread is the amount of time that elapses between the first arriving path and the last arriving (non-negligible) path. As seen in Figure 3.13, the delay spread can be found by inspecting $A_\tau(\Delta \tau) \triangleq A_\tau(\Delta \tau)$, that is, by setting $\Delta t = 0$ in the channel autocorrelation function. $A_\tau(\Delta \tau)$ is often referred to as the *multipath intensity profile, or power-delay profile*. If $A_\tau(\Delta \tau)$ has non-negligible values from $(0, \tau_{max})$, the maximum delay spread is τ_{max}. Intuitively, this is an important definition because it specifies how many taps v will be needed in the discrete representation of the channel impulse response, since

$$v \approx \frac{\tau_{max}}{T_s},$$ (3.29)

where T_s is the sampling time. But this definition is not rigorous, since it is not clear what "non-negligible" means mathematically. More quantitatively, the average and RMS delay spread are often used instead of τ_{max} and are defined as follows:

$$\mu_\tau = \frac{\int_0^\infty \Delta \tau A_\tau(\Delta \tau) d(\Delta \tau)}{\int_0^\infty A_\tau(\Delta \tau) d(\Delta \tau)}$$ (3.30)

$$\tau_{RMS} = \sqrt{\frac{\int_0^\infty (\Delta\tau - \mu_\tau)^2 A_\tau(\Delta\tau) d(\Delta\tau)}{\int_0^\infty A_\tau(\Delta\tau) d(\Delta\tau)}} . \tag{3.31}$$

Intuitively, τ_{RMS} gives a measure of the width, or spread, of the channel response in time. A large τ_{RMS} implies a highly dispersive channel in time and a long channel impulse response (large v), whereas a small τ_{RMS} indicates that the channel is not very dispersive and hence might require only a few taps to accurately characterize. A general rule of thumb is that $\tau_{max} \approx 5\tau_{RMS}$.

Table 3.2 shows some typical values for the RMS delay spread and the associated channel coherence bandwidth for two candidate WiMAX frequency bands. This table demonstrates that longer-range channels have more frequency-selective fading.

The channel coherence bandwidth B_c is the frequency-domain dual of the channel delay spread. The coherence bandwidth gives a rough measure for the maximum separation between a frequency f_1 and a frequency f_2 where the channel frequency response is correlated. That is:

$$|f_1 - f_2| \leq B_c \quad \Rightarrow \quad H(f_1) \approx H(f_2)$$
$$|f_1 - f_2| > B_c \quad \Rightarrow \quad H(f_1) \text{ and } H(f_2) \text{ are uncorrelated}$$

Just as τ_{max} is a ballpark value describing the channel duration, B_c is a ballpark value describing the range of frequencies over which the channel stays constant. Given the channel delay spread, it can be shown that

$$B_c \approx \frac{1}{5\tau_{RMS}} \approx \frac{1}{\tau_{max}} . \tag{3.32}$$

Exact relations can be found between B_c and τ_{RMS} by arbitrarily defining notions of coherence, but the important and prevailing feature is that B_c and τ are inversely related.

3.4.2 Doppler Spread and Coherence Time

Whereas the power-delay profile gave the statistical power distribution of the channel over time for a signal transmitted for only an instant, the *Doppler power spectrum* gives the statistical power distribution of the channel versus frequency for a signal transmitted at one exact frequency, generally normalized as $f = 0$ for convenience. Whereas the power-delay profile was caused by multipath between the transmitter and the receiver, the Doppler power spectrum is caused by *motion* between the transmitter and receiver. The Doppler power spectrum is the Fourier transform of $A_t(\Delta t)$, that is:

$$\rho_t(\Delta f) = \int_{-\infty}^\infty A_t(\Delta t) e^{-\Delta f \cdot \Delta t} (d\Delta t) . \tag{3.33}$$

Table 3.2 Some Typical RMS Delay Spread and Approximate Coherence Bandwidths for Various WiMAX Applications

Environment	f_c(GHz)	RMS Delay τ_{RMS} (ns)	Coherence Bandwidth $B_c \approx \dfrac{1}{5\tau_{RMS}}$ (MHz)	Reference
Urban	9.1	1,300	0.15	[22]
Rural	9.1	1,960	0.1	[22]
Indoor	9.1	270	0.7	[22]
Urban	5.3	44	4.5	[36]
Rural	5.3	66	3.0	[36]
Indoor	5.3	12.4	16.1	[36]

Unlike the power-delay profile, the Doppler power spectrum is nonzero strictly for $\Delta f \in (-f_D, f_D)$, where f_D is called the maximum Doppler, or Doppler spread. That is, $\rho_t(\Delta f)$ is strictly bandlimited. The Doppler spread is

$$f_D = \frac{vf_c}{c}, \tag{3.34}$$

where v is the maximum speed between the transmitter and the receiver, f_C is the carrier frequency, and c is the speed of light. As can be seen, over a large bandwidth, the Doppler will change, since the frequency over the entire bandwidth is *not* f_C. However, as long as the communication bandwidth $B \ll f_c$, the Doppler power spectrum can be treated as approximately constant. This generally is true for all but ultrawideband (UWB) systems.

Owing to the time/frequency uncertainty principle,[7] since $\rho_t(\Delta f)$ is strictly bandlimited, its time/frequency dual $A_t(\Delta t)$ cannot be strictly time-limited. Since $A_t(\Delta t)$ gives the correlation of the channel over time, the channel, strictly speaking, exhibits nonzero correlation between any two time instants. In practice, however, it is possible to define a channel coherence time T_C, which similarly to coherence bandwidth, gives the period of time over which the channel is significantly correlated. Mathematically:

$$|t_1 - t_2| \leq T_c \quad \Rightarrow \quad \mathbf{h}(t_1) \approx \mathbf{h}(t_2)$$
$$|t_1 - t_2| > t_c \quad \Rightarrow \quad \mathbf{h}(t_1) \text{ and } \mathbf{h}(t_2) \text{ are uncorrelated}$$

The coherence time and Doppler spread are also inversely related:

---

7. The time/frequency uncertainty principle mandates that no waveform can be perfectly isolated in both time and frequency.

Table 3.3 Summary of Broadband Fading Parameters, with Rules of Thumb

Quantity	If "Large"?	If "Small"?	WiMAX Design Impact
Delay spread, τ	If $\tau \gg T$, frequency selective	If $\tau \ll T$, frequency flat	The larger the delay spread relative to the symbol time, the more severe the ISI.
Coherence band-width, B_c	If $\dfrac{1}{B_c} \ll T$, frequency flat	If $\dfrac{1}{B_c} \gg T$, frequency selective	Provides a guideline to subcarrier width $B_{sc} \approx B_c/10$ and hence number of subcarriers needed in OFDM: $L \geq 10B/B_c$.
Doppler spread, $f_D = \dfrac{f_c v}{c}$	If $f_c v \gg c$, fast fading	If $f_c v \leq c$, slow fading	As f_D/B_{sc} becomes non-negligible, subcarrier orthogonality is compromised.
Coherence time, T_c	If $T_c \gg T$, slow fading	If $T_c \leq T$, fast fading	T_c small necessitates frequent channel estimation and limits the OFDM symbol duration but provides greater time diversity.
Angular spread, θ_{RMS}	NLOS channel, lots of diversity	Effectively LOS channel, not much diversity	Multiantenna array design, beamforming versus diversity.
Coherence distance, D_c	Effectively LOS channel, not much diversity	NLOS channel, lots of diversity	Determines antenna spacing.

$$T_c \approx \frac{1}{f_D}. \tag{3.35}$$

This makes intuitive sense: If the transmitter and the receiver are moving fast relative to each other and hence the Doppler is large, the channel will change much more quickly than if the transmitter and the receiver are stationary.

Table 3.4 gives some typical values for the Doppler spread and the associated channel coherence time for two candidate WiMAX frequency bands. This table demonstrates one of the reasons that mobility places extra constraints on the system design. At high frequency and mobility, the channel changes completely around 500 times per second, placing a large burden on channel-estimation algorithms and making the assumption of accurate transmitter channel knowledge questionable. Subsequent chapters (especially 5–7) discuss why accurate channel knowledge is important in WiMAX. Additionally, the large Doppler at high mobility and frequency can also degrade the OFDM subcarrier orthogonality, as discussed in Chapter 4.

Table 3.4 Some Typical Doppler Spreads and Approximate Coherence Times for Various WiMAX Applications

f_c (GHz)	Speed (kmph)	Speed (mph)	Maximum Doppler, f_D (Hz)	Coherence Time, $T_c \approx \dfrac{1}{f_D}$ (msec)
2.5	2	1.2	4.6	200
2.5	45	27.0	104.2	10
2.5	100	60.0	231.5	4
5.8	2	1.2	10.7	93
5.8	45	27.0	241.7	4
5.8	100	60.0	537.0	2

3.4.3 Angular Spread and Coherence Distance

So far, we have focused on how the channel response varies over time and how to quantify its delay and correlation properties. However, channels also vary over space. We do not attempt to rigorously treat all aspects of spatial/temporal channels but will summarize a few important points.

The RMS angular spread of a channel can be denoted as θ_{RMS} and refers to the statistical distribution of the angle of the arriving energy. A large θ_{RMS} implies that channel energy is coming in from many directions; a small θ_{RMS} implies that the received channel energy is more focused. A large angular spread generally occurs when there is a lot of local scattering, which results in more statistical diversity in the channel; more focused energy results in less statistical diversity.

The dual of angular spread is coherence distance, D_c. As the angular spread increases, the coherence distance decreases, and vice versa. A coherence distance of d means that any physical positions separated by d have an essentially uncorrelated received signal amplitude and phase. An approximate rule of thumb [8] is

$$D_c \approx \frac{.2\lambda}{\theta_{RMS}}. \tag{3.36}$$

The case of Rayleigh fading, discussed in Section 3.5.1, assumes a uniform angular spread; the well-known relation is

$$D_c \approx \frac{9\lambda}{16\pi}. \tag{3.37}$$

An important trend to note from the preceding relations is that the coherence distance increases with the carrier wavelength λ. Thus, higher-frequency systems have shorter coherence distances.

Angular spread and coherence distance are particularly important in multiple-antenna systems. The coherence distance gives a rule of thumb for how far apart antennas should be spaced in order to be statistically independent. If the coherence distance is very small, antenna arrays can be effectively used to provide rich diversity. The importance of diversity is introduced in Section 3.6. On the other hand, if the coherence distance is large, space constraints may make it impossible to take advantage of spatial diversity. In this case, it would be preferable to have the antenna array cooperate and use beamforming. The trade-offs between beamforming and linear array processing are discussed in Chapter 5.

3.5 Modeling Broadband Fading Channels

In order to design and benchmark wireless communication systems, it is important to develop channel models that incorporate their variations in time, frequency, and space. Models are classified as either *statistical* or *empirical*. Statistical models are simpler and are useful for analysis and simulations. Empirical models are more complicated but usually represent a specific type of channel more accurately.

3.5.1 Statistical Channel Models

As we have noted, the received signal in a wireless system is the superposition of numerous reflections, or multipath components. The reflections may arrive very closely spaced in time—for example, if there is local scattering around the receiver—or at relatively longer intervals. Figure 3.11 showed that when the reflections arrive at nearly the same time, constructive and destructive interference between the reflections causes the envelope of the aggregate received signal $r(t)$ to vary substantially.

In this section, we summarize statistical methods for characterizing the amplitude and power of $r(t)$ when all the reflections arrive at about the same time. First, we consider the special case of the multipath intensity profile, where $A_\tau(\Delta\tau) \approx 0$ for $\Delta\tau \neq 0$. That is, we concern ourselves only with the scenario in which all the received energy arrives at the receiver at the same instant: step 1 in our pedagogy. In practice, this is true only when the symbol time is much greater than the delay spread—$T \gg \tau_{max}$—so these models are often said to be valid for narrowband fading channels. In addition to assuming a negligible multipath delay spread, we first consider just a snapshot value of $r(t)$ and provide statistical models for its amplitude and power under various assumptions. We then consider how these statistical values are correlated in time, frequency, and space: step 2. Finally, we relax all the assumptions and consider how wideband fading channels evolve in time, frequency, and space: step 3.

3.5.1.1 Rayleigh Fading

Suppose that the number of scatterers is large and that the angles of arrival between them are uncorrelated. From the Central Limit Theorem, it can be shown that the in-phase (cosine) and

---

Sidebar 3.4 A Pedagogy for Developing Statistical Models

Our pedagogy for developing statistical models of wireless channels consists of three steps discussed in the sections noted.

1. **Section 3.5.1:** First, consider a single channel sample corresponding to a single principal path between the transmitter and the receiver:

$$h(\tau,t) \to h_0 \delta(\tau,t) \ ..$$

   Attempt to quantify: How is the value of $|h_0|$ statistically distributed?

2. **Section 3.5.2:** Next, consider how this channel sample h_0 evolves over time:

$$h(\tau,t) \to h_0(t) \delta(\tau) \ ..$$

   Attempt to quantify: How does the value $|h_0(t)|$ change over time? That is, how is $h_0(t)$ correlated with some $h_0(t + \Delta t)$?

3. **Section 3.5.2 and Section 3.5.3:** Finally, represent $h(\tau,t)$ as a general time-varying function. One simple approach is to model $h(\tau,t)$ as a general multipath channel with $v + 1$ tap values. The channel sample value for each of these taps is distributed as determined in step 1, and evolves over time as specified by step 2.

---

quadrature (sine) components of $r(t)$, denoted as $r_I(t)$ and $r_Q(t)$, follow two independent time-correlated Gaussian random processes.

Consider a snapshot value of $r(t)$ at time $t = 0$, and note that $r(0) = r_I(0) + r_Q(0)$. Since the values $r_I(0)$ and $r_Q(0)$ are Gaussian random variables, it can be shown that the distribution of the envelope amplitude $|r| = \sqrt{r_I^2 + r_Q^2}$ is Rayleigh and that the received power $|r|^2 = r_I^2 + r_Q^2$ is exponentially distributed. Formally,

$$f_{|r|}(x) = \frac{2x}{P_r} e^{-x^2/P_r}, \ x \geq 0, \tag{3.38}$$

and

$$f_{|r|^2}(x) = \frac{1}{P_r} e^{-x/P_r}, \ x \geq 0, \tag{3.39}$$

where P_r is the average received power owing to shadowing and pathloss, as described, for example, in Equation (3.10). The pathloss and shadowing determine the mean received power—assuming they are fixed over some period of time—and the total received power fluctuates around this mean, owing to the fading (see Figure 3.13). It can also be noted that in this setup,

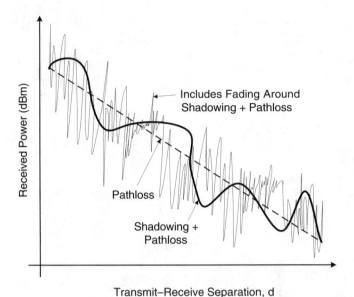

Figure 3.13 Plot showing the three major trends: pathloss, shadowing ,and fading all on the same plot: empirical, simulated, or a good CAD drawing

the Gaussian random variables r_I and r_Q each have zero mean and variance $\sigma^2 = P_r/2$. The phase of $r(t)$ is defined as

$$\theta_r = \tan^{-1}\left(\frac{r_Q}{r_I}\right), \tag{3.40}$$

which is uniformly distributed from 0 to 2π, or equivalently from $[-\pi, \pi]$ any other contiguous full period of the carrier signal.[8]

3.5.1.2 LOS Channels: Ricean distribution

An important assumption in the Rayleigh fading model is that all the arriving reflections have a mean of zero. This will not be the case if there is a dominant path—for example, a LOS path—between the transmitter and the receiver. For a LOS signal, the received envelope distribution is more accurately modeled by a Ricean [24] distribution, which is given by

$$f_{|r|}(x) = \frac{x}{\sigma^2} e^{-(x^2+\mu^2)/2\sigma^2} I_0\left(\frac{x\mu}{\sigma^2}\right), \ x \geq 0, \tag{3.41}$$

---

8. Strictly, Equation (3.40) will give only values from $[0, \pi]$, but it is conventional that the sign of r_I and r_Q determines the quadrant of the phase. For example, if r_I and r_Q are negative, $\theta_r \in [\pi, 3\pi/2]$.

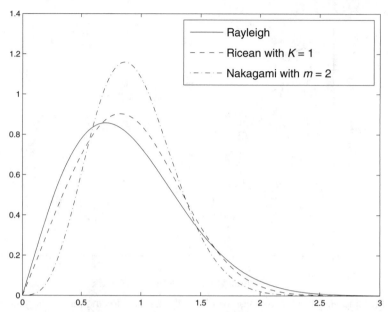

Figure 3.14 Probability distributions $f_{|r|}(x)$ for Rayleigh, Ricean with $K = 1$, and Nakagami with $m = 2$ and average received power $P_r = 1$ for all

where μ^2 is the power of the LOS component and I_0 is the 0th-order, modified Bessel function of the first kind. Although more complicated than a Rayleigh distribution, this expression is a generalization of the Rayleigh distribution. This can be confirmed by observing that

$$\mu = 0 \Rightarrow I_0 (\frac{x\mu}{\sigma^2}) = 1 ,$$

so the Ricean distribution reduces to the Rayleigh distribution in the absence of a LOS component. Except in this special case, the Ricean phase distribution θ_r is not uniform in $[0, 2\pi]$ and is not described by a straightforward expression.

Since the Ricean distribution depends on the LOS component's power μ^2, a common way to characterize the channel is by the relative strengths of the LOS and scattered paths. This factor, K, is quantified as

$$K = \frac{\mu^2}{2\sigma_2} \tag{3.42}$$

and is a natural description of how strong the LOS component is relative to the NLOS components. For $K = 0$, the Ricean distribution again reduces to Rayleigh, and as $K \to \infty$, the physical meaning is that there is only a single LOS path and no other scattering. Mathematically, as K grows large, the Ricean distribution is quite Gaussian about its mean μ with decreasing variance, physically meaning that the received power becomes increasingly deterministic.

The average received power under Ricean fading is the combination of the scattering power and the LOS power: $P_r = 2\sigma^2 + \mu^2$. Although it is not straightforward to directly find the Ricean power distribution $f_{|r|^2}(x)$, the Ricean envelope distribution in terms of K can be found by subbing $\mu^2 = KP_r/(K+1)$ and $2\sigma^2 = P/(K+1)$ into Equation (3.41).

Although its simplicity makes the Rayleigh distribution more amenable to analysis than the Ricean distribution, the Ricean distribution is usually a more accurate depiction of wireless broadband systems, which typically have one or more dominant components. This is especially true of fixed wireless systems, which do not experience fast fading and often are deployed to maximize LOS propagation.

3.5.1.3 A More General Model: Nakagami-m Fading

The last statistical fading model that we discuss is the Nakagami-m fading distribution [18]. The PDF (probability density function) of Nakagami fading is parameterized by m and is given as

$$f_{|r|}(x) = \frac{2m^m x^{2m-1}}{\Gamma(m)P_r^m} e^{-mx^2/P_r}, m \geq 0.5.$$ (3.43)

Although this expression appears to be just as—or even more— ungainly as the Ricean distribution, the dependence on x is simpler; hence the Nakagami distribution can in many cases be used in tractable analysis of fading channel performance [30]. Additionally, it is more general, as $m = (K + 1)^2/(2K +1)$ gives an approximate Ricean distribution, and $m = 1$ gives a Rayleigh. As $m \to \infty$, the receive power tends to a constant, P_r. The power distribution for Nakagami fading is

$$f_{|r|^2}(x) = (\frac{m}{P_r})^m \frac{x^{m-1}}{\Gamma(m)} e^{-mx/P_r}, m \geq 0.5.$$ (3.44)

Similarly, the power distribution is also amenable to integration.

3.5.2 Statistical Correlation of the Received Signal

The statistical methods in the previous section discussed how *samples* of the received signal are statistically distributed. We considered the Rayleigh, Ricean, and Nakagami-m statistical models and provided the PDFs that giving the likelihoods of the received signal envelope and power at a given time instant (Figure 3.14). What is of more interest, though, is how to link those statistical models with the channel autocorrelation function, $A_c(\Delta\tau,\Delta t))$, in order to understand how the envelope signal $r(t)$ evolves over time or changes from one frequency or location to another.

For simplicity and consistency, we use Rayleigh fading as an example distribution here, but the concepts apply equally for any PDF. We first discuss correlation in different domains separately but conclude with a brief discussion of how the correlations in different domains interact.

3.5.2.1 Time Correlation

In the time domain, the channel $h(\tau = 0, t)$ can intuitively be thought of as consisting of approximately one new sample from a Rayleigh distribution every T_c seconds, with the values in between interpolated. But, it will be useful to be more rigorous and accurate in our description of the fading envelope. As discussed in Section 3.4, the autocorrelation function $A_t(\Delta t)$ describes how the channel is correlated in time. Similarly, its frequency-domain Doppler power spectrum $\rho_t(\Delta f)$ provides a band-limited description of the same correlation, since it is simply the Fourier transform of $A_t(\Delta t)$. In other words, the power-spectral density of the channel $h(\tau = 0, t)$ should be $\rho_t(\Delta f)$. Since uncorrelated random variables have a flat power spectrum, a sequence of independent complex Gaussian random numbers can be multiplied by the desired Doppler power spectrum $\rho_t(\Delta f)$; then, by taking the inverse fast fourier transform, a correlated narrowband sample signal $h(\tau = 0, t)$ can be generated. The signal will have a time correlation defined by $\rho_t(\Delta f)$ and be Rayleigh, owing to the Gaussian random samples in frequency.

For the specific case of uniform scattering [16], it can been shown that the Doppler power spectrum becomes

$$\rho_t(\Delta f) = \begin{cases} \dfrac{P_r}{4\pi} \dfrac{1}{f_D \sqrt{1 - (\dfrac{\Delta f}{f_D})^2}}, & |\Delta f| \leq f_D \\ 0, & \Delta_f > f_D \end{cases} \tag{3.45}$$

A plot of this realization of $\rho_t(\Delta f)$ is shown in Figure 3.15. It is well known that the inverse Fourier transform of this function is the 0th order Bessel function of the first kind, which is often used to model the time autocorrelation function, $A_c(\delta t)$, and hence predict the time-correlation properties of narrowband fading signals. A specific example of how to generate a Rayleigh fading signal envelope with a desired Doppler f_D, and hence channel coherence time $T_c \approx f_D^{-1}$, is provided in Matlab (see Sidebar 3.4).

3.5.2.2 Frequency Correlation

Similarly to time correlation, a simple intuitive notion of fading in frequency is that the channel in the frequency domain, $H(f, t = 0)$, can be thought of as consisting of approximately one new random sample every B_c Hz, with the values in between interpolated. The Rayleigh fading model assumes that the received quadrature signals in time are complex Gaussian. Similar to the development in the previous section where by complex Gaussian values in the frequency domain can be converted to a correlated Rayleigh envelope in the time domain, complex Gaussian values in the time domain can likewise be converted to a correlated Rayleigh frequency envelope $|H(f)|$.

The correlation function that maps from uncorrelated time-domain (τ domain) random variables to a correlated frequency response is the multipath intensity profile, $A_\tau(\Delta \tau)$. This makes sense: Just as $\rho_t(\Delta f)$ describes the channel time correlation in the frequency domain, $A_\tau(\Delta \tau)$ describes the channel frequency correlation in the time domain. Note that in one familiar special

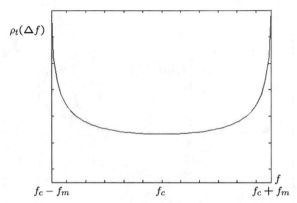

Figure 3.15 The spectral correlation owing to Doppler, $\rho_t(\Delta f)$ for uniform scattering: Equation (3.45)

case, there is only one arriving path, in which case $A_\tau(\Delta\tau) = \delta(\Delta\tau)$. Hence, the values of $|H(f)|$ are correlated over all frequencies since the Fourier transform of $\delta(\Delta\tau)$ is a constant over all frequency. This scenario is called flat fading; in practice, whenever $A_\tau(\Delta\tau)$ is narrow ($\tau_{max} \ll T$), the fading is approximately flat.

If the arriving quadrature components are approximately complex Gaussian, a correlated Rayleigh distribution might be a reasonable model for the gain $|H(f)|$ on each subcarrier of a typical OFDM system. These gain values could also be generated by a suitably modified version of the provided simulation, where in particular, the correlation function used changes from that in Equation (3.45) to something like an exponential or uniform distribution or any function that reasonably reflects the multipath intensity profile $A_\tau(\Delta\tau)$.

3.5.2.3 The Selectivity/Dispersion Duality

Two quite different effects from fading are *selectivity* and *dispersion*. By *selectivity*, we mean that the signal's received value is changed by the channel over time or frequency. By *dispersion*, we mean that the channel is dispersed, or spread out, over time or frequency. Selectivity and dispersion are time/frequency duals of each other: Selectivity in time causes dispersion in frequency, and selectivity in frequency causes dispersion in time—or vice versa (see Figure 3.17).

For example, the Doppler effect causes dispersion in frequency, as described by the Doppler power spectrum $\rho_t(\Delta f)$. This means that frequency components of the signal received at a specific frequency f_0 will be dispersed about f_0 in the frequency domain with a probability distribution function described by $\rho_t(\Delta f)$. As we have seen, this dispersion can be interpreted as a time-varying amplitude, or selectivity, in time.

Similarly, a dispersive multipath channel that causes the paths to be received over a period of time τ_{max} causes selectivity in the frequency domain, known as frequency-selective fading. Because symbols are traditionally sent one after another in the time domain, time dispersion

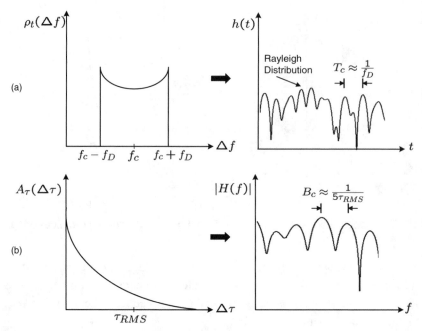

Figure 3.16 (a) The shape of the Doppler power spectrum $\rho_t(\Delta f)$ determines the correlation envelope of the channel in time. (b) Similarly, the shape of the multipath intensity profile $A_\tau(\Delta \tau)$ determines the correlation pattern of the channel frequency response.

usually causes much more damaging interference than frequency dispersion does, since adjacent symbols are smeared together.

3.5.2.4 Multidimensional Correlation

In order to present the concepts as clearly as possible, we have thus far treated time, frequency, and spatial correlations separately. In reality, signals are correlated in all three domains.

A broadband wireless data system with mobility and multiple antennas is an example of a system in which all three types of fading will play a significant role. The concept of doubly selective (in time and frequency) fading channels [25] has received recent attention for OFDM. The combination of these two types of correlation is important because in the context of OFDM, they appear to compete with each other. On one hand, a highly frequency-selective channel—resulting from a long multipath channel as in a wide area wireless broadband network—requires a large number of potentially closely spaced subcarriers to effectively combat the intersymbol interference and small coherence bandwidth. On the other hand, a highly mobile channel with a large Doppler causes the channel to fluctuate over the resulting long symbol period, which degrades the subcarrier orthogonality. In the frequency domain, the Doppler frequency shift can cause significant inter carrier interference as the carriers become more closely spaced. Although the mobility and multipath delay spread must reach fairly severe levels before this doubly selective effect becomes significant, this problem facing mobile WiMAX systems does not have a comparable

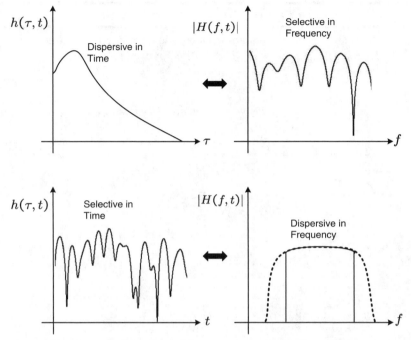

Figure 3.17 The dispersion/electivity duality: Dispersion in time causes frequency selectivity; dispersion in frequency causes time selectivity.

precedent. The scalable nature of the WiMAX physical layer—notably, variable numbers of subcarriers and guard intervals—will allow custom optimization of the system for various environments and applications.

3.5.3 Empirical Channel Models

The parametric statistical channel models discussed thus far in the chapter do not take into account specific wireless propagation environments. Although exactly modeling a wireless channel requires complete knowledge of the surrounding scatterers, such as buildings and plants, the time and computational demands of such a methodology are unrealistic, owing to the near-infinite number of possible transmit/receive locations and the fact that objects are subject to movement. Therefore, empirical and semiempirical wireless channel models have been developed to accurately estimate the pathloss, shadowing, and small-scale fast fading. Although these models are generally not analytically tractable, they are very useful for simulations and to fairly compare competing designs. Empirical models are based on extensive measurement of various propagation environments, and they specify the parameters and methods for modeling the typical propagation scenarios in various wireless systems. Compared to parametric channel models, the empirical channel models take into account such realistic factors as angle of arrival (AoA),

Sidebar 3.5 A Rayleigh Fading Simulation in Matlab

The following Matlab function generates a stochastic correlated Rayleigh fading envelope with effective Doppler frequency f_D. See Figure 3.18 for example-generated envelopes.

```
function [Ts, z_dB] = rayleigh_fading(f_D, t, f_s)
% Inputs
%    f_D : [Hz] Doppler frequency
%    t   : simulation time interval length, time interval [0,t]
%    f_s : [Hz] sampling frequency, set to 1000 if smaller.
% Outputs
%    Ts   : [Sec] [1xN double] time instances for the Rayleigh signal
%    z_dB : [dB] [1xN double] Rayleigh fading signal
% Required parameters
if f_s < 1000, f_s = 1000; end  % [Hz} Min. required sampling rate
N = ceil(t*f_s);       % Number of samples
Ts = linspace(0,t,N);
if mod(N,2) == 1, N = N+1; end   % Use even number of samples
f = linspace(-f_s,f_s,N);
% Generate I & Q complex Gaussian samples in frequency domain
Gfi_p = randn(2,N/2); Gfq_p = randn(2,N/2);
CGfi_p = Gfi_p(1,:)+i*Gfi_p(2,:); CGfq_p = Gfq_p(1,:)+i*Gfq_p(2,:);
CGfi = [fliplr(CGfi_p)' CGfi_p ]; CGfq = [fliplr(CGfq_p)' CGfq_p ];
% Generate fading spectrum for shaping Gaussian line spectra
P_r = 1; % normalize average received envelope to 0dB
S_r = P_r/(4*pi)./(f_D*sqrt(1-(f/f_D).^2)); %Doppler spectra
% Set samples outside the Doppler frequency range to 0
idx1 = find(f>f_D); idx2 = find(f<-f_D);
S_r(idx1) = 0; S_r(idx2) = 0;
% Generate r_I(t) and r+Q(t) using inverse FFT:
r_I = N*ifft(CGfi.*sqrt(S_r));
r_Q = -i*N*ifft(CGfq.*sqrt(S_r));
% Finally, generate the Rayleigh distributed signal envelope
z = sqrt(abs(r_I).^2+abs(r_Q).^2);
z_dB = 20*log10(z);
z_dB = z_dB(1:length(Ts)); % Return correct number of points
```

angle of departure (AoD), antenna array fashion, angular spread (AS), and antenna array gain pattern.

Different empirical channel models exist for different wireless scenarios, such as suburban macro-, urban macro-, and urbanmicro cells. For channels experienced in different wireless standards, the empirical channel models are also different. Here, we briefly introduce the common physical parameters and methodologies used in several major empirical channel models. These models are also applicable to the multiple-antenna systems described in Chapter 6.

3.5.3.1 3GPP

The 3GPP channel model is widely used in modeling the outdoor macro- and microcell wireless environments. The empirical channel models for other systems, such as 802.11n and 802.20, are

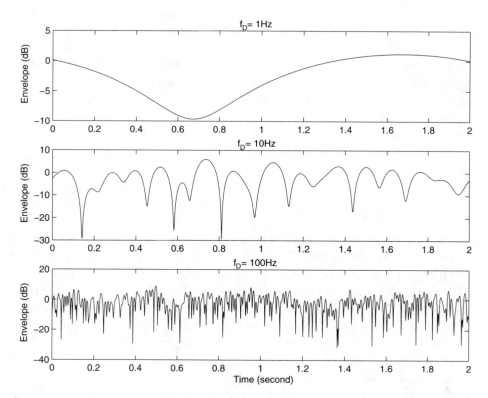

Figure 3.18 Sample channel gains in dB from the provided Rayleigh fading Matlab function for Doppler frequencies of $f_D = $ 1Hz, 10Hz, and 100Hz

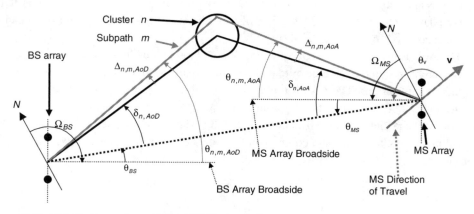

Figure 3.19 3GPP channel model for MIMO simulations

similar in most aspects, with subtle differences in the terminology and specific parameters. The 3GPP channel model is commonly used in WiMAX performance modeling.

1. First, we need to specify the environment in which an empirical channel model is used: suburban macro-, urban macro-, or urban microenvironment. The BS-to-BS distance is typically larger than 3 km for a macroenvironment and less than 1 km for an urban microenvironment.

2. The pathloss is specified by empirical models for various scenarios. For the 3GPP macro-cell environment, the pathloss is given as

$$PL[dB] \;=\; (44.9 - 6.55\log_{10}(h_{bs})\log_{10}(\frac{d}{1000}) + 45.5 + (35.46 - 1.1h_{ms})\log_{10}(f_c)$$
$$- 13.82\log_{10}(h_{bs}) + 0.7h_{ms} + C, \qquad (3.46)$$

   where h_{bs} is the BS antenna height in meters, h_{ms} is the MS antenna height in meters, f_c is the carrier frequency inMHz, d is the distance in meters between the BS and the MS, and C is a constant factor ($C = 0$ dB for suburban macro and $C = 3$ dB for urban macro).

3. The received signal at the mobile receiver consists of N time-delayed versions of the transmitted signal. The N paths are characterized by powers and delays that are chosen according to the channel-generation procedures. The number of paths N ranges from 1 to 20 and is dependent on the specific channel models. For example, the 3GPP channel model has $N = 6$ multipath components. The power distribution normally follows the exponential profile, but other power profiles are also supported.

4. Each multipath component corresponds to a cluster of M subpaths, each of which characterizes the incoming signal from a scatterer. The M subpaths define a cluster of adjacent scatterers and therefore have the same multipath delay. The M subpaths have random phases and subpath gains, specified by the given procedure in different stands. For 3GPP, the phases are random variables uniformly distributed from 0 to $360°$, and the subpath gains are given by Equation (3.47).

5. The AoD is usually within a narrow range in outdoor applications owing to the lack of scatterers around the BS transmitter and is often assumed to be uniformly distributed in indoor applications. The AoA is typically assumed to be uniformly distributed, owing to the abundance of local scattering around the mobile receiver.

6. The final channel is created by summing up the M subpath components. In the 3GPP channel model, the nth multipath component from the uth transmit antenna to the sth receive antenna is given as

$$
h_{u,s,n}(t) \;=\; \sqrt{\frac{P_n \sigma_s}{M}} \sum_{m=1}^{M}
\left(
\begin{array}{l}
\sqrt{G_{BS}\left(\theta_{n,m,AoD}\right)}\exp\!\left(j\!\left[kd_s \sin\!\left(\theta_{n,m,AoD}+\Phi_{n,m}\right)\right]\right)\times \\[6pt]
\sqrt{G_{BS}\left(\theta_{n,m,AoA}\right)}\exp\!\left(jkd_u \sin\!\left(\theta_{n,m,AoA}\right)\right)\times \\[6pt]
\exp\!\left(jk\,\|\mathbf{v}\|\cos\!\left(\theta_{n,m,AoA}-\theta_v\right)t\right)
\end{array}
\right),
\tag{3.47}
$$

where

P_n is the power of the nth path, following exponential distribution.

σ_s is the lognormal shadow fading, applied as a bulk parameter to the n paths. The shadow fading is determined by the delay spread (DS), angle spread (AS), and shadowing parameters, which are correlated random variables generated with specific procedures.

M is the number of subpaths per path.

$\theta_{n,m,AoD}$ is the the AoD for the mth subpath of the nth path.

$\theta_{n,m,AoA}$ is the the AoA for the mth subpath of the nth path.

$G_{BS}\left(\theta_{n,m,AoD}\right)$ is the BS antenna gain of each array element.

$G_{BS}\left(\theta_{n,m,AoA}\right)$ is the MS antenna gain of each array element.

k is the wave number $\dfrac{2\pi}{\lambda}$, where λ is the carrier wavelength in meters.

d_s is the distance in meters from BS antenna element s from the reference ($s = 1$) antenna.

d_u is the distance in meters from MS antenna element u from the reference ($u = 1$) antenna.

$\Phi_{n,m}$ is the phase of the mth subpath of the nth path, uniformly distributed between 0 and $360°$.

$\|\mathbf{v}\|$ is the magnitude of the MS velocity vector, which consists of the velocity of the MS array elements.

θ_v is the angle of the MS velocity vector.

3.5.3.2 Semiempirical Channel Models

The preceding empirical channel models provide a very thorough description of the propagation environments. However, the sheer number of parameters involved makes constructing a fully empirical channel model relatively time consuming and computationally intensive. Alternatives are semiempirical channel models, which provide the accurate inclusion of the practical parameters in a real wireless system while maintaining the simplicity of statistical channel models.

Examples of the simpler empirical channel models include 3GPP2 pedestrian A, pedestrian B, vehicular A, and vehicular B models, suited for low-mobility pedestrian mobile users and higher-mobility vehicular mobile users. The multipath profile is determined by the number of multipath taps and the power and delay of each multipath component. Each multipath component is modeled as independent Rayleigh fading with a potentially different power level, and the correlation in the time domain is created according to a Doppler spectrum corresponding to the specified speed. The pedestrian A is a flat-fading model corresponding to a single Rayleigh fading component with a speed of 3 kmph; the pedestrian B model corresponds to a multipath profile with four paths of delays [0. 11. 19. 41] microseconds and the power profile given as [1 0. 1071 0.0120 0.0052]. For the vehicular A model, the mobile speed is specified at 30 kmph. Four multipath components exist, each with delay profile [0 0.11 0.19 0.41] microseconds and power profile [1 0.1071 0.0120 0.0052]. For the vehicular B model, the mobile speed is 30 kmph, with six multipath components, delay profile [0 0.2 0.8 1.2 2.3 3.7] microseconds, and power profile [1 0.813 0.324 0.158 0.166 0.004].

Another important empirical channel model for the 802.16 WiMAX fixed broadband wireless system is the Stanford University Interim (SUI) channel model. This model provides six typical channels for the typical terrain types of the continental United States: SUI1 to SUI6 channels. Each of these models addresses a specific channel scenario with low or high Doppler spread, small or large delay spread, different LOS factors, different spatial correlations at the transmitter, and receiver antenna array. For all six models, the channel consists of three multipath fading taps whose delay and power profiles are different.

These empirical channel models follow the fundamental principles of the statistical parametric models discussed previously in this chapter, while considering empirical measurement results. As such, semiempirical channel models are suitable for link-level simulations and performance evaluation in real-world broadband wireless environments.

3.6 Mitigation of Fading

The fading characteristic of wireless channels is perhaps the most important difference between wireless and wired communication system design.[9] Since frequency-selective fading is more prominent in wideband channels—since a wideband channel's bandwidth is usually much greater than the coherence bandwidth—we refer to channels with significant time dispersion or frequency selectivity as *broadband fading* and to channels with only frequency dispersion or time selectivity as *narrowband fading*. We now briefly review and differentiate between narrowband and broadband fading. The next several chapters of the book are devoted to in-depth exploration of techniques that overcome or exploit fading.

---

9. The other most notable differentiating factors for wireless are that all users nominally interfere with one another in the shared wireless medium and that portability puts severe power constraints on the mobile transceivers.

3.6.1 Narrowband (Flat) Fading

Many different techniques are used to overcome narrowband fading, but most can be collectively referred to as *diversity*. Because the received signal power is random, if several (mostly) uncorrelated versions of the signal can be received, chances are good that at least one of the versions has adequate power. Without diversity, high-data-rate wireless communication is virtually impossible. Evidence of this is given in Figure 3.20, which shows the effect of unmitigated fading in terms of the received average bit error rate (BER). The BER probability for QAM systems in additive white Gaussian noise (AWGN) can accurately be approximated by the following bound [11]:

$$P_b \leq 0.2 e^{-1.5\gamma/(M-1)}, \tag{3.48}$$

where $M \geq 4$ is the M QAM alphabet size.[10] Note that the probability of error decreases very rapidly (exponentially) with the SNR, so decreasing the SNR linearly causes the BER to increase exponentially. In a fading channel, then, the occasional instances when the channel is in a deep fade dominate the BER, particularly when the required BER is very low. From observing the Rayleigh distribution in Equation (3.39), we can see that it requires dramatically increased P_r to continually decrease the probability of a deep fade. This trend is captured plainly in Figure 3.20, where we see that at reasonable system BERs, such as $10^{-5} - 10^{-6}$, the required SNR is over 30 dB higher in fading! Clearly, it is not desirable, or even possible, to increase the power by over a factor of 1,000. Furthermore, in an interference-limited system, increasing the power will not significantly raise the effective SINR.

Although BER is a more analytically convenient measure, since it is directly related to the SINR—for example, via Equation (3.38), a more common and relevant measure in WiMAX is the packet error rate (PER), or equivalently block error rate (BLER) or frame error rate (FER). All these measures refer to the probability that at least one bit is in error in a block of L bits. This is the more relevant measure, since the detection of a single bit error in a packet by the cyclic redundancy check (CRC) causes the packet to be discarded by the receiver. An expression for PER is

$$PER \leq 1 - (1 - P_b)^L, \tag{3.49}$$

where P_b is the BER and L is the packet length. This expression is true with equality when all bits are equally likely to be in error. If the bit errors are correlated, the PER improves. It is clear that PER and BER are directly related, so reducing PER and BER are roughly equivalent objectives.

Diversity is the key to overcoming the potentially devastating performance loss from fading channels and to improving PER and BER.

---

10. For example, $M = 4$ is QPSK, $M = 16$ is 16 QAM, and so on.

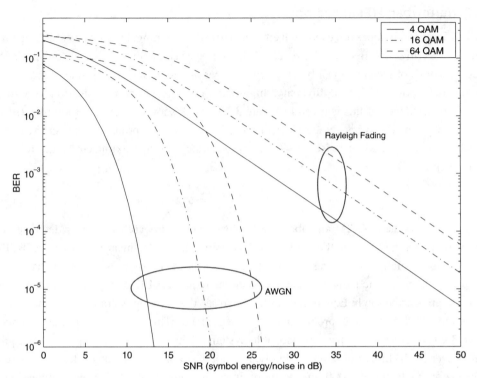

Figure 3.20 Flat fading causes a loss of at least 20 dB–30 dB at reasonable BER values.

3.6.1.1   Time Diversity

Two important forms of time diversity are coding/interleaving and adaptive modulation. Coding and interleaving techniques intelligently introduce redundancy in the transmitted signal so that each symbol is likely to have its information spread over a few channel coherence times. This way, after appropriate decoding, a deep fade affects all the symbols just a little bit rather than completely knocking out the symbols that were unluckily transmitted during the deep fade. Transmitters with adaptive modulation must have knowledge of the channel. Once they do, they usually choose the modulation technique that will achieve the highest possible data rate while still meeting a BER requirement. For example, in Equation (3.48), as the constellation alphabet size M increases, the BER also increases. Since the data rate is proportional to $\log_2 M$, we would like to choose the largest alphabet size M such that the required BER is met. If the channel is in a very deep fade, no symbols may be sent, to avoid making errors. Adaptive modulation and coding are an integral part of the WiMAX standard and are discussed further in Chapters 5 and 9.

3.6.1.2   Spatial Diversity

Spatial diversity, another extremely common and powerful form of diversity, is usually achieved by having two or more antennas at the receiver and/or the transmitter. The simplest form of space diversity consists of two receive antennas, where the stronger of the two signals is selected. As long as the antennas are spaced sufficiently, the two received signals will undergo approximately uncorrelated fading. This type of diversity is sensibly called *selection diversity* and is illustrated in Figure 3.21. Even though this simple technique completely discards half of the received signal, most of the deep fades can be avoided, and the average SNR is also increased. More sophisticated forms of spatial diversity include receive antenna arrays (two or more antennas) with maximal ratio combining, transmit diversity using spacetime codes, and combinations of transmit and receive diversity. Spatial-signaling techniques are expected to be crucial to achieving high spectral efficiency in WiMAX and are discussed in detail in Chapter 5.

3.6.1.3 Frequency Diversity

It is usually not straightforward to achieve frequency diversity unless the signal is transmitted over a large bandwidth. But in this case, the signal undergoes increasingly severe time dispersion.[11] Techniques that achieve frequency diversity while maintaining robustness to time dispersion are discussed in Section 3.6.2.

3.6.1.4 Diversity-Types Interactions

The use of diversity in one domain can decrease the utility of diversity in another domain. For example, imagine what the dark line in Figure 3.21 will look like as the number of branches (antennas) becomes large. Naturally, the selected signal will become increasingly flat in time, since at each instant, the best signal is selected. Hence, in this example, the *gain* from using time diversity, such as coding and interleaving, will not be as great as if no spatial diversity was used. Put simply, the total diversity gain is less than the sum of the two individual gains. So, although the overall performance is maximized by using all the forms of diversity, as diversity causes the effective channel to get closer to an AWGN channel, additional sources of diversity achieve diminishing returns.

3.6.2 Broadband Fading

As we have emphasized, frequency-selective fading causes dispersion in time, which causes adjacent symbols to interfere with each other unless $T \gg \tau_{max}$. Since the data rate R is proportional to $1/T$, high-data-rate systems almost invariably have a substantial multipath delay spread, $T \ll \tau_{max}$, and experience very serious intersymbol interference as a result. Choosing a technique to effectively combat it is a central design decision for any high-data-rate system. Increasingly, OFDM is the most popular choice for combatting ISI. OFDM is discussed in detail in the next chapter; here, let's briefly consider the other notable techniques for ISI mitigation.

---

11. An exception to this is frequency hopping, whereby a narrowband signal hops from one frequency slot to another in a large bandwidth. For frequency diversity, the frequency slot size would preferably be on the order of B_c.

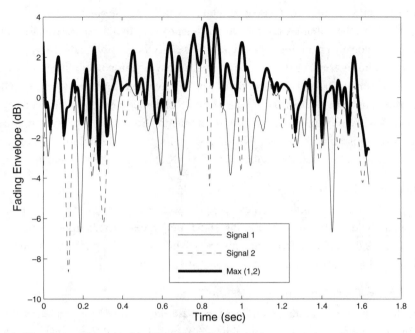

Figure 3.21 Simple two-branch selection diversity eliminates most deep fades.

3.6.3 Spread Spectrum and Rake Receivers

Somewhat counterintuitively, speeding up the *transmission* rate can help combat multipath fading, assuming that the *data* rate is kept the same. Since speeding up the transmission rate for a narrowband data signal results in a wideband transmission, this technique is called *spread spectrum*. Spread-spectrum techniques are generally broken into two quite different categories: direct sequence and frequency hopping. Direct-sequence spread spectrum, also known as code division multiple access (CDMA), is used widely in cellular voice networks and is effective at multiplexing a large number of variable-rate users in a cellular environment. Frequency hopping is used in some low-rate wireless local area networks (LANs) such as Bluetooth, and also for its interference-averaging properties in GSM cellular networks.

Some of WiMAX's natural competitors for wireless broadband data services have grown out of the CDMA cellular voice networks—notably 1xEV-DO and HSDPA/HSUPA—as discussed in Chapter 1. However, CDMA is not an appropriate technology for high data rates, and 1xEV-DO and HSDPA are CDMA in name only.[12] Essentially, for both types of spread spectrum, a large bandwidth is used to send a relatively small data rate. This is a reasonable approach for low-data-rate communications, such as voice, whereby a large number of users can be statistically multiplexed to yield a high overall system performance. For high-data-rate systems, each

---

12. In 1xEV-DO and HSDPA, users are multiplexed in the time rather than the code domain, and the spreading factor is very small.

user must use several codes simultaneously, which generally results in self-interference. Although this self-interference can be corrected with an equalizer (see Section 3.6.4), this largely defeats the purpose of using spread spectrum to help with intersymbol interference.

In short, spread-spectrum is not a natural choice for wireless broadband networks, since by definition, the data rate of a spread-spectrum system is less than its bandwidth. The same trend has been observed in wireless LANs: Early wireless LANs (802.11 and 802.11b) were spread spectrum[13] and had relatively low spectral efficiency; later wireless LANs (802.11a and 802.11g) used OFDM for multipath suppression and achieved much higher data rates in the same bandwidth.

3.6.4 Equalization

Equalizers are the most logical alternative for ISI suppression to OFDM, since they don't require additional antennas or bandwidth and have moderate complexity. Equalizers are implemented at the receiver and attempt to reverse the distortion introduced by the channel. Generally, equalizers are broken into two classes: linear and decision directed (nonlinear).

A *linear equalizer* simply runs the received signal through a filter that roughly models the inverse of the channel. The problem with this approach is that it inverts not only the channel but also the received noise. This noise enhancement can severely degrade the receiver performance, especially in a wireless channel with deep frequency fades. Linear receivers are relatively simple to implement but achieve poor performance in a time-varying and severe-ISI channel.

A *nonlinear equalizer* uses previous symbol decisions made by the receiver to cancel out their subsequent interference and so are often called *decision-feedback equalizers* (DFEs). Recall that the problem with multipath is that many separate paths are received at different time offsets, so prior symbols cause interference with later symbols. If the receiver knows the prior symbols, it can subtract out their interference. One problem with this approach is that it is common to make mistakes about what the prior symbols were, especially at low SNR, which causes error propagation. Also, nonlinear equalizers pay for their improved performance relative to linear receivers with sophisticated training and increased computational complexity.

Maximum-likelihood sequence detection (MLSD) is the optimum method of suppressing ISI but has complexity that scales like $O(M^v)$, where M is the constellation size and v is the channel delay. Therefore, MLSD is generally impractical on channels with a relatively long delay spread or high data rate but is often used in some low-data-rate outdoor systems, such as GSM. For a high-data-rate broadband wireless channel, MLSD is not expected to be practical in the foreseeable future, although suboptimal approximations, such as delayed-decision-feedback sequence estimation (DDFSE),

---

13. Note that the definition of spread spectrum is somewhat loose. The FCC has labeled even the 11Mbps in 20MHz 802.11b system as "spread spectrum," but this is generally inconsistent with its historical definition that the bandwidth be much larger than the data rate. See, for example, [26, 31, 33] and the references therein.

which is a hybrid of MLSD and decision-feedback equalization [7] and reduced-state sequence estimation (RSSE) [9] are reasonable suboptimal approximations for MLSD in practical scenarios [12].

3.6.5 The Multicarrier Concept

The philosophy of multicarrier modulation is that rather than fighting the time-dispersive ISI channel, why not use its diversity? For this, a large number of subcarriers (L) are used in parallel, so that the symbol time for each goes from $T \rightarrow LT$. In other words, rather than sending a single signal with data rate R and bandwidth B, why not send L signals at the same time, each having bandwidth B/L and data rate R/L? In this way, if $B/L \ll B_c$, each signal will undergo approximately flat fading, and the time dispersion for each signal will be negligible. As long as the number of subcarriers L is large enough, the condition $B/L \ll B_c$ can be met. This elegant idea is the basic principle of orthogonal frequency division multiplexing (OFDM). In the next chapter, we take a close look at this increasingly popular modulation technique, discussing its theoretical basis and implementation challenges.

3.7 Summary and Conclusions

In this chapter, we attempted to understand and characterize the challenging and multifaceted broadband wireless channel.

- The average value of the channel power can be modeled based simply on the distance between the transmitter and the receiver, the carrier frequency, and the pathloss exponent.
- The large-scale perturbations from this average channel can be characterized as *lognormal shadowing*.
- Cellular systems must contend with severe interference from neighboring cells; this interference can be reduced through sectoring and frequency-reuse patterns.
- The small-scale channel effects are known collectively as *fading*. Broadband wireless channels have autocorrelation functions that tell us a lot about their behavior.
- Realistic models for time, frequency, and spatial correlation can be developed from popular statistical channel models, such as Rayleigh, Ricean, and Nakagami.
- A number of diversity-achieving techniques are available for both narrowband and broadband fading.

3.8 Bibliography

[1] J. G. Andrews. Interference cancellation for cellular systems: A contemporary overview. *IEEE Wireless Communications Magazine*, 12(2):19–29, April 2005.

[2] S. Catreux, P. Driessen, and L. Greenstein. Attainable throughput of an interference-limited multiple-input multiple-output (MIMO) cellular system. *IEEE Transactions on Communications*, 49(8):1307–1311, August 2001.

[3] W. Choi and J. G. Andrews. Base station cooperatively scheduled transmission in a cellular MIMO TDMA system. In *Proceedings, Conference on Information Sciences and Systems (CISS)*, March 2006.

[4] W. Choi and J. G. Andrews. Downlink Performance and Capacity of Distributed Antenna Systems in a Multicell Environment. *IEEE Transactions on Wireless Communications*, 6(1), January 2007.

[5] W. Choi and J. Y. Kim. Forward-link capacity of a DS/CDMA system with mixed multirate sources. *IEEE Transactions on Vehicular Technology*, 50(3):737–749, May 2001.

[6] H. Dai, A. Molisch, and H. V. Poor. Downlink capacity of interference-limited MIMO systems with joint detection. *IEEE Transactions on Wireless Communications*, 3(2):442–453, March 2004.

[7] A. Duel-Hallen and C. Heegard. Delayed decision-feedback sequence estimation. *IEEE Transactions on Communications*, 37:428–436, May 1989.

[8] G. Durgin. *Space-Time Wireless Channels*. Prentice Hall, 2003.

[9] M. V. Eyuboglu and S. U. Qureshi. Reduced-state sequence estimation with set partitioning and decision feedback. *IEEE Transactions on Communications*, 36:13–20, January 1988.

[10] L. F. Fenton. The sum of log-normal probability distributions in scatter transmission systems. *IRE Transactions Communications*, 8:57–67, March 1960.

[11] G. Foschini and J. Salz. Digital communications over fading radio channels. *Bell Systems Technical Journal*, pp. 429–456, February 1983.

[12] W. H. Gerstacker and R. Schober. Equalization concepts for EDGE. *IEEE Transactions on Wireless Communications*, 1(1):190–199, January 2002.

[13] A. J. Goldsmith. *Wireless Communications*. Cambridge University Press, 2005.

[14] R. Hasegawa, M. Shirakabe, R. Esmailzadeh, and M. Nakagawa. Downlink performance of a CDMA system with distributed base station. In *Proceedings, IEEE Vehicular Technology Conference*, pp. 882–886, October 2003.

[15] S. Jafar, G. Foschini, and A. Goldsmith. PhantomNet: Exploring optimal multicellular multiple antenna systems. In *Proceedings, IEEE Vehicular Technology Conference*, pp. 24–28, September 2002.

[16] W. C. Jakes. *Microwave Mobile Communications*. Wiley-Interscience, 1974.

[17] G. J. M. Janssen, P. A. Stigter, and R. Prasad. Wideband indoor channel measurements and BER analysis of frequency selective multipath channels at 2.4, 4.75, and 11.5GHz. *IEEE Transactions on Communications*, 44(10):1272–1288, October 1996.

[18] M. Nakagami. The m-distribution: A general formula of intensity distribution of rapid fading. *Statistical Methods in Radio Wave Propagation*, Pergamon, Oxford, England, pp. 3–36, 1960.

[19] T. Okumura, E. Ohmori, and K. Fukuda. Field strength and its variability in VHF and UHF land mobile service. *Review Electrical Communication Laboratory*, pp. 825–873, September 1968.

[20] P. Papazian. Basic transmission loss and delay spread measurements for frequencies between 430 and 5750MHz. *IEEE Transactions on Antennas and Propagation*, 53(2):694–701, February 2005.

[21] D. Parsons. *The Mobile Radio Propagation Channel*. Wiley, 1992.

[22] T. S. Rappaport.*Wireless Communications: Principles and Practice*, 2d ed. Prentice-Hall, 2002.

[23] W. Rho and A. Paulraj. Performance of the distributed antenna systems in a multicell environment. In *Proceedings, IEEE Vehicular Technology Conference*; pp. 587–591, April 2003.

[24] S. O. Rice. Statistical properties of a sine wave plus random noise. *Bell Systems Technical Journal*, 27:109–157, January 1948.

[25] P. Schniter. Low-complexity equalization of OFDM in doubly-selective channels. *IEEE Transactions on Signal Processing*, 52(4):1002–1011, April 2004.

[26] R. Scholtz. The origins of spread-spectrum communications. *IEEE Transactions on Communications*, 30(5):822–854, May 1982.

[27] S. Schwartz and Y. S. Yeh. On the distribution function and moments of power sums with log-normal components. *Bell Systems Technical Journal*, 61:1441–1462, September 1982.

[28] S. Y. Seidel, T. Rappaport, S. Jain, M. Lord, and R. Singh. Path loss, scattering and multipath delay statistics in four European cities for digital cellular and microcellular radiotelephone. *IEEE Transactions on Vehicular Technology*, 40(4):721–730, November 1990.

[29] S. Shamai and B. Zaidel. Enhancing the cellular downlink capacity via co-processing at the transmitting end. In *Proceedings, IEEE Vehicular Technology Conference*, pp. 1745–1749, May 2001.

[30] M. K. Simon and M. S. Alouini. *Digital Communication over Generalized Fading Channels: A Unified Approach to the Performance Analysis*. Wiley, 2000.

[31] M. K. Simon, J. K. Omura, R. A. Scholtz, and B. K. Levitt. *Spread Spectrum Communications Handbook*, rev. ed. McGraw-Hill, 1994.

[32] G. L. Stuber. *Principles of Mobile Communication*, 2d ed. Kluwer, 2001.

[33] A. J. Viterbi. *CDMA—Principles of Spread Spectrum Communication*. Addison-Wesley, 1995.

[34] M. D. Yacoub. *Foundations of Mobile Radio Engineering*. CRC Press, 1993.

[35] H. Zhang and H. Dai. Co-channel interference mitigation and cooperative processing in downlink multicell multiuser MIMO networks. *European Journal on Wireless Communications and Networking*, 4th quarter, 2004.

[36] X. Zhao, J. Kivinen, and P. Vainikainen. Propagation Characteristics for Wideband Outdoor Mobile Communications at 5.3GHz. *IEEE Journal on Selected Areas in Communications*, 20(3):507–514, April 2002.

[37] K. Karakayli, G. J. Foschini, and R.A. Valenzuala. Network coordination for spectrally efficient communications in cellular systems. *IEEE Wireless Communications Magazine,* 13(4): pp. 56–61, August 2006.

[38] T. C. Ng and W. Yu. Joint optimaization of relay strategies and power allocation in a cooperative cellular network. To appear, *IEEE Journal of Selective Areas of Communications*.

[39] S. S. (Shitz), O. Somekh, and B. M. Zaidel. Multi-cell communications: An information theoretic perspective. In *Joint Workshop on Communications and Coding (JWCC)*, Florence, Italy, October 2004.

CHAPTER **4**

Orthogonal Frequency Division Multiplexing

Orthogonal frequency division multiplexing (OFDM) is a multicarrier modulation technique that has recently found wide adoption in a widespread variety of high-data-rate communication systems, including digital subscriber lines, wireless LANs (802.11a/g/n), digital video broadcasting, and now WiMAX and other emerging wireless broadband systems such as the proprietary Flash-OFDM developed by Flarion (now QUALCOMM), and 3G LTE and fourth generation cellular systems. OFDM's popularity for high-data-rate applications stems primarily from its efficient and flexible management of intersymbol interference (ISI) in highly dispersive channels.

As emphasized in Chapter 3, as the channel delay spread τ becomes an increasingly large multiple of the symbol time T_s, the ISI becomes very severe. By definition, a high-data-rate system will generally have $\tau \gg T_s$, since the number of symbols sent per second is high. In a non-line of sight (NLOS) system, such as WiMAX, which must transmit over moderate to long distances, the delay spread will also frequently be large. In short, wireless broadband systems of all types will suffer from severe ISI and hence will require transmitter and/or receiver techniques that overcome the ISI. Although the 802.16 standards include single-carrier modulation techniques, the vast majority of, if not all, 802.16-compliant systems will use the OFDM modes, which have also been selected as the preferred modes by the WiMAX Forum.

To develop an understanding of how to use OFDM in a wireless broadband system, this chapter:

- Explains the elegance of multicarrier modulation and how it works in theory
- Emphasizes a practical understanding of OFDM system design, covering such key concepts as the cyclic prefix, frequency equalization, and synchronization[1]

---

1. Channel estimation for OFDM is covered in Chapter 5 in the context of MIMO-OFDM.

• Discusses implementation issues for WiMAX systems, such as the peak-to-average ratio, and provides illustrative examples related to WiMAX.

4.1 Multicarrier Modulation

The basic idea of multicarrier modulation is quite simple and follows naturally from the competing desires for high data rates and ISI-free channels. In order to have a channel that does not have ISI, the symbol time T_s has to be larger—often significantly larger—than the channel delay spread τ. Digital communication systems simply cannot function if ISI is presents; an error floor quickly develops, and as T_s approaches or falls below τ, the bit error rate becomes intolerable. As noted previously, for wideband channels that provide the high data rates needed by today's applications, the desired symbol time is usually much smaller than the delay spread, so intersymbol interference is severe.

In order to overcome this problem, multicarrier modulation divides the high-rate transmit bit stream into L lower-rate substreams, *each* of which has $T_s/L \gg \tau$ and is hence effectively ISI free. These individual substreams can then be sent over L parallel subchannels, maintaining the total desired data rate. Typically, the subchannels are orthogonal under ideal propagation conditions, in which case multicarrier modulation is often referred to as orthogonal frequency division multiplexing (OFDM). The data rate on each of the subchannels is much less than the total data rate, so the corresponding subchannel bandwidth is much less than the total system bandwidth. The number of substreams is chosen to ensure that each subchannel has a bandwidth less than the coherence bandwidth of the channel, so the subchannels experience relatively flat fading. Thus, the ISI on each subchannel is small. Moreover, in the digital implementation of OFDM, the ISI can be completely eliminated through the use of a cyclic prefix.

Example 4.1 A certain wideband wireless channel has a delay spread of 1μ sec. We assume that in order to overcome ISI, $T_s \geq 10\tau$.
1. What is the maximum bandwidth allowable in this system?
2. If multicarrier modulation is used and we desire a 5MHz bandwidth, what is the required number of subcarriers?

For question 1, if it is assumed that $T_s = 10\tau$ in order to satisfy the ISI-free condition, the maximum bandwidth would be $1/T_s = .1/\tau = 100$ KHz, far below the intended bandwidths for WiMAX systems.

In question 2, if multicarrier modulation is used, the symbol time goes to $T = LT_s$. The delay-spread criterion mandates that the new symbol time is still bounded to 10 percent of the delay spread: $(LT_s)^{-1} = 100$ Khz. But the 5MHz bandwidth requirement gives $(T_s)^{-1} = 5$ MHz Hence, $L \geq 50$ allows the full 5MHz bandwidth to be used with negligible ISI.

In its simplest form, multicarrier modulation divides the wideband incoming data stream into L narrowband substreams, each of which is then transmitted over a different orthogonal-frequency subchannel. As in Example 4.1, the number of substreams L is chosen to make the symbol time on each substream much greater than the delay spread of the channel or, equivalently, to make the substream bandwidth less than the channel-coherence bandwidth. This ensures that the substreams will not experience significant ISI.

A simple illustration of a multicarrier transmitter and receiver is given in Figure 4.1, Figure 4.2, and Figure 4.3. Essentially, a high data rate signal of rate R bps and with a passband bandwidth B is broken into L parallel substreams, each with rate R/L and passband bandwidth B/L. After passing through the channel $H(f)$, the received signal would appear as shown in Figure 4.3, where we have assumed for simplicity that the pulse shaping allows a perfect spectral shaping so that there is no subcarrier overlap.[2] As long as the number of subcarriers is sufficiently large to allow the subcarrier bandwidth to be much less than the coherence bandwidth, that is, $B/L \ll B_c$, it can be ensured that each subcarrier experiences approximately flat fading. The mutually orthogonal signals can then be individually detected, as shown in Figure 4.2.

Hence, the multicarrier technique has an interesting interpretation in both the time and frequency domains. In the time domain, the symbol duration on each subcarrier has increased to $T = LT_s$, so letting L grow larger ensures that the symbol duration exceeds the channel-delay spread, $T \gg \tau$, which is a requirement for ISI-free communication. In the frequency domain, the subcarriers have bandwidth $B/L \ll B_c$, which ensures flat fading, the frequency-domain equivalent to ISI-free communication.

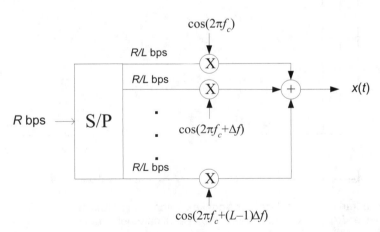

Figure 4.1 A basic multicarrier transmitter: A high-rate stream of R bps is broken into L parallel streams, each with rate R/L and then multiplied by a different carrier frequency.

2. In practice, there would be some roll-off factor of β, so the actual consumed bandwidth of such a system would be $(1 + \beta)B$. As we will see, however, OFDM avoids this inefficiency by using a cyclic prefix.

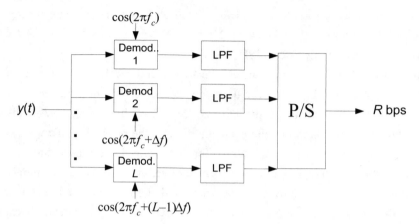

Figure 4.2 A basic multicarrier receiver: Each subcarrier is decoded separately, requiring L independent receivers.

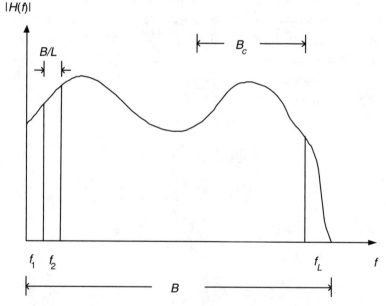

Figure 4.3 The transmitted multicarrier signal experiences approximately flat fading on each subchannel, since $B/L \ll B_c$, even though the overall channel experiences frequency-selective fading: $B > B_c$.

Although this simple type of multicarrier modulation is easy to understand, it has several crucial shortcomings. First, in a realistic implementation, a large bandwidth penalty will be inflicted, since the subcarriers can't have perfectly rectangular pulse shapes and still be time limited. Additionally, very high quality (and hence, expensive) low-pass filters will be required to maintain the orthogonality of the subcarriers at the receiver. Most important, this scheme requires L independent RF units and demodulation paths. In Section 4.2, we show how OFDM overcomes these shortcomings.

4.2 OFDM Basics

In order to overcome the daunting requirement for L RF radios in both the transmitter and the receiver, OFDM uses an efficient computational technique, discrete Fourier transform (DFT), which lends itself to a highly efficient implementation commonly known as the fast Fourier transform (FFT). The FFT and its inverse, the IFFT, can create a multitude of orthogonal subcarriers using a single radio.

4.2.1 Block Transmission with Guard Intervals

We begin by grouping L data symbols into a block known as an *OFDM symbol*. An OFDM symbol lasts for a duration of T seconds, where $T = LT_s$. In order to keep each OFDM symbol independent of the others after going through a wireless channel, it is necessary to introduce a guard time between OFDM symbols:

This way, after receiving a series of OFDM symbols, as long as the guard time T_g is larger than the delay spread of the channel τ, each OFDM symbol will interfere only with itself:

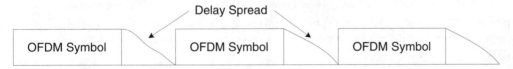

Put simply, OFDM transmissions allow ISI *within* an OFDM symbol. But by including a sufficiently large guard band, it is possible to guarantee that there is no interference *between* subsequent OFDM symbols.

4.2.2 Circular Convolution and the DFT

Now that subsequent OFDM symbols have been rendered orthogonal with a guard interval, the next task is to attempt to remove the ISI *within* each OFDM symbol. As described in Chapter 3, when an input data stream $x[n]$ is sent through a linear time-invariant Finite Impulse Response (FIR) channel $h[n]$, the output is the linear convolution of the input and the channel: $y[n] = x[n] * h[n]$. However, let's imagine computing $y[n]$ in terms of a *circular* convolution:

$$y[n] = x[n] \circledast h[n] = h[n] \circledast x[n], \tag{4.1}$$

where

$$x[n] \circledast h[n] = h[n] \circledast x[n] \triangleq \sum_{k=0}^{L-1} h[k]x[n-k]_L, \tag{4.2}$$

and the circular function $x[n]_L = x[n \bmod L]$ is a *periodic* version of $x[n]$ with period L. In other words, each value of $y[n] = h[n] \circledast x[n]$ is the sum of the product of L terms.[3]

In this case of circular convolution, it would then be possible to take the DFT of the channel output $y[n]$ to get

$$\text{DFT}\{y[n]\} = \text{DFT}\{h[n] \circledast x[n]\}, \tag{4.3}$$

which yields in the frequency domain

$$Y[m] = H[m]X[m]. \tag{4.4}$$

Note that the duality between circular convolution in the time domain and simple multiplication in the frequency domain is a property unique to the DFT. The L point DFT is defined as

$$\text{DFT}\{x[n]\} = X[m] \triangleq \frac{1}{\sqrt{L}} \sum_{n=0}^{L-1} x[n] e^{-j\frac{2\pi nm}{L}}, \tag{4.5}$$

whereas its inverse, the IDFT, is defined as

$$\text{IDFT}\{X[m]\} = x[n] \triangleq \frac{1}{\sqrt{L}} \sum_{m=0}^{L-1} X[m] e^{j\frac{2\pi nm}{L}}. \tag{4.6}$$

Referring to Equation (4.4), this innocent formula describes an ISI-free channel in the frequency domain, where each input symbol $X[m]$ is simply scaled by a complex value $H[m]$. So, given knowledge of the channel-frequency response $H[m]$ at the receiver, it is trivial to recover the input symbol by simply computing

$$\hat{X}[m] = \frac{Y[m]}{H[m]}, \tag{4.7}$$

where the estimate $\hat{X}[m]$ will generally be imperfect, owing to additive noise, cochannel interference, imperfect channel estimation, and other imperfections. Nevertheless, in principle, the ISI, which is the most serious form of interference in a wideband channel, has been mitigated.

A natural question to ask at this point is, Where does this circular convolution come from? After all, nature provides a linear convolution when a signal is transmitted through a linear channel. The answer is that this circular convolution can be faked by adding a specific prefix, the *cyclic prefix* (CP), onto the transmitted vector.

---

3. For a more thorough tutorial on circular convolution, see [35] or the Connexions Web resource http://cnx.rice.edu/.

4.2.3 The Cyclic Prefix

The key to making OFDM realizable in practice is the use of the FFT algorithm, which has low complexity. In order for the IFFT/FFT to create an ISI-free channel, the channel must appear to provide a circular convolution, as seen in Equation (4.4). Adding cyclic prefix to the transmitted signal, as is shown in Figure 4.4, creates a signal that appears to be $x[n]_L$, and so $y[n] = x[n] \circledast h[n]$.

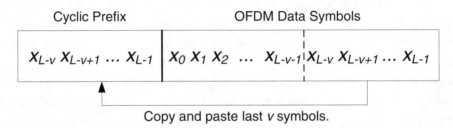

Figure 4.4 The OFDM cyclic prefix

Let's see how this works. If the maximum channel delay spread has a duration of $v+1$ samples, adding a guard band of at least v samples between OFDM symbols makes each OFDM symbol independent of those coming before and after it, and so only a single OFDM symbol can be considered. Representing such an OFDM symbol in the time domain as a length L vector gives

$$\mathbf{x} = [x_1 \ x_2 \dots x_L].\tag{4.8}$$

After applying a cyclic prefix of length v, the transmitted signal is

$$\mathbf{x}_{cp} = [\underbrace{x_{L-v} \ x_{L-v+1} \dots x_{L-1}}_{\text{Cyclic Prefix}} \ \underbrace{x_0 \ x_1 \dots x_{L-1}}_{\text{Original Data}}].\tag{4.9}$$

The output of the channel is by definition $\mathbf{y}_{cp} = \mathbf{h} * \mathbf{x}_{cp}$, where $\mathbf{h}$ is a length $v+1$ vector describing the impulse response of the channel during the OFDM symbol.[4] The output $\mathbf{y}_{cp}$ has $(L+v)+(v+1)-1 = L+2v$ samples. The first v samples of $\mathbf{y}_{cp}$ contain interference from the preceding OFDM symbol and so are discarded. The last v samples disperse into the subsequent OFDM symbol, so also are discarded. This leaves exactly L samples for the desired output $\mathbf{y}$, which is precisely what is required to recover the L data symbols embedded in $\mathbf{x}$.

Our claim is that these L samples of $\mathbf{y}$ will be equivalent to $\mathbf{y} = \mathbf{x} \circledast \mathbf{h}$. Various proofs are possible; the most intuitive is a simple inductive argument. Consider y_0, the first element in $\mathbf{y}$. As shown in Figure 4.5, owing to the cyclic prefix, y_0 depends on x_0 and the circularly wrapped values $x_{L-v} \dots x_{L-1}$. That is:

---

4. It can generally be reasonably assumed that the channel remains constant over an OFDM symbol, since the OFDM symbol time T is usually much less than the channel coherence time, T_c.

$$
\begin{aligned}
y_0 &= h_0 x_0 + h_1 x_{L-1} + \ldots + h_v x_{L-v} \\
y_1 &= h_0 x_1 + h_1 x_0 + \ldots + h_v x_{L-v+1} \\
&\vdots \\
y_{L-1} &= h_0 x_{L-1} + h_1 x_{L-2} + \ldots + h_v x_{L-v-1}.
\end{aligned}
$$

$$(4.10)$$

From inspecting Equation (4.2), we see that this is exactly the value of $y_0, y_1, \ldots, y_{L-1}$ resulting from $\mathbf{y} = \mathbf{x} \circledast \mathbf{h}$. Thus, by mimicking a circular convolution, a cyclic prefix that is at least as long as the channel duration allows the channel output $\mathbf{y}$ to be decomposed into a simple multiplication of the channel frequency response $\mathbf{H} = \text{DFT}\{\mathbf{h}\}$ and the channel frequency domain input, $\mathbf{X} = \text{DFT}\{\mathbf{x}\}$.

The cyclic prefix, although elegant and simple, is not entirely free. It comes with both a bandwidth and power penalty. Since v redundant symbols are sent, the required bandwith for OFDM increases from B to $(L+v/L)B$. Similarly, an additional v symbol must be counted against the transmit-power budget. Hence, the cyclic prefix carries a power penalty of $10\log_{10}(L+v/L)$ dB in addition to the bandwidth penalty. In summary, the use of the cyclic prefix entails data rate and power losses that are both

$$
\text{Rate Loss} = \text{Power Loss} = \frac{L}{L+v}.
$$

The "wasted" power has increased importance in an interference-limited wireless system, causing interference to neighboring users. One way to reduce the transmit-power penalty is noted in Sidebar 4.1.

It can be noted that for $L \gg v$, the inefficiency owing to the cyclic prefix can be made arbitrarily small by increasing the number of subcarriers. However, as the later parts of this chapter explain, numerous other important sacrifices must be made as L grows large. As with most system design problems, desirable properties, such as efficiency, must be traded off against cost and required tolerances.

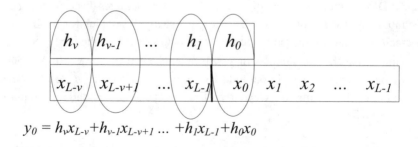

$$
y_0 = h_v x_{L-v} + h_{v-1} x_{L-v+1} \ldots + h_1 x_{L-1} + h_0 x_0
$$

Figure 4.5 Circular convolution created by OFDM cyclic prefix

Example 4.2 In this example, we will find the minimum and maximum date rate loss due to the cyclic prefix in WiMAX. We will consider a 10MHz channel bandwidth, where the maximum delay spread has been determined to be $\tau = 5\,\mu$sec. From Table 8.3, it can be seen that the choices for guard band size in WiMAX are $G = \{1/4, 1/8, 1/16, 1/32\}$ and the number of subcarriers must be one of $L = \{128, 256, 512, 1024, 2048\}$.

At a symbol rate of 10MHz, a delay spread of $5\,\mu$ sec affects 50 symbols, so we require a CP length of at least $v = 50$.

The minimum overhead will be for the largest number of subcarriers, so this yields $L = 2048$. In this case, $v/L = 50/2048 = 1/40.96$ so the minimum guard band of $1/32$ will suffice. Hence, the data rate loss is only 1/32 in this case. The maximum overhead occurs when the number of subcarriers is small.

If $L = 128$, then $v/L = 50/128$, so even an overhead of $1/4$ won't be sufficient to preserve subcarrier orthogonality. More subcarriers are required. For $L = 256$, $v/L < 1/4$, so in this case ISI-free operation is possible, but at a data rate loss of 1/4.

Sidebar 4.1 An Alternative Prefix

One alternative to the cyclic prefix is to use a zero prefix, which constitutes a null guard band. One commercial system that proposes this is the Multiband OFDM system that has been standardized for ultrawideband (UWB) operation by the WiMedia Alliance.[a] As shown in Figure 4.6, the multiband OFDM transmitter simply sends a prefix of null data so that there is no transmitter-power penalty. At the receiver, the "tail" can be added back in, which recreates the effect of a cyclic prefix, so the rest of the OFDM system can function as usual.

Why wouldn't every OFDM system use a zero prefix, then, since it reduces the transmit power by $10\log_{10}((L+v)/L)$ dB? There are two reasons. First, the zero prefix generally increases the receiver power by $10\log_{10}((L+v)/L)$ dB, since the tail now needs to be received, whereas with a cyclic prefix, it can be ignored. Second, additional noise from the received tail symbols is added back into the signal, causing a higher noise power $\sigma^2 \rightarrow ((L+v)/L)\sigma^2$. The designer must weigh these trade-offs to determine whether a zero or a cyclic prefix is preferable. WiMAX systems use a cyclic prefix.

a. This was originally under the context of the IEEE 802.15.3 subcommittee, which has since disbanded.

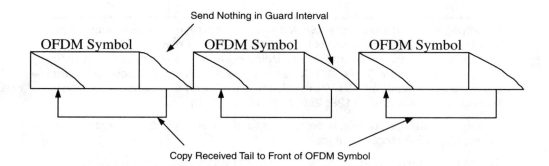

Figure 4.6 The OFDM zero prefix allows the circular channel to be recreated at the receiver.

4.2.4 Frequency Equalization

In order for the received symbols to be estimated, the complex channel gains for each subcarrier must be known, which corresponds to knowing the amplitude and phase of the subcarrier. For simple modulation techniques, such as QPSK, that don't use the amplitude to transmit information, only the phase information is sufficient.

After the FFT is performed, the data symbols are estimated using a one-tap frequency-domain equalizer, or FEQ, as

$$\hat{X}_l = \frac{Y_l}{H_l} \, , \tag{4.11}$$

where H_l is the *complex* response of the channel at the frequency $f_c + (l-1)\Delta f$, and therefore it both corrects the phase and equalizes the amplitude before the decision device. Note that although the FEQ inverts the channel, there is no problematic noise enhancement or coloring, since both the signal and the noise will have their powers directly scaled by $|1/H_l|^2$.

4.2.5 An OFDM Block Diagram

Let us now briefly review the key steps in an OFDM communication system (Figure 4.7). In OFDM, the encoding and decoding are done in the frequency domain, where $\mathbf{X}$, $\mathbf{Y}$, and $\hat{\mathbf{X}}$ contain the L transmitted, received, and estimated data symbols.

1. The first step is to break a wideband signal of bandwidth B into L narrowband signals (subcarriers), each of bandwidth B/L. This way, the aggregate symbol rate is maintained, but each subcarrier experiences flat fading, or ISI-free communication, as long as a cyclic prefix that exceeds the delay spread is used. The L subcarriers for a given OFDM symbol are represented by a vector $\mathbf{X}$, which contains the L current symbols.

2. In order to use a single wideband radio instead of L independent narrowband radios, the subcarriers are modulated using an IFFT operation.

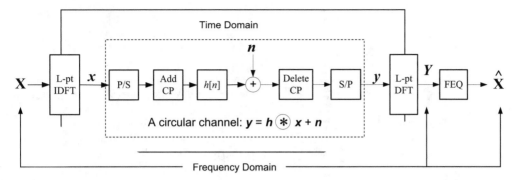

Figure 4.7 An OFDM system in vector notation.

3. In order for the IFFT/FFT to decompose the ISI channel into orthogonal subcarriers, a cyclic prefix of length v must be appended after the IFFT operation. The resulting $L + v$ symbols are then sent in serial through the wideband channel.

4. At the receiver, the cyclic prefix is discarded, and the L received symbols are demodulated, using an FFT operation, which results in L data symbols, each of the form $Y_l = H_l X_l + N_l$ for subcarrier l.

5. Each subcarrier can then be equalized via an FEQ by simply dividing by the complex channel gain $H[i]$ for that subcarrier. This results in $\hat{X}_l = X_l + N_l / H_l$.

We have neglected a number of important practical issues thus far. For example, we have assumed that the transmitter and the receiver are perfectly synchronized and that the receiver perfectly knows the channel, in order to perform the FEQ. In the next section, we present the implementation issues for OFDM in WiMAX.

4.3 An Example: OFDM in WiMAX

To gain an appreciation for the time- and frequency-domain interpretations of OFDM, WiMAX systems can be used as an example. Although simple in concept, the subtleties of OFDM can be confusing if each signal-processing step is not understood. To ground the discussion, we consider a passband OFDM system and then give specific values for the important system parameters.

Figure 4.8 shows a passband OFDM modulation engine. The inputs to this figure are L independent QAM symbols (the vector **X**), and these L symbols are treated as separate subcarriers. These L data-bearing symbols can be created from a bit stream by a symbol mapper and serial-to-parallel convertor (S/P). The L-point IFFT then creates a time-domain L-vector **x** that is cyclic extended to have length $L(1 + G)$, where G is the fractional overhead. This longer vector is then parallel-to-serial (P/S) converted into a wideband digital signal that can be amplitude modulated with a single radio at a carrier frequency of $f_c = \omega_c / 2\pi$.

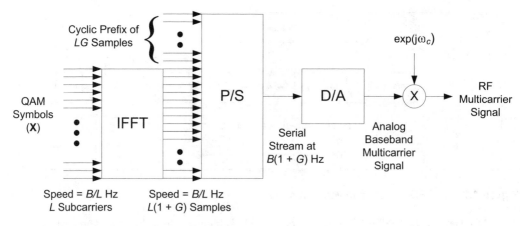

Figure 4.8 Closeup of the OFDM baseband transmitter

This procedure appears to be relatively straightforward, but in order to be a bit less abstract, we now use some plausible values for the parameters. (Chapter 8 enumerates all the legal values for the OFDM parameters B, L, L_d, and G.) The key OFDM parameters are summarized in Table 4.1, along with some potential numerical values for them. As an example, if 16 QAM modulation were used ($M = 16$), the raw (neglecting coding) data rate of this WiMAX system would be

$$R = \frac{B}{L} \frac{L_d \log_2(M)}{1+G} \tag{4.12}$$

$$= \frac{10^7 \text{MHz}}{1024} \frac{768 \log_2(16)}{1.125} = 24 \text{Mbps.} \tag{4.13}$$

In words, each L_d data-carrying subcarriers of bandwidth B/L carries $\log_2(M)$ bits of data. An additional overhead penalty of $(1+G)$ must be paid for the cyclic prefix, since it consists of redundant information and sacrifices the transmission of actual data symbols.

4.4 Timing and Frequency Synchronization

In order to demodulate an OFDM signal, the receiver needs to perform two important synchronization tasks. First, the timing offset of the symbol and the optimal timing instants need to be determined. This is referred to as *timing synchronization*. Second, the receiver must align its carrier frequency as closely as possible with the transmitted carrier frequency. This is referred to as *frequency synchronization*. Compared to single-carrier systems, the timing-synchronization requirements for OFDM are in fact somewhat relaxed, since the OFDM symbol structure naturally accommodates a reasonable degree of synchronization error. On the other hand, frequency-synchronization requirements are significantly more stringent, since the orthogonality of the data symbols is reliant on their being individually discernible in the frequency domain.

Table 4.1 Summary of OFDM Parameters

Symbol	Description	Relation	Example WiMAX value
$B*$	Nominal bandwidth	$B = 1/T_s$	10MHz
$L*$	Number of subcarriers	Size of IFFT/FFT	1024
$G*$	Guard fraction	% of L for CP	1/8
L_d*	Data subcarriers	L–pilot/null subcarriers	768
T_s	Sample time	$T_s = 1/B$	1 μ sec
N_g	Guard symbols	$N_g = GL$	128
T_g	Guard time	$T_g = T_s N_g$	12.8 μ sec
T	OFDM symbol time	$T = T_s(L + N_g)$	115.2 μ sec
B_{sc}	Subcarrier bandwidth	$B_{sc} = B/L$	9.76 KHz

* Denotes WiMAX-specified parameters; the other OFDM parameters can all be derived from these values.

Figure 4.9 shows an OFDM symbol in time (a) and frequency (b). In the time domain, the IFFT effectively modulates each data symbol onto a unique carrier frequency. In Figure 4.9, only two of the carriers are shown: The transmitted signal is the superposition of all the individual carriers. Since the time window is $T = 1\mu sec$ and a rectangular window is used, the frequency response of each subcarrier becomes a "sinc" function with zero crossings every $1/T = 1$MHz. This can be confirmed using the Fourier transform $\mathcal{F}\{\cdot\}$, since

$$\mathcal{F}\{\cos(2\pi f_c) \cdot \text{rect}(t/T)\} = \mathcal{F}\{\cos(2\pi f_c)\} * \mathcal{F}\{\text{rect}(2t/T)\} \tag{4.14}$$

$$= \text{sinc}\left(T(f - f_c)\right), \tag{4.15}$$

where $\text{rect}(x) = 1, x \in (-0.5, 0.5)$, and zero elsewhere. This frequency response is shown for $L = 8$ subcarriers in Figure 4.9b.

The challenge of timing and frequency synchronization can be appreciated by inspecting these two figures. If the timing window is slid to the left or the right, a unique phase change will be introduced to each of the subcarriers. In the frequency domain, if the carrier frequency synchronization is perfect, the receiver samples at the peak of each subcarrier, where the desired subcarrier amplitude is maximized, and the intercarrier interference (ICI) are zero. However, if

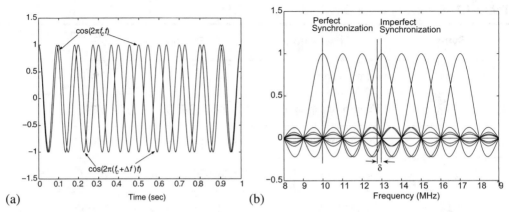

Figure 4.9 OFDM synchronization in (a) time and (b) frequency. Here, two subcarriers in the time domain and eight subcarriers in the frequency domain are shown, where $f_c = 10$ MHz, and the subcarrier spacing $\Delta f = 1$ Hz.

the carrier frequency is misaligned by some amount δ, some of the desired energy is lost, and more significantly, intercarrier interference is introduced.

The following two subsections examine timing and frequency synchronization. Although the development of good timing and frequency synchronization algorithms for WiMAX systems is the responsibility of each equipment manufacturer, we give some general guidelines on what is required of a synchronization algorithm and discuss the penalty for imperfect synchronization. It should be noted that synchronization is one of the most challenging problems in OFDM implementation, and the development of efficient and accurate synchronization algorithms presents an opportunity for technical differentiation and intellectual property.

4.4.1 Timing Synchronization

The effect of timing errors in symbol synchronization is somewhat relaxed in OFDM owing to the presence of a cyclic prefix. In Section 4.2.3, we assumed that only the L time-domain samples after the cyclic prefix were used by the receiver. Indeed, this corresponds to "perfect" timing synchronization, and in this case, even if the cyclic prefix length N_g is equivalent to the length of the channel impulse response v, successive OFDM symbols can be decoded ISI free.

If perfect synchronization is not maintained, it is still possible to tolerate a timing offset of τ seconds without any degradation in performance, as long as $0 \leq \tau \leq T_m - T_g$, where T_g is the guard time (cyclic prefix duration), and T_m is the maximum channel delay spread. Here, $\tau < 0$ corresponds to sampling earlier than at the ideal instant, whereas $\tau > 0$ is later than the ideal instant. As long as $0 \leq \tau \leq T_m - T_g$, the timing offset can be included by the channel estimator in the complex gain estimate for each subchannel, and the appropriate phase shift can be applied by the FEQ without any loss in performance—at least in theory. This acceptable range of τ is referred to as the timing-synchronization margin and is shown in Figure 4.10.

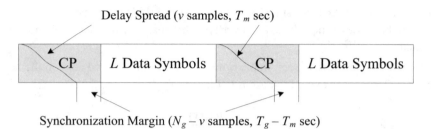

Figure 4.10 Timing-synchronization margin

On the other hand, if the timing offset τ is not within this window $0 \le \tau \le T_m - T_g$, intersymbol interference occurs regardless of whether the phase shift is appropriately accounted for. This can be confirmed intuitively for the scenario that $\tau > 0$ and for $\tau < T_m - T_g$. For the case $\tau > 0$, the receiver loses some of the desired energy, since only the delayed version of the early samples $x_0, x_1, \ldots$ is received, and incorporates undesired energy from the subsequent symbol. Similarly for $\tau < T_m - T_g$: Desired energy is lost while interference from the preceding symbol is included in the receive window. For both of these scenarios, the SNR loss can be approximated by

$$\Delta SNR(\tau) \approx -2 \left(\frac{\tau}{LT_s} \right)^2, \tag{4.16}$$

which makes intuitive sense and has been shown more rigorously in the literature on synchronization for OFDM [40]. Important observations from this expression follow.

- SNR decreases quadratically with the timing offset.
- Longer OFDM symbols are increasingly immune from timing offset; that is, more subcarriers help.
- Since in general $\tau \ll LT_s$, timing-synchronization errors are not that critical as long as the induced phase change is corrected.

In summary, to minimize SNR loss owing to imperfect timing synchronization, the timing errors should be kept small compared to the guard interval, and a small margin in the cyclic prefix length is helpful.

4.4.2 Frequency Synchronization

OFDM achieves a higher degree of bandwidth efficiency than do other wideband systems. The subcarrier packing is extremely tight compared to conventional modulation techniques, which require a guard band on the order of 50 percent or more, in addition to special transmitter architectures, such as the Weaver architecture or single-sideband modulation, that suppress the redundant negative-frequency portion of the passband signal. The price to be paid for this bandwidth

efficiency is that the multicarrier signal shown in Figure 4.9 is very sensitive to frequency offsets, owing to the fact that the subcarriers overlap rather than having each subcarrier truly spectrally isolated.

The form of the subcarriers seen in the right side of Figure 4.9, and also on the book cover,
are called "sinc" forms. The sinc function is defined as

$$\text{sinc}(x) = \frac{\sin(\pi x)}{\pi x}.$$

With this definition, it can be confirmed that $\text{sinc}(0) = 1$ and that zero crossings occur at ± 1, ± 2,
± 3, ... Sinc functions occur commonly because they are the frequency response of a rectangular
function. Since the sine waves existing in each OFDM symbol are truncated every T seconds, the
width of the main lobe of the subcarrier sinc functions is $2/T$, i.e., there are zero crossings every
$1/T$ Hz. Therefore, N subcarriers can be packed into a bandwidth of N/T Hz, with the tails of the
subcarriers trailing off on either side, as can be seen in the right side of Figure 4.9.

Since the zero crossings of the frequency domain sinc pulses all line up as seen in
Figure 4.9, as long as the frequency offset $\delta = 0$, there is no interference between the subcarriers.
One intuitive interpretation for this is that since the FFT is essentially a frequency-sampling
operation, if the frequency offset is negligible, the receiver simply samples **y** at the peak points
of the sinc functions, where the ICI is zero from all the neighboring subcarriers.

In practice, of course, the frequency offset is not always zero. The major causes for this are
mismatched oscillators at the transmitter and the receiver and Doppler frequency shifts owing to
mobility. Since precise crystal oscillators are expensive, tolerating some degree of frequency offset is essential in a consumer OFDM system such as WiMAX. For example, if an oscillator is
accurate to 0.1 parts per million (ppm), $f_{\text{offset}} \approx (f_c)(0.1\text{ppm})$. If $f_c = 3\text{GHz}$ and the Doppler is
100Hz, $f_{\text{offset}} = 300 + 100\text{Hz}$, which will degrade the orthogonality of the received signal, since
now the received samples of the FFT will contain interference from the adjacent subcarriers. We
now analyze this intercarrier interference in order to better understand its effect on OFDM
performance.

The matched filter receiver corresponding to subcarrier l can be simply expressed for the
case of rectangular windows, neglecting the carrier frequency, as

$$x_l(t) = X_l e^{j\frac{2\pi l t}{LT_s}}, \tag{4.17}$$

where $1/LT_s = \Delta f$, and again LT_s is the duration of the data portion of the OFDM symbol:
$T = T_g + LT_s$. An interfering subcarrier m can be written as

$$x_{l+m}(t) = X_m e^{j\frac{2\pi(l+m)t}{LT_s}}. \tag{4.18}$$

If the signal is demodulated with a fractional frequency offset of δ, $|\delta| \leq \frac{1}{2}$

$$\hat{x}_{l+m}(t) = X_m e^{j\frac{2\pi(l+m+\delta)t}{LT_s}}. \tag{4.19}$$

The ICI between subcarriers l and $l + m$ using a matched filter, the FFT, is simply the inner product between them:

$$I_m = \int_0^{LT_s} x_l(t)\hat{x}_{l+m}(t)dt = \frac{LT_s X_m \left(1 - e^{-j2\pi(\delta+m)}\right)}{j2\pi(m+\delta)}. \tag{4.20}$$

It can be seen that in Equation (4.20), $\delta = 0 \Rightarrow I_m = 0$, and $m = 0 \Rightarrow I_m = 0$, as expected. The total average ICI energy per symbol on subcarrier l is then

$$ICI_l = E[\sum_{m \neq l} | I_m |^2] \approx C_0 (LT_s \delta)^2 \mathcal{E}_x, \tag{4.21}$$

where C_0 is a constant that depends on various assumptions, and $\mathcal{E}_x$ is the average symbol energy [27, 37]. The approximation sign is used because this expression assumes that there are an infinite number of interfering subcarriers. Since the interference falls off quickly with m, this assumption is very accurate for subcarriers near the middle of the band and is pessimistic by a factor of 2 at either end of the band.

The SNR loss induced by frequency offset is given by

$$\Delta SNR = \frac{\mathcal{E}_x / N_o}{\mathcal{E}_x / \left(N_o + C_0 (LT_s \delta)^2\right)} \tag{4.22}$$

$$= 1 + C_0 (LT_s \delta)^2 SNR. \tag{4.23}$$

Important observations from the ICI expression (Equation (4.23)) and Figure 4.11 follow.

- SNR decreases quadratically with the frequency offset.
- SNR decreases quadratically with the number of subcarriers.
- The loss in SNR is also proportional to the SNR itself.
- In order to keep the loss negligible—say, less than 0.1 dB, the relative frequency offset needs to be about 1 percent to 2 percent of the subcarrier spacing or even lower, to preserve high SNRs.
- Therefore, this is a case in which reducing the CP overhead by increasing the number of subcarriers causes an offsetting penalty, introducing a trade-off.

In order to further reduce the ICI for a given choice of L, nonrectangular windows can also be used [31, 38].

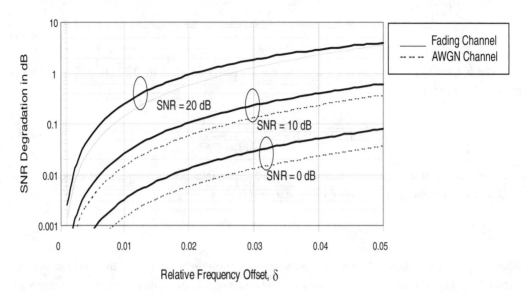

Figure 4.11 SNR loss as a function of the frequency offset δ, relative to the subcarrier spacing.

4.4.3 Obtaining Synchronization in WiMAX

The preceding two sections discussed the consequences of imperfect time and frequency synchronization. Many synchronization algorithms have been developed in the literature; a partial list includes [6, 16, 28, 40, 43]. Generally, the methods can be categorized as based on either pilot symbol or blind—cyclic prefix.

In the first category, known pilot symbols are transmitted. Since the receiver knows what was transmitted, attaining quick and accurate time and frequency synchronization is easy, but at the cost of surrendering some throughput. In the WiMAX downlink, the preamble consists of a known OFDM symbol that can be used to attain initial synchronization. In the WiMAX uplink, the periodic ranging (described in Chapter 8) can be used to synchronize. Since WiMAX is an OFDM-based system with many competing users, each user implements the synchronization at the mobile. This requires the base station to communicate the frequency offset to the MS.

Blind means that pilot symbols are not available to the receiver, so in the second category, the receiver must do the best it can without explicitly being able to determine the effect of the channel. In the absence of pilot symbols, the cyclic prefix, which contains redundancy, can also be used to attain time and frequency synchronization [40]. This technique is effective when the number of subcarriers is large or when the offsets are estimated over a number of consecutive symbols. The principal benefit of CP-based methods is that pilot symbols are not needed, so the data rate can nominally be increased. In WiMAX, accurate synchronization, and especially channel estimation, are considered important enough to warrant the use of pilot symbols, so the blind techniques are not usually used for synchronization. They could be used to track the channel in between preambles or ranging signals, but the frequency and timing offsets generally vary slowly enough that this is not required.

4.5 The Peak-to-Average Ratio

OFDM signals have a higher peak-to-average ratio (PAR)—often called a peak-to-average-power ratio (PAPR)—than single-carrier signals do. The reason is that in the time domain, a multicarrier signal is the sum of many narrowband signals. At some time instances, this sum is large and at other times is small, which means that the peak value of the signal is substantially larger than the average value. This high PAR is one of the most important implementation challenges that face OFDM, because it reduces the efficiency and hence increases the cost of the RF power amplifier, which is one of the most expensive components in the radio. In this section, we quantify the PAR problem, explain its severity in WiMAX, and briefly offer some strategies for reducing the PAR.

4.5.1 The PAR Problem

When transmitted through a nonlinear device, such as a high-power amplifier (HPA) or a digital-to-analog converter (DAC) a high peak signal, generates out-of-band energy (spectral regrowth) and in-band distortion (constellation tilting and scattering). These degradations may affect the system performance severely. The nonlinear behavior of an HPA can be characterized by amplitude modulation/amplitude modulation (AM/AM) and amplitude modulation/phase modulation (AM/PM) responses. Figure 4.12 shows a typical AM/AM response for an HPA, with the associated input and output back-off regions (IBO and OBO, respectively).

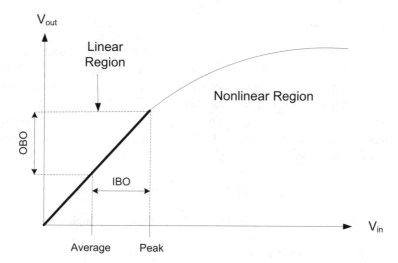

Figure 4.12 A typical power amplifier response. Operation in the linear region is required in order to avoid distortion, so the peak value must be constrained to be in this region, which means that on average, the power amplifier is underutilized by a back-off amount.

To avoid such undesirable nonlinear effects, a waveform with high peak power must be transmitted in the linear region of the HPA by decreasing the average power of the input signal. This is called (input) *backoff* (IBO) and results in a proportional output backoff (OBO). High backoff reduces the power efficiency of the HPA and may limit the battery life for mobile applications. In addition to inefficiency in terms of power, the coverage range is reduced, and the cost of the HPA is higher than would be mandated by the average power requirements.

The input backoff is defined as

$$IBO = 10\log_{10}\frac{P_{inSat}}{\bar{P}_{in}},\qquad(4.24)$$

where P_{inSat} is the saturation power, above which is the nonlinear region, and $\bar{P}_{in}$ is the average input power. The amount of backoff is usually greater than or equal to the PAR of the signal.

The power efficiency of an HPA can be increased by reducing the PAR of the transmitted signal. For example, the efficiency of a class A amplifier is halved when the input PAR is doubled or the operating point (average power) is halved [5, 13]. The theoretical efficiency limits for two classes of HPAs are shown in Figure 4.13. Clearly, it would be desirable to have the average and peak values be as close together as possible in order to maximize the efficiency of the power amplifier.

In addition to the large burden placed on the HPA, a high PAR requires high resolution for both the transmitter's DAC and the receiver's ADC, since the dynamic range of the signal is proportional to the PAR. High-resolution D/A and A/D conversion places an additional complexity, cost, and power burden on the system.

4.5.2 Quantifying the PAR

Since multicarrier systems transmit data over a number of parallel-frequency channels, the resulting waveform is the superposition of L narrowband signals. In particular, each of the L output samples from an L-pt IFFT operation involves the sum of L complex numbers, as can be seen in Equation (4.6). Because of the Central Limit Theorem, the resulting output values $\{x_1, x_2,\ldots,x_L\}$ can be accurately modeled, particularly for large L, as complex Gaussian random variables with zero mean and variance $\sigma^2 = \mathcal{E}_x/2$; that is the real and imaginary parts both have zero mean and variance $\sigma^2 = \mathcal{E}_x/2$. The amplitude of the output signal is

$$|x[n]| = \sqrt{(Re\{x[n]\})^2 + (Im\{x[n]\})^2},\qquad(4.25)$$

which is Rayleigh distributed with parameter σ^2. The output power is therefore

$$|x[n]|^2 = (Re\{x[n]\})^2 + (Im\{x[n]\})^2,\qquad(4.26)$$

which is exponentially distributed with mean $2\sigma^2$. The important thing to note is that the output amplitude and hence power are *random*, so the PAR is not a deterministic quantity, either.

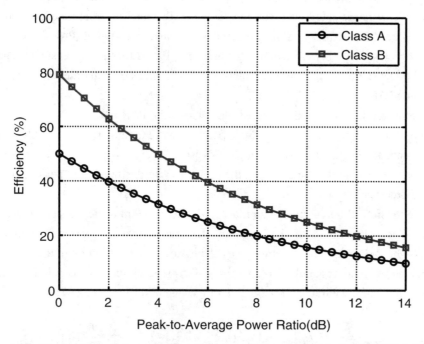

Figure 4.13 Theoretical efficiency limits of linear amplifiers [26]. A typical OFDM PAR is in the 10 dB range, so the power amplifier efficiency is 50% to 75% lower than in a single-carrier system.

The PAR of the transmitted analog signal can be defined as

$$PAR \triangleq \frac{\max\limits_{t} |x(t)|^2}{E[|x(t)|^2]}, \qquad (4.27)$$

where naturally, the range of time to be considered has to be bounded over some interval. Generally, the PAR is considered for a single OFDM symbol, which consists of $L + N_g$ samples, or a time duration of T, as this chapter has explained. Similarly, the discrete-time PAR can be defined for the IFFT output as

$$PAR \triangleq \frac{\max\limits_{l \in (0, L+N_g)} |x_l|^2}{E[|x_l|^2]} = \frac{\mathcal{E}_{max}}{\mathcal{E}_x} \qquad (4.28)$$

It is important to recognize, however, that although the average energy of IFFT outputs $x[n]$ is the same as the average energy of the inputs $X[m]$ and equal to $\mathcal{E}_x$, the analog PAR is *not* generally the same as the PAR of the IFFT samples, owing to the interpolation performed by the D/A convertor. Usually, the analog PAR is higher than the digital (Nyquist sampled)[5] PAR. Since the PA is by definition analog, the analog PAR is what determines the PA performance.

Similarly, digital-signal processing (DSP) techniques developed to reduce the digital PAR may not always have the anticipated effect on the analog PAR, which is what matters. In order to bring the analog PAR expression in Equation (4.27) and the digital PAR expression in Equation (4.28) closer together, oversampling can be considered for the digital signal. That is, a factor M additional samples can be used to interpolate the digital signal in order to better approximate its analog PAR.

It can be proved that the maximum possible value of the PAR is L, which occurs when all the subcarriers add up constructively at a single point. However, although it is possible to choose an input sequence that results in this very high PAR, such an expression for PAR is misleading. For independent binary inputs, for example, the probability of this maximum peak value occurring is on the order of 2^{-L}.

Since the theoretical maximum (or similar) PAR value seldom occurs, a statistical description of the PAR is commonly used. The complementary cumulative distribution function (CCDF = 1 – CDF) of the PAR is the most commonly used measure. The distribution of the OFDM PAR has been studied by many researchers [3, 32, 33, 44]. Among these, van Nee and de Wild [44] introduced a simple and accurate approximation of the CCDF for large $L(\geq 64)$:

$$\mathrm{CCDF}(L,\mathcal{E}_{\max}) = 1 - G(L,\mathcal{E}_{\max}) = 1 - F(L,\mathcal{E}_{\max})^{\beta L} = 1 - \left(1 - \exp(-\frac{\mathcal{E}_{\max}}{2\sigma^2})\right)^{\beta L}, \qquad (4.29)$$

where $\mathcal{E}_{\max}$ is the peak power level and β is a pseudoapproximation of the oversampling factor, which is given empirically as $\beta = 2.8$. Note that the PAR is $\mathcal{E}_{\max}/2\sigma^2$ and $F(L,\mathcal{E}_{\max})$ is the cummulative distribution function (CDF) of a single Rayleigh-distributed subcarrier with parameter σ^2. The basic idea behind this approximation is that unlike a Nyquist-sampled signal, the samples of an oversampled OFDM signal are correlated, making it difficult to derive an exact peak distribution. The CDF of the Nyquist-sampled signal power can be obtained by

$$G(L,\mathcal{E}_{\max}) = P(\max \| x(t) \| \leq \mathcal{E}_{\max}) = F(L,\mathcal{E}_{\max})^L, \qquad (4.30)$$

With this result as a baseline, the oversampled case can be approximated in a heuristic way by regarding the oversampled signal as generated by βL Nyquist-sampled subcarriers. Note, however, that β is not equal to the oversampling factor M. This simple expression is quite effective for generating accurate PAR statistics for various scenarios, and sample results are displayed in Figure 4.14. As expected, the approximation is accurate for large L, and the PAR of OFDM system increases with L but not nearly linearly.

---

5. *Nyquist sampling* means the minimum allowable sampling frequency without irreversible information loss, that is, no oversampling is performed.

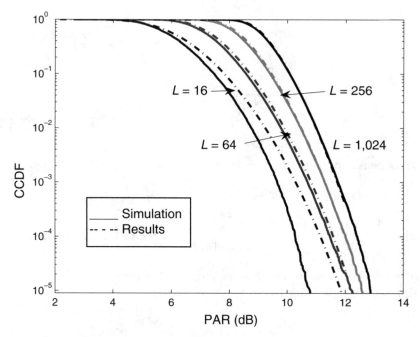

Figure 4.14 CCDF of PAR for QPSK OFDM system: $L = 16, 64, 256, 1,024$

4.5.3 Clipping: Living with a High PAR

In order to avoid operating the Power Amplifier (PA) in the nonlinear region, the input power can be reduced by an amount about equal to the PAR. However, two important facts related to this IBO amount can be observed from Figure 4.14. First, since the highest PAR values are uncommon, it might be possible to simply "clip" off the highest peaks in order to reduce the IBO and effective PAR, at the cost of some, ideally, minimal distortion of the signal. Second, and conversely, it can be seen that even for a conservative choice of IBO—say, 10dB—there is still a distinct possibility that a given OFDM symbol will have a PAR that exceeds the IBO and causes clipping. See Sidebar 4.2 for more discussion of how to predict the required backoff amount.

Clipping, sometimes called soft limiting, truncates the amplitude of signals that exceed the clipping level as

$$\tilde{x}_L[n] = \begin{cases} Ae^{j\angle x[n]}, & f \mid x[n] \mid > A \\ x[n], & f \mid x[n] \mid \le A \end{cases} \tag{4.31}$$

where $x[n]$ is the original signal, and $\tilde{x}[n]$ is the output after clipping. The soft limiter can be equivalently thought of as a peak cancellation technique, like that shown in Figure 4.15. The soft-limiter output can be written in terms of the original signal and a canceling, or clipping, signal as

$$\tilde{x}[n] = x[n] + c[n], \ for \ n = 0,\ldots,L-1 \tag{4.32}$$

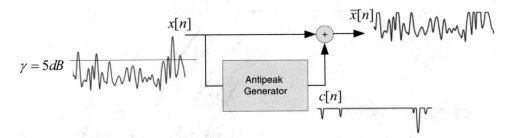

Figure 4.15 A peak cancellation as a model of soft limiter when $\gamma = 5dB$

where $c[n]$ is the clipping signal defined by

$$c[n] = \begin{cases} |A - |x[n]|| e^{j\theta[n]}, & \text{if } |x[n]| > A \\ 0, & \text{if } |x[n]| \le A \end{cases} \quad \text{for } n = 0, ..., NL - 1 - 3pt \quad (4.33)$$

where $\theta[n] = \arg(-x[n])$; that is, the phase of $c[n]$ is out of phase with $x[n]$ by $180°$, and A is the *clipping level,* which is defined as

$$\gamma \triangleq \frac{A}{\sqrt{E\{|x[n]|^2\}}} = \frac{A}{\sqrt{\mathcal{E}_x}} . \quad (4.34)$$

In such a peak-cancellation strategy, an antipeak generator estimates peaks greater than clipping level. The clipped signal $\tilde{x}[n]$ can be obtained by adding a time-shifted and scaled signal $c[n]$ to the original signal $x[n]$. The exact clipping signal $c[n]$ can be generated to reduce PAR, using a variety of techniques.

Obviously, clipping reduces the PAR at the expense of distorting the signal by the additive signal $c[n]$. The two primary drawbacks from clipping are (1) spectral regrowth—frequency-domain leakage—which causes unacceptable interference to users in neighboring RF channels, and (2) distortion of the desired signal. We now consider these two effects separately.

4.5.3.1 Spectral Regrowth

The clipping noise can be expressed in the frequency domain through the use of the DFT. The resulting clipped frequency-domain signal $\tilde{X}$ is

$$\tilde{X}_k = X_k + C_k, \quad k = 0, ..., L - 1, \quad (4.35)$$

where C_k represents the clipped-off signal in the frequency domain. In Figure 4.16, the power-spectral density of the original (X), clipped ($\tilde{X}$), and clipped-off (C) signals are plotted for different clipping ratios γ of 3 dB, 5 dB, and 7 dB. The following deleterious effects are observed. First, the clipped-off signal C_k is strikingly increased as the clipping ratio is lowered from 7 dB to 3 dB. This increase shows the correlation between X_k and C_k inside the desired band at low clipping ratios, and causes the in-band signal to be attenuated as the clipping ratio is lowered.

Sidebar 4.2 Quantifying PAR: The Cubic Metric

Although the PAR gives a reasonable estimate of the amount of PA backoff required, it is not precise. That is, backing off on the output power by 3 dB may not reduce the effects of nonlinear distortion by 3 dB. Similarly, the penalty associated with the PAR does not necessarily follow a dB-for-dB relationship. A typical PA gain can be reasonably modeled as

$$v_{out}(t) = c_1 v_{in}(t) + c_2 (v_{in}(t))^3, \qquad (4.36)$$

where c_1 and c_2 are amplifier-dependent constants. The cubic term in the equation causes several types of distortion, including both in- and out-of-band distortion. Therefore, Motorola [29] proposed a "cubic metric" for estimating the amount of amplifier backoff needed in order to reduce the distortion effects by a prescribed amount. The cubic metric (CM) is defined as

$$CM = \frac{20 \log_{10}[\bar{v}^3]_{\text{rms}} - 20 \log_{10}[\bar{v}_{\text{ref}}^3]_{\text{rms}}}{c_3}, \qquad (4.37)$$

where $\bar{v}$ is the signal of interest normalized to have an RMS value of 1, and $\bar{v}_{\text{ref}}$ is a low-PAR reference signal, usually a simple BPSK voice signal, also normalized to have an RMS value of 1. The constant c_3 is found empirically through curve fitting; it was found that $c_3 \approx 1.85$ in [29].

The advantage of the cubic metric is that initial studies show that it very accurately predicts—usually within 0.1 dB—the amount of backoff required by the PA in order to meet distortion constraints.

Second, it can be seen that the out-of-band interference caused by the clipped signal $\tilde{X}$ is determined by the shape of clipped-off signal C_k. Even the seemingly conservative clipping ratio of 7 dB violates the specification for the transmit spectral mask of IEEE 802.16e-2005, albeit barely.

4.5.3.2 In-Band Distortion

Although the desired signal and the clipping signal are clearly correlated, it is possible, based on the Bussgang Theorem, to model the in-band distortion owing to the clipping process as the combination of uncorrelated additive noise and an attenuation of the desired signal [14, 15, 34]:

$$\tilde{x}[n] = \alpha x[n] + d[n], \quad for \ n = 0, 1, \ldots, L-1. \qquad (4.38)$$

Now, $d[n]$ is uncorrelated with the signal $x[n]$, and the attenuation factor α is obtained by

$$\alpha = 1 - e^{-\gamma^2} + \frac{\sqrt{\pi}\gamma}{2} erfc(\gamma). \qquad (4.39)$$

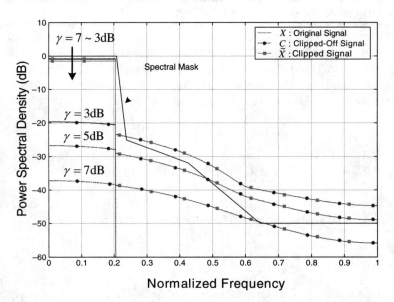

Figure 4.16 Power-spectral density of the unclipped (original) and clipped (nonlinearly distorted) OFDM signals with 2,048 block size and 64 QAM when clipping ratio (γ) is 3 dB, 5 dB, and 7 dB in soft limiter

The attenuation factor α is plotted in Figure 4.17 as a function of the clipping ratio γ. The attenuation factor α is negligible when the clipping ratio γ is greater than 8dB, so high clipping ratios, the correlated clipped-off signal $c[n]$ in Equation (4.33), can be approximated by uncorrelated noise $d[n]$. That is, $c[n] \approx d[n]$ as $\gamma \uparrow$. The variance of the uncorrelated clipping noise can be expressed assuming a stationary Gaussian input $x[n]$ as

$$\sigma_d^2 = \mathcal{E}_x(1 - \exp(-\gamma^2) - \alpha^2). \tag{4.40}$$

In WiMAX, the error vector magnitude (EVM) is used as a means to estimate the accuracy of the transmit filter and D/A converter, as well as the PA, nonlinearity. The EVM is essentially the average error vector relative to the desired constellation point and can be caused by a degradation in the system. The EVM over an OFDM symbol is defined as

$$\text{EVM} = \sqrt{\frac{\frac{1}{N}\sum_{k=1}^{L}(\Delta I_k^2 + \Delta Q_k^2)}{S_{\max}^2}} = \sqrt{\frac{\sigma_d^2}{S_{\max}^2}}, \tag{4.41}$$

where $S_{\max}$ is the maximum constellation amplitude. The concept of EVM is illustrated in Figure 4.18. In the case of clipping, a given EVM specification can be easily translated into an SNR requirement by using the variance σ_d^2 of the clipping noise.

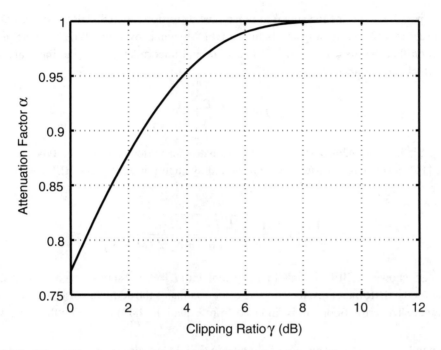

Figure 4.17 Attenuation factor α as a function of the clipping ratio γ

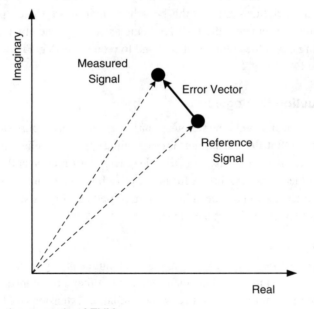

Figure 4.18 Illustrative example of EVM

It is possible to define the signal-to-noise-plus-distortion ratio (SNDR) of one OFDM symbol in order to estimate the impact of clipped OFDM signals over an AWGN channel under the assumption that the distortion $d[n]$ is Gaussian and uncorrelated with the input and channel noise, which has variance $N_0/2$:

$$SNDR = \frac{\alpha^2 \mathcal{E}_x}{\sigma_d^2 + N_0/2}.$$ (4.42)

The bit error probability (BEP) can be evaluated for various modulation types by using the SNDR [14]. In the case of Multilevel-QAM and average power $\mathcal{E}_x$, the BEP can be approximated as

$$P_b \approx \frac{4}{\log_2 M}\left(1 - \frac{1}{\sqrt{M}}\right) Q\left(\sqrt{\frac{3\mathcal{E}_x \alpha^2}{(\sigma_d^2 + N_0/2)(M-1)}}\right).$$ (4.43)

Figure 4.19 shows the BER for an OFDM system with $L = 2,048$ subcarriers and 64 QAM modulation. As the SNR increases, the clipping error dominates the additive noise, and an error floor is observed. The error floor can be inferred from Equation (4.43) by letting the noise variance $N_0/2 \to 0$.

A number of additional studies of clipping in OFDM systems have been completed in recent years [3, 4, 14, 32, 39]. In some cases, clipping may be acceptable, but in WiMAX systems, the margin for error is quite tight, as much of this book has emphasized. Hence, more aggressive and innovative techniques for reducing the PAR are being actively pursued by the WiMAX community in order to bring down the component cost and to reduce the degradation owing to the nonlinear effects of the PA.

4.5.4 PAR-Reduction Strategies

To alleviate the nonlinear effects, numerous approaches have been pursued. The first plan of attack is to reduce PAR at the transmitter, through either peak cancellation or signal mapping [20]. Another set of techniques focuses on OFDM signal reconstruction at the receiver in spite of the introduced nonlinearities [23, 42]. A further approach is to attempt to predistort the analog signal so that it will appear to have been linearly amplified [8]. In this section, we look at techniques that attempt PAR reduction at the transmitter.

4.5.4.1 Peak Cancellation

This class of PAR-reduction techniques applies an antipeak signal to the desired signal. Although clipping is the most obvious such technique, other important peak-canceling techniques include tone reservation (TR) and active constellation extension (ACE).

Clipping can be improved by an iterative process of clipping and filtering, since the filtering can be used to subdue the spectral regrowth and in band distortion [2]. After several iterations of

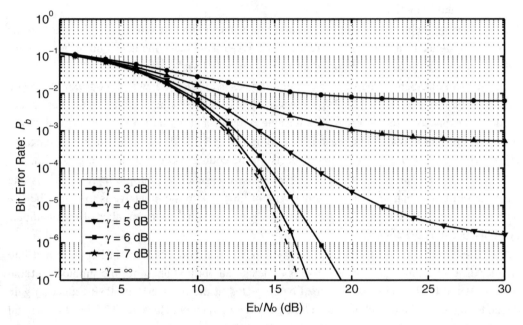

Figure 4.19 Bit error rate probability for a clipped OFDM signal in AWGN with various clipping ratios

clipping and filtering, the residual in-band distortion can be restored by iterative estimation and cancellation of the clipping noise [9].

Tone reservation reduces PAR by intelligently adding power to unused carriers, such as the null subcarriers specified by WiMAX. The reduced signal can be expressed as

$$\tilde{x}[n] = x[n] + c[n] = Q_L(\mathbf{X} + \mathbf{C}) , \tag{4.44}$$

where Q_L is the IDFT matrix of size $L \times L$, $\mathbf{X} = \{X_k, k \in \mathcal{R}^c\}$ is a complex symbol vector, $\mathbf{C} = \{C_k, k \in \mathcal{R}\}$ is a complex tone-reservation vector, and $\mathcal{R}$ is the reserved-tones set. There is no distortion from TR, because the reserved-tone carriers and the data carriers are orthogonal. To reduce PAR using TR, a simple gradient algorithm and a Fourier projection algorithm have been proposed in [41] and [18], respectively. However, these techniques converge slowly. For faster convergence, an active-set approach is used in [25].

Another peak-canceling technique is active constellation extension [24]. Essentially, the corner points of an M-QAM constellation can be extended without any loss of SNR, and this property can be used to decrease the PAR without negatively affecting the performance, as long as ACE is allowed only when the minimum distance is guaranteed. Unfortunately, the gain in PAR reduction is inversely proportional to the constellation size in M-QAM.

4.5.4.2   Signal Mapping

Signal-mapping techniques share in common that some redundant information is added to the transmitted signal in a manner that reduces the PAR. This class includes coding techniques, selected mapping (SLM), and partial transmit sequence (PTS).

The main idea behind the various coding schemes is to select a low PAR codeword based on the desired transmit symbols [22, 36]. However, most of the decoding techniques for these codes require an exhaustive search and so are feasible only for a small number of subcarriers. Moreover, it is difficult to maintain a reasonable coding rate in OFDM when the number of subcarriers grows large. The implementation prospects for the coding-based techniques appear dim.

In selected mapping, one OFDM symbol is used to generate multiple representations that have the same information as the original symbol [30]. The basic objective is to select the one with minimum PAR; the gain in PAR reduction is proportional to the number of the candidate symbols, but so is the complexity.

PTS is similar to SLM; however, the symbol in the frequency domain is partitioned into smaller disjoint subblocks. The objective is to design an optimal phase for the subblock set that minimizes the PAR. The phase can then be corrected at the receiver. The PAR-reduction gain depends on the number of subblocks and the partitioning method. However, PTS has exponential search complexity with the number of subblocks.

SLM and PTS are quite flexible and effective, but their principal drawbacks are that the receiver structure must be changed, and transmit overhead (power and symbols) is required to send the needed information for decoding. Hence, these techniques, in contrast to peak-cancellation techniques, would require explicit support by the WiMAX standard.

4.6 OFDM's Computational Complexity Advantage

One of its principal advantages relative to single-carrier modulation with equalization is that OFDM requires much lower computational complexity for high-data-rate communication. In this section, we compare the computational complexity of an equalizer with that of a standard IFFT/FFT implementation of OFDM.

An equalizer operation consists of a series of multiplications with several delayed versions of the signal. The number of delay taps in an equalizer depends on the symbol rate of the system and the delay spread in the channel. To be more precise, the number of equalizer taps is proportional to the bandwidth-delay-spread product $T_m/T_s \approx BT_m$. We have been calling this quantity v, or the number of ISI channel taps. An equalizer with v taps performs v complex multiply-and-accumulate (CMAC) operations per received symbol. Therefore, the complexity of an equalizer is of the order

$$O(v \cdot B) = O(B^2 T_m). \tag{4.45}$$

In an OFDM system, the IFFT and FFT are the principal computational operations. It is well known that the IFFT and FFT each have a complexity of $O(L\log_2 L)$, where L is the FFT block size. In the case of OFDM, L is the number of subcarriers. As this chapter has shown, for a fixed cyclic prefix overhead, the number of subcarriers L must grow linearly with the bandwidth-delay-spread product $v = BT_m$. Therefore, the computational complexity for each OFDM symbol is of the order $O(BT_m \log_2 BT_m)$. There are B/L OFDM symbols sent each second. Since $L \propto BT_m$, this means that there are order $O(1/T_m)$ OFDM symbols per second, so the computational complexity in terms of CMACs for OFDM is

$$O(BT_m \log_2 BT_m)O(1/T_m) = O(B\log_2 BT_m). \tag{4.46}$$

Clearly, the complexity of an equalizer grows as the square of the data rate, since both the symbol rate and the number of taps increase linearly with the data rate. For an OFDM system, the increase in complexity grows with the data rate only slightly faster than linearly. This difference is dramatic for very large data rates, as shown in Figure 4.20.

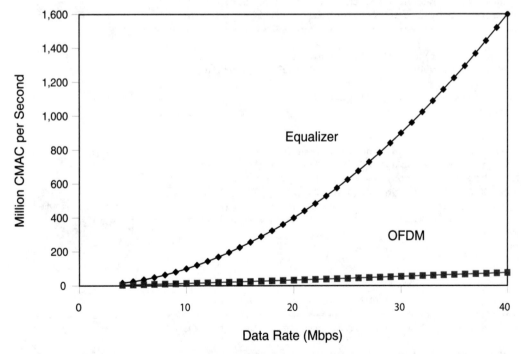

Figure 4.20 OFDM has an enormous complexity advantage over equalization for broadband data rates. The delay spread is $T_m = 2\,\mu sec$, the OFDM symbol period is $T = 20\,\mu sec$, and the considered equalizer is a DFE.

4.7 Simulating OFDM Systems

In this section, we provide some resources for getting started on simulating an OFDM system. The popular LabVIEW simulation package from National Instruments can be used to develop virtual instruments (VI's) that implement OFDM in a graphical user interface. See also [1]. Other communication system building blocks and multicarrier modulation tools are available in the National Instruments Modulation Toolkit.

OFDM functions can also be developed in Matlab.[6] Here, we provide Matlab code for a baseband OFDM transmitter and receiver for QPSK symbols. These functions can be modified to transmit and receive M-QAM symbols or passband—complex baseband—signals.

```
function x=OFDMTx(N, bits, num, nu)
%=============================================================
% x=OFDMTx( N, bits, num, nu,zero_tones)
%
% APSK transmitter for OFDM
%
% N is the FFT size
% bits is a {1,-1} stream of length (N/2-1)*2*num (baseband, zero DC)
% num is the number of OFDM symbols to produce
% nu is the cyclic prefix length
%=============================================================

if length(bits) ~= (N/2-1)*2*num
 error('bits vector of improper length -- Aborting');
end x=[];
real_index = -1; %initial values
imag_index = 0;

for a=1:num
real_index = max(real_index)+2:2:max(real_index)+N-1;
imag_index = max(imag_index)+2:2:max(imag_index)+N-1;
X = (bits(real_index) + j*bits(imag_index))/2;
X=[0 X 0]; % zero nyquist and DC
x_hold=sqrt(N)*ifft([[X,conj(fliplr(X(2:length(X)-1)))]]); %Baseband so symmeteic
x_hold=[x_hold(length(x_hold)-nu+1:length(x_hold)),x_hold]; %Add CP
x=[x,x_hold];
end
x=real(x);


function bits_out = OFDMRx(N,y,h,num_symbols,nu)
%=============================================================
% bits_out = OFDMRx(M,N,y,num_symbols,nu,zero_tones)
%
% N is the FFT size
% y is received channel output for all symbols (excess delay spread removed)
% h is the (estimated) channel impulse response
```

---

6. These functions also can be used in MathScript for LabVIEW.

```
% num_symbols is the # of symbols (each of N*sqrt(M) bits and N+nu samples)
% nu is the cyclic prefix length
%===========================================================
if length(y) ~= (N + nu)*num_symbols
 error('received vector of improper length -- Aborting');
end bits_out=[]; bits_out_cur = zeros(1,N-2);

for a=1:num_symbols
 y_cur = y((a-1)*(N+nu)+1+nu:a*(N+nu)); % Get current OFDM symbol, strip CP
 X_hat = 1/sqrt(N)*fft(y_cur)./fft(h,N); %FEQ
 X_hat = X_hat(1:N/2);
 real_index = 1:2:N-2-1; % Don't inc. X_hat (1) because is DC, zeroed
 imag_index = 2:2:N-2;
 bits_out_cur(real_index) = sign(real(X_hat(2:length(X_hat))));
 bits_out_cur(imag_index) = sign(imag(X_hat(2:length(X_hat))));
 bits_out=[bits_out,bits_out_cur];
end
```

4.8 Summary and Conclusions

This chapter has covered the theory of OFDM, as well as the important design and implementation-related issues.

- OFDM overcomes even severe intersymbol interference through the use of the IFFT and a cyclic prefix.
- OFDM is the core modulation strategy used in WiMAX systems.
- Two key details of OFDM implementation are synchronization and management of the peak-to-average ratio.
- In order to aid WiMAX engineers and students, examples relating OFDM to WiMAX, including simulation code, were provided.

4.9 Bibliography

[1] J. G. Andrews and T. Kim. MIMO-OFDM design using LabVIEW. www.ece.utexas.edu/~jandrews/molabview.html.html.

[2] J. Armstrong. Peak-to-average power reduction for OFDM by repeated clipping and frequency domain filtering. *Electronics Letters*, 38(8):246–247, February 2002.

[3] A. Bahai, M. Singh, A. Goldsmith, and B. Saltzberg. A new approach for evaluating clipping distortion in multicarrier systems. *IEEE Journal on Selected Areas in Communications*, 20(5):1037–1046, 2002.

[4] P. Banelli and S. Cacopardi. Theoretical analysis and performance of OFDM signals in nonlinear AWGN channels. *IEEE Transactions on Communications*, 48(3):430–441, March 2000.

[5] R. Baxley and G. Zhou. Power savings analysis of peak-to-average power ratio in OFDM. *IEEE Transactions on Consumer Electronics*, 50(3):792–798, 2004.

[6] H. Bolcskei. Blind estimation of symbol timing and carrier frequency offset in wireless OFDM systems. *IEEE Transactions on Communications*, 49:988–99, June 2001.

[7] R. W. Chang. Synthesis of band-limited orthogonal signals for multichannel data transmission. *Bell Systems Technical Journal*, 45:1775–1796, December 1966.

[8] S. Chang and E. J. Powers. A simplified predistorter for compensation of nonlinear distortion in OFDM systems. *IEEE Globecom*, pp. 3080–3084, San Antonio, TX, November 2001.

[9] H. Chen and A. Haimovich. Iterative estimation and cancellation of clipping noise for OFDM signals. *IEEE Communications Letters*, 7(7):305– 307, July 2003.

[10] L. J. Cimini. Analysis and simulation of a digital mobile channel using orthogonal frequency division multiplexing. *IEEE Transactions on Communications*, 33(7):665–675, July 1985.

[11] J. M. Cioffi. *Digital Communications, Chapter 4: Multichannel Modulation*. Course notes. www.stanford.edu/class/ee379c/.

[12] J. M. Cioffi. A multicarrier primer. Stanford University/Amati T1E1 contribution, I1E1.4/91–157, November 1991.

[13] S. C. Cripps. *RF Power Amplifiers for Wireless Communications*. Artech House, 1999.

[14] D. Dardari, V. Tralli, and A. Vaccari. A theoretical characterization of nonlinear distortion effects in OFDM systems. *IEEE Transactions on Communications*, 48(10):1755–1764, October 2000.

[15] M. Friese. On the degradation of OFDM-signals due to peak-clipping in optimally predistorted power amplifiers. In *Proceedingseedings IEEE Globecom*, pp. 939–944, November 1998.

[16] T. Fusco. *Synchronization techniques for OFDM systems*. PhD thesis, Universita di Napoli Federico II, 2005.

[17] R. G. Gallager. *Information Theory and Reliable Communications*. Wiley, 1968. 33.

[18] A. Gatherer and M. Polley. Controlling clipping probability in DMT transmission. In *Proceedings of the Asilomar Conference on Signals, Systems and Computers*, pp. 578–584, November 1997.

[19] A. J. Goldsmith. *Wireless Communications*. Cambridge University Press, 2005.

[20] S. H. Han and J. H. Lee. An overview of peak-to-average power ratio reduction techniques for multi-carrier transmission. *IEEE Wireless Communications*, 12(2):56–65, 2005.

[21] J. L. Holsinger. Digital communication over fixed time-continuous channels with memory, with special application to telephone channels. PhD thesis, Massachusetts Institute of Technology, 1964.

[22] A. E. Jones, T. A. Wilkinson, and S. K. Barton. Block coding scheme for reduction of peak to mean envelope power ratio of multicarrier transmission schemes. *Electronics Letters*, 30(25):2098–2099, December 1994.

[23] D. Kim and G. Stuber. Clipping noise mitigation for OFDM by decision-aided reconstruction. *IEEE Communications Letters*, 3(1):4–6, January 1999.

[24] B. Krongold and D. Jones. PAR reduction in OFDM via active constellation extension. *IEEE Transactions on Broadcasting*, 49(3):258–268, 2003.

[25] B. Krongold and D. Jones. An active-set approach for OFDM PAR reduction via tone reservation. *IEEE Transactions on Signal Processing*, 52(2):495–509, February 2004.

[26] S. Miller and R. O'Dea. Peak power and bandwidth efficient linear modulation. *IEEE Transactions on Communications*, 46(12):1639–1648, December 1998.

[27] P. Moose. A technique for orthogonal frequency division multiplexing frequency offset correction. *IEEE Transactions on Communications*, 42(10):2908–2914, October 1994.

[28] M. Morelli and U. Mengali. An improved frequency offset estimator for OFDM applications. *IEEE Communications Letters*, 3(3), March 1999.

[29] Motorola. Comparison of PAR and cubic metric for power de-rating. TSG-RAN WG1#37 Meeting, Montreal, Canada, Document # R1-040522, May 2004.

[30] S. Müller and J. Huber. A comparison of peak power reduction schemes for OFDM. In *Proceedings, IEEE Globecom*, pp. 1–5, November 1997.

[31] C. Muschallik. Improving an OFDM reception using an adaptive nyquist windowing. *IEEE Transactions on Consumer Electronics*, 42(3):259–269, August 1996.

[32] H. Nikopour and S. Jamali. On the performance of OFDM systems over a Cartesian clipping channel: A theoretical approach. *IEEE Transactions on Wireless Communications*, 3(6):2083–2096, 2004.

[33] H. Ochiai and H. Imai. On the distribution of the peak-to-average power ratio in OFDM signals. *IEEE Transactions on Communications*, 49(2):282–289, 2001.

[34] H. Ochiai and H. Imai. Performance analysis of deliberately clipped OFDM signals. *IEEE Transactions on Communications*, 50(1):89–101, January 2002.

[35] A. V. Oppenheim and R. W. Schafer. *Discrete-Time Signal Processing*. Prentice Hall, 1989.

[36] K. G. Paterson and V. Tarokh. On the existence and construction of good codes with low peak-to-average power ratios. *IEEE Transactions on Information Theory*, 46(6):1974–1987, September 2000.

[37] T. Pollet, M. V. Bladel, and M. Moeneclaey. BER sensitivity of OFDM systems to carrier frequency offset and Wiener phase noise. *IEEE Transactions on Communications*, 43(234):191–193, February/March/April 1995.

[38] A. Redfern. Receiver window design for multicarrier communication systems. *IEEE Journal on Selected Areas in Communications*, 20(5):1029–1036, June 2002.

[39] G. Santella and F. Mazzenga. A hybrid analytical-simulation procedure for performance evaluation in M-QAMOFDM schemes in presence of nonlinear distortions. *IEEE Transactions on Vehicular Technology*, 47(1):142–151, February 1998.

[40] T. M. Schmidl and D. C. Cox. Robust frequency and timing synchronization for OFDM. *IEEE Transactions on Communications*, 45(12):1613–1621, December 1997.

[41] J. Tellado. *Multicarrier Modulation with low PAR: Applications to DSL and wireless*. Kluwer, 2000.

[42] J. Tellado, L. Hoo, and J. Cioffi. Maximum-likelihood detection of nonlinearly distorted multicarrier symbols by iterative decoding. *IEEE Transactions on Communications*, 51(2):218–228, February 2003.

[43] J. van de Beek, M. Sandell, and P. Borjesson. ML estimation of time and frequency offset in OFDM systems. *IEEE Transactions on Signal Processing*, 45:1800–1805, July 1997.

[44] R. van Nee and A. de Wild. Reducing the peak-to-average power ratio of OFDM. In *Vehicular Technology Conference, 1998. VTC 98. 48th IEEE*, 3:2072–2076, 1998.

[45] S. Weinstein and P. Ebert. Data transmission by frequency-division multiplexing using the discrete Fourier transform. *IEEE Transactions on Communications*, 19(5):628–634, October 1971.

CHAPTER 5

Multiple-Antenna Techniques

T he use of multiple antennas allows independent channels to be created in space and is one of the most interesting and promising areas of recent innovation in wireless communications. Chapter 4 explained how WiMAX systems are able to achieve *frequency* diversity through the use of multicarrier modulation. The focus of this chapter is *spatial* diversity, which can be created without using the additional bandwidth that time and frequency diversity both require. In addition to providing spatial diversity, antenna arrays can be used to focus energy (beamforming) or create multiple parallel channels for carrying unique data streams (spatial multiplexing). When multiple antennas are used at both the transmitter and the receiver, these three approaches are often collectively referred to as multiple/input multiple output (MIMO) communication[1] and can be used to

1. Increase the system reliability (decrease the bit or packet error rate)

2. Increase the achievable data rate and hence system capacity

3. Increase the coverage area

4. Decrease the required transmit power

---

1. Use of the term MIMO (pronounced "My-Moe") generally assumes multiple antennas are at *both* the transmitter and the receiver. SIMO (single input/multiple output) and MISO (multiple input/ single output) refer, respectively, to only a single antenna at the transmitter or the receiver. Without further qualification, MIMO is often assumed to mean specifically the spatial multiplexing approach, since spatial multiplexing transmits multiple independent data streams and hence has multiple inputs and outputs.

However, these four desirable attributes usually compete with one another; for example, an increase in data rate often will require an increase in either the error rate or transmit power. The way in which the antennas are used generally reflects the relative value attached by the designer to each of these attributes, as well as such considerations as cost and space. Despite the cost associated with additional antenna elements and their accompanying RF chains, the gain from antenna arrays is so enormous that there is little question that multiple antennas will play a critical role in WiMAX systems. Early WiMAX products will likely be conservative in the number of antennas deployed at both the base station (BS) and the mobile station (MS), and also are likely to value system reliability (diversity) over aggressive data rates (spatial multiplexing). We expect that in the medium to long term, though, WiMAX systems will need to aggressively use many of the multiple-antenna techniques discussed in this chapter in order to meet the WiMAX vision for a mobile broadband Internet experience.

This chapter begins with receive diversity, which is the most well-established form of spatial diversity. Transmit diversity, which requires quite a different approach, is discussed next. Beamforming is then summarized and contrasted with diversity. Spatial multiplexing, which is the most contemporary and unproven of the MIMO techniques, is then considered, emphasizing the shortcomings of MIMO theory in the context of cellular systems. We then look at how the channel can be acquired at the receiver and the transmitter, reviewing first MIMO-OFDM channel estimation and then channel feedback techniques. We conclude with a discussion of advanced MIMO techniques that may find a future role in the WiMAX standard. The performance improvement that we forecast for WiMAX systems due to MIMO techniques is detailed in Chapters 11 and 12.

5.1 The Benefits of Spatial Diversity

As demonstrated in Chapter 3 and repeated in Figure 5.1, even two appropriately spaced antennas appear to be sufficient to eliminate most deep fades, which paints a promising picture for the potential benefits of spatial diversity. One main advantage of spatial diversity relative to time and frequency diversity is that no additional bandwidth or power is needed in order to take advantage of spatial diversity. The cost of each additional antenna, its RF chain, and the associated signal processing required to modulate or demodulate multiple spatial streams may not be negligible, but this trade-off is often very attractive for a small number of antennas, as we demonstrate in this chapter.

We now briefly summarize the main advantages of spatial diversity, which will be explored in more depth in the subsequent sections of this chapter.

5.1.1 Array Gain

When multiple antennas are present at the receiver, two forms of gain are available: *diversity* gain and *array* gain. Diversity gain results from the creation of multiple independent channels between the transmitter and the receiver and is a product of the statistical richness of those channels. Array gain, on the other hand, does not rely on statistical diversity between the channels

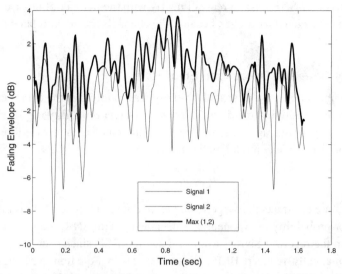

Figure 5.1 Simple two-branch selection diversity eliminates most deep fades.

and instead achieves its performance enhancement by coherently combining the energy received by each of the antennas. Even if the channels are completely correlated, as might happen in a line-of-sight system, the received SNR increases linearly with the number of receive antennas, N_r, owing to the array gain.

For a $N_t \times N_r$ system, the array gain is N_r, which can be seen for a $1 \times N_r$ as follows. In correlated flat fading, each antenna $i \in (1, N_r)$ receives a signal that can be characterized as

$$y_i = h_i x + n_i = hx + n_i,\tag{5.1}$$

where $h_i = h$ for all the antennas, since they are perfectly correlated. Hence, the SNR on a single antenna is

$$\gamma_i = \frac{|h|^2}{\sigma^2},\tag{5.2}$$

where the noise power is σ^2 and we assume unit signal energy ($\mathcal{E}_x = E |x|^2 = 1$). If all the receive antenna paths are added, the resulting signal is

$$y = \sum_{i=1}^{N_r} y_i = N_r hx + \sum_{i=1}^{N_r} n_i,\tag{5.3}$$

and the combined SNR, assuming that just the noise on each branch is uncorrelated, is

$$\gamma_\Sigma = \frac{|N_r h|^2}{N_r \sigma^2} = \frac{N_r |h|^2}{\sigma^2}.\tag{5.4}$$

Hence, the received SNR also increases linearly with the number of receive antennas even if those antennas are correlated. However, because the channels are all correlated in this case, there is no diversity gain.

5.1.2 Diversity Gain and Decreased Error Rate

Traditionally, the main objective of spatial diversity was to improve the communication reliability by decreasing the sensitivity to fading. The physical-layer reliability is typically measured by the outage probability, or average bit error rate. In additive noise, the bit error probability (BEP) can be written for virtually any modulation scheme as

$$P_b \approx c_1 e^{-c_2 \gamma}, \tag{5.5}$$

where c_1 and c_2 are constants that depend on the modulation type, and γ is the received SNR. Because the error probability is exponentially decreasing with SNR, the few instances in a fading channel when the received SNR is low dominate the BEP, since even modestly higher SNR values have dramatically reduced BEP, as can be seen in Equation (5.5). In fading, without diversity, the average BEP can be written, analogous to Equation (5.5), as

$$\bar{P}_b \approx c_3 \gamma^{-1}. \tag{5.6}$$

This simple inverse relationship between SNR and BEP is *much* weaker than a decaying exponential, which is why it was observed in Figure 3.22 that the BEP with fading is dramatically worse than without fading.

If sufficiently spaced[2] N_t transmit antennas and N_r receive antennas are added to the system, it is said that the *diversity order* is $N_d = N_r N_t$, since that is the number of uncorrelated channel paths between the transmitter and the receiver. Since the probability of all the N_d uncorrelated channels having low SNR is very small, the diversity order has a dramatic effect on the system reliability. With diversity, the average BEP improves to

$$\bar{P}_b \approx c_4 \gamma^{-N_d}, \tag{5.7}$$

which is an enormous improvement. On the other hand, if only an array gain was possible—for example, if the antennas are not sufficiently spaced or the channel is LOS—the average BEP would decrease only from Equation (5.6) to

$$\bar{P}_b \approx c_5 (N_d \gamma)^{-1}, \tag{5.8}$$

since the array gain provides only a linear increase in SNR. The difference between Equation (5.7) and Equation (5.8) is quite dramatic as γ and N_d increase and is shown in Figure 5.2, where it is assumed that the constants $c_i = 1$, which is equivalent to normalizing the

---

2. Recall from Chapter 3 that, generally, about half a wavelength is sufficient for the antenna elements to be sufficiently uncorrelated.

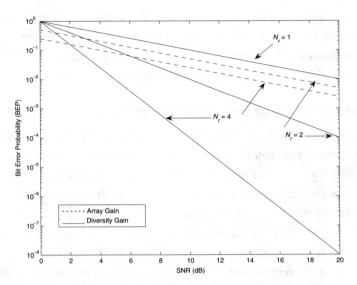

Figure 5.2 BEP trends for $N_r = [1 \ 2 \ 4]$. Here, the BEP (0 dB) is normalized to 1 for each technique. Statistical diversity has a dramatic impact on BEP, whereas the impact from the array gain is only incremental.

BEP to 1 for $\gamma = 1$. The trend is clear: Sufficient spacing for the antennas is critical for increasing the system reliability.

5.1.3 Increased Data Rate

Diversity techniques are very effective at averaging out fades in the channel and thus increasing the system reliability. Receive-diversity techniques also increase the average received SNR at best linearly, owing to the array gain. The Shannon capacity formula gives the maximum achievable data rate of a single communication link in additive white Gaussian noise (AWGN) as

$$C = B\log_2(1+\gamma), \tag{5.9}$$

where C is the capacity, or maximum error-free data rate; B is the bandwidth of the channel; and γ is again the SNR (or SINR). Owing to advances in coding, and with sufficient diversity, it may be possible to approach the Shannon limit in some wireless channels.

Since antenna diversity increases the SNR linearly, diversity techniques increase the capacity only *logarithmically* with respect to the number of antennas. In other words, the data rate benefit rapidly diminishes as antennas are added. However, it can be noted that when the SNR is low, the capacity increase is close to linear with SNR, since $\log(1+x) \approx x$ for small x. Hence in low-SNR channels, diversity techniques increase the capacity about linearly, but the overall throughput is generally still poor owing to the low SNR.

In order to get a more substantial data rate increase at higher SNRs, the multiantenna channel can instead be used to send multiple independent streams. Spatial multiplexing has the ability to achieve a *linear* increase in the data rate with the number of antennas at moderate to high SINRs through the use of sophisticated signal-processing algorithms. Specifically, the capacity can be increased as a multiple of $\min(N_t, N_r)$; that is, capacity is limited by the minimum of the number of antennas at either the transmitter or the receiver.

5.1.4 Increased Coverage or Reduced Transmit Power

The benefits of diversity can also be harnessed to increase the coverage area and to reduce the required transmit power, although these gains directly compete with each other, as well as with the achievable reliability and data rate. We first consider the increase in coverage area due to spatial diversity. For simplicity, assume that there are N_r receive antennas and just one transmit antenna. Due to simply the array gain, the average SNR is approximately $N_r\gamma$, where γ is the average SNR per branch. From the simplified pathloss model of Chapter 3, $P_r = P_t P_o d^{-\alpha}$, it can be found that the increase in coverage range is $N_r^{1/\sigma}$, and so the coverage area improvement is $N_r^{2/\sigma}$, without even considering the diversity gain. Hence, the system reliability would be greatly enhanced even with this range extension. Similar reasoning can be used to show that the required transmit power can be reduced by $10\log_{10}N_r$ dB while maintaining a diversity gain of $N_t \times N_r$.

5.2 Receive Diversity

The most prevalent form of spatial diversity is receive diversity, often with only two antennas. This type of diversity is nearly ubiquitous on cellular base stations and wireless LAN access points. Receive diversity places no particular requirements on the transmitter but requires a receiver that processes the N_r received streams and combines them in some fashion (Figure 5.3). In this section, we overview two of the widely used combining algorithms: selection combining (SC) and maximal ratio combining (MRC). Although receive diversity is highly effective in both flat fading and frequency-selective fading channels, we focus on the flat-fading scenario, in which the signal received by each of the N_r antennas is uncorrelated and has the same average power.

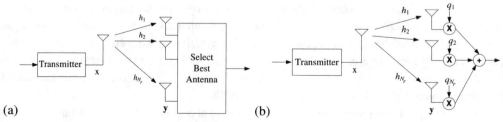

(a)   (b)

Figure 5.3 Receive diversity: (a) selection combining and (b) maximal ratio combining

5.2.1 Selection Combining

Selection combining is the simplest type of combiner, in that it simply estimates the instantaneous strengths of each of the N_r streams and selects the highest one. Since it ignores the useful energy on the other streams, SC is clearly suboptimal, but its simplicity and reduced hardware requirements make it attractive in many cases.

The diversity gain from using selection combining can be confirmed quite quickly by considering the outage probability, defined as the probability that the received SNR drops below some required threshold, $P_{out} = P[\gamma < \gamma_o] = p$. Assuming N_r uncorrelated receptions of the signal,

$$
\begin{aligned}
P_{out} &= P[\gamma_1 < \gamma_o, \gamma_2 < \gamma_o, \ldots, \gamma_{N_r} < \gamma_o], \\
&= P[\gamma_1 < \gamma_o]P[\gamma_2 < \gamma_o]\ldots P[\gamma_{N_r} < \gamma_o], \\
&= p^{N_r}.
\end{aligned}
\tag{5.10}
$$

For a Rayleigh fading channel,

$$
p = 1 - e^{-\gamma_o / \overline{\gamma}},
\tag{5.11}
$$

where $\overline{\gamma}$ is the average received SNR at that location—for example, owning to pathloss. Thus, selection combining dramatically decreases the outage probability to

$$
P_{out} = (1 - e^{-\gamma_o / \overline{\gamma}})^{N_r}.
\tag{5.12}
$$

The average received SNR for N_r branch SC can be derived in Rayleigh fading to be

$$
\begin{aligned}
\overline{\gamma}_{sc} &= \overline{\gamma} \sum_{i=1}^{N_r} \frac{1}{i}, \\
&= \overline{\gamma}(1 + \frac{1}{2} + \frac{1}{3} + \ldots + \frac{1}{N_r}).
\end{aligned}
\tag{5.13}
$$

Hence, although each added (uncorrelated) antenna does increase the average SNR,[3] it does so with rapidly diminishing returns. The average BEP can be derived by averaging (integrating) the appropriate BEP expression in AWGN against the exponential distribution. Plots of the BEP with different amounts of selection diversity are shown in Figure 5.4, and although the performance improvement with increasing N_r diminishes, the improvement from the first few antennas is dramatic. For example, at a target BEP of 10^{-4}, about 15 dB of improvement is

---

3. It can be noted that Equation (5.13) does not in fact converge, that is, $N_r \to \infty \Rightarrow \gamma_{sc} \to \infty$, because the tail of the exponential function allows arbitrarily high SNR. In practice, this is impossible, since the number of colocated uncorrelated antennas could rarely exceed single digits, and the SNR of a single branch never does approach infinity.

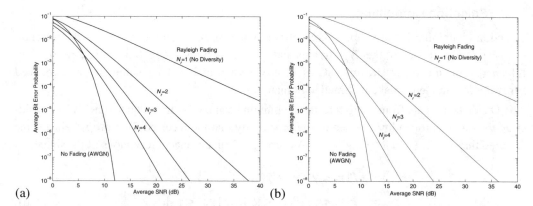

Figure 5.4 Average bit error probability for (a) selection combining and (b) maximal ratio combinings using coherent BPSK. Owing to its array gain, MRC typically achieves a few dB better SNR than does SC.

achieved by adding a single receive antenna, and the improvement increases to 20 dB with an additional antenna.

5.2.2 Maximal Ratio Combining

Maximal ratio combining combines the information from all the received branches in order to maximize the ratio of signal-to-noise power, which gives it its name. MRC works by weighting each branch with a complex factor $q_i = |q_i| e^{j\phi_i}$ and then adding up the N_r branches, as shown in Figure 5.4. The received signal on each branch can be written as $x(t)h_i$, assuming that the fading is flat with a complex value of $h_i = |h_i| e^{j\theta_i}$ on the ith branch.

The combined signal can then be written as S

$$y(t) = x(t) \sum_{i=1}^{N_r} |q_i| \|h_i| \exp\{j(\phi_i + \theta_i)\}. \tag{5.14}$$

If we let the phase of the combining coefficient $\phi_i = -\theta_i$ for all the branches, the signal-to-noise ratio of $y(t)$ can be written as

$$\gamma_{MRC} = \frac{\mathcal{E}_x (\sum_{i=1}^{N_r} |q_i| \|h_i|)^2}{\sigma^2 \sum_{i=1}^{N_r} |q_i|^2}, \tag{5.15}$$

where $\mathcal{E}_x$ is the transmit signal energy. Maximizing this expression by taking the derivative with respect to $|q_i|$ gives the maximizing combining values as $|q_i^*|^2 = |h_i|^2 / \sigma^2$; that is, each branch is multiplied by its SNR. In other words, branches with better signal energy should be enhanced, whereas branches with lower SNRs should be given relatively less weight. The resulting signal-to-noise ratio can be found to be

$$\gamma_{MRC} = \frac{\mathcal{E}_x \sum_{i=1}^{N_r} |h_i|^2}{\sigma^2} = \sum_{i=1}^{N_r} \gamma_i. \tag{5.16}$$

MRC is intuitively appealing: The total SNR is achieved by simply adding up the branch SNRs when the appropriate weighting coefficients are used. It should be noted that although MRC does in fact maximize SNR and generally performs well, it may not be optimal in many cases since it ignores interference powers the statistics of which may differ from branch to branch.

Equal gain combining (EGC), which corrects only the phase and hence as the name of the technique suggests uses $|q_i| = 1$ and $\phi_i = -\theta_i$ for all the combiner branches, achieves a post-combining SNR of

$$\gamma_{EGC} = \frac{\mathcal{E}_x \sum_{i=1}^{N_r} |h_i|^2}{N_r \sigma^2} \tag{5.17}$$

The most notable difference between Equation (5.17) and Equation (5.16) is that EGC incurs a noise penalty in trade for not requiring channel gain estimation. EGC is hence subopti-mal compared to MRC, assuming that the MRC combiner has accurate knowledge of $|h_i|$, par-ticularly when the noise variance is high and there are several receive branches. For an interference-limited cellular system, such as WiMAX, MRC would be strongly preferred to either EGC or SC, despite the fact that the latter techniques are somewhat simpler. The BEP per-formance of MRC is shown in Figure 5.4 for $N_r = 1$. Although the BEP slopes are similar to selection combining, since the techniques have the same diversity order, the SNR gain is several dB owing to its array gain, which may be especially significant at the SINR operating points expected in interference-limited WiMAX systems—usually less than 10 dB. An additional important advantage of MRC in frequency-selective fading channels is that all the frequency diversity can be used, whereas an RF antenna-selection algorithm would simply select the best *average* antenna and then must live with the potentially deep fades at certain frequencies.

5.3 Transmit Diversity

Transmit spatial diversity is a newer phenomenon than receive diversity and has become widely implemented only in the early 2000s. Because the signals sent from different transmit antennas interfere with one another, processing is required at both the transmitter and the receiver in order to achieve diversity while removing or at least attenuating the spatial interference. Transmit diversity is particularly attractive for the downlink of infrastructure-based systems such as WiMAX, since it shifts the burden for multiple antennas to the transmitter, which in this case is a base station, thus greatly benefitting MSs that have severe power, size, and cost constraints.

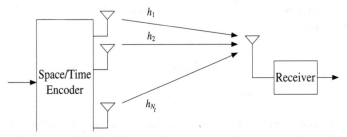

Figure 5.5 Open-loop transmit diversity

Additionally, if the multiple antennas are already at the base station for uplink receive diversity, the incremental cost of using them for transmit diversity is very low.

Multiple-antenna transmit schemes—both transmit diversity and spatial multiplexing—are often categorized as either *open loop* or *closed loop*. Open-loop systems do *not* require knowledge of the channel at the transmitter. On the contrary, closed-loop systems require channel knowledge at the transmitter, thus necessitating either channel reciprocity—same uplink and downlink channel, possible in TDD—or more commonly a feedback channel from the receiver to the transmitter.

5.3.1 Open-Loop Transmit Diversity

The most popular open-loop transmit-diversity scheme is space/time coding, whereby a code known to the receiver is applied at the transmitter. Although the receiver must know the channel to decode the space/time code, this is not a large burden, since the channel must be known for other decoding operations anyway. Space/time coding was first suggested in the early 1990s before generating intense interest in the late 1990s. Of the many types of space/time codes, we focus here on space/time block codes (STBCs), which lend themselves to easy implementation and are defined for transmit diversity in WiMAX systems.

A key breakthrough in the late 1990s was a space/time block code referred to as either the Alamouti code—after its inventor [1]—or the orthogonal space/time block code (OSTBC). This simple code has become the most popular means of achieving transmit diversity, owing to its ease of implementation—linear at both the transmitter and the receiver—and its optimality with regards to diversity order.

The simplest STBC corresponds to two transmit antennas and a single receive antenna. If two symbols to be transmitted are s_1 and s_2, the Alamouti code sends the following over two symbol times:

	Antenna 1	2
Time 0	s_1	s_2
1	$-s_2^*$	s_1^*

The 2×4 Alamouti STBC is referred to as a rate 1 code, since the data rate is neither increased nor decreased; two symbols are sent over two time intervals. Rather than directly increasing the data rate, the goal of space/time coding is to harness the spatial diversity of the channel.

Assuming a flat-fading channel, $h_1(t)$ is the complex channel gain from antenna 1 to the receive antenna, and $h_2(t)$ is from antenna 2. An additional assumption is that the channel is constant over two symbol times; that is, $h_1(t = 0) = h_1(t = T) = h_1$. This is a reasonable assumption if $f_D T \ll 1$, which is usually true.[4]

The received signal $r(t)$ can be written as

$$
\begin{aligned}
r(0) &= h_1 s_1 + h_2 s_2 + n(0), \\
r(T) &= -h_1 s_2^* + h_2 s_1^* + n(T),
\end{aligned}
\tag{5.18}
$$

where $n(\cdot)$ is a sample of white Gaussian noise. The following diversity-combining scheme can then be used, assuming that the channel is known at the receiver:

$$
\begin{aligned}
y_1 &= h_1^* r(0) + h_2 r^*(T), \\
y_2 &= h_2^* r(0) - h_1 r^*(T).
\end{aligned}
\tag{5.19}
$$

Hence, for example, it can be seen that

$$
\begin{aligned}
y_1 &= h_1^*(h_1 s_1 + h_2 s_2 + n(0)) + h_2(-h_1^* s_2 + h_2^* s_1 + n^*(T)), \\
&= (|h_1|^2 + |h_2|^2)s_1 + h_1^* n(0) + h_2 n^*(T),
\end{aligned}
\tag{5.20}
$$

and, proceeding similarly, that

$$
y_2 = (|h_1|^2 + |h_2|^2)s_2 + h_2^* n(0) - h_1 n^*(T).
\tag{5.21}
$$

Hence, this very simple decoder that linearly combines the two received samples $r(0)$ and $r^*(T)$ is able to eliminate all the spatial interference. The resulting SNR can be computed as

$$
\begin{aligned}
\gamma_\Sigma &= \frac{(|h_1|^2 + |h_2|^2)^2}{|h_1|^2 \sigma^2 + |h_2|^2 \sigma^2} \frac{\mathcal{E}_x}{2}, \\
&= \frac{(|h_1|^2 + |h_2|^2)}{\sigma^2} \frac{\mathcal{E}_x}{2}, \\
&= \frac{\sum_{i=1}^{2} |h_i|^2}{\sigma^2} \frac{\mathcal{E}_x}{2}.
\end{aligned}
\tag{5.22}
$$

---

4. Owing to the flat-fading assumption, the STBC in an OFDM system is generally performed in the frequency domain, where each subcarrier experiences flat fading. However, this leads to a long symbol time T and may cause the channel-invariance assumption to be compromised, resulting in a modest performance loss in the event of high mobility. See, for example, [40] and [53].

Referring to Equation (5.16), we can see that this is similar to the gain from MRC. However, in order to keep the transmit power the same as in the MRC case, each transmit antenna must halve its transmit power so that the total transmit energy per actual data symbol is $\mathcal{E}_x$ for both cases. That is, for STBC, $E \mid s_1 \mid^2 = E \mid s_2 \mid^2 = \mathcal{E}_x/2$, since each is sent twice.

In summary, the 2×1 Alamouti code achieves the same diversity order and data rate as a 1×2 receive diversity system with MRC but with a 3 dB penalty, owing to the redundant transmission required to remove the spatial interference at the receiver. The linear decoder used here is the maximum-likelihood decoder—in zero mean noise—so is optimum as well as simple.

Space/time trellis codes introduce memory and achieve better performance than orthogonal STBCs—about 2 dB in many cases—but have decoding complexity that scales as $O(M^{\min\{N_t, N_r\}})$, where M is again the constellation order. Orthogonal STBCs, on the other hand, have complexity that scales only as $O(\min\{N_t, N_r\})$, so the complexity reduction is quite considerable for high-spectral-efficiency systems with many antennas at both the transmitter and the receiver.

It should be noted that in WiMAX or any OFDM-based system, the space/time coding can be implemented as space/frequency block codes (SFBC) [6], where adjacent *subcarriers*, rather than time slots, are coded over. This assumes that adjacent subcarriers have the same amplitude and phase, which is typically approximately true in practice. All the other development is identical. If space/time coding is used in OFDM, the STBC is implemented over two OFDM symbols. Since OFDM symbols can be quite long in duration, care must be taken to make sure that the channel is constant over subsequent OFDM symbols. Details for how STBCs and SFBCs are implemented in WiMAX are given in Chapter 8.

5.3.2 $N_t \times N_r$ Transmit Diversity

It would be desirable to achieve the gains of both MRC and STBC simultaneously, and that is indeed possible in several cases. In general, however, orthogonal STBCs, such as the 2×1 Alamouti code, do not exist for most combinations of transmit and receive antennas. As a result, a substantial amount of research has proposed various techniques for achieving transmit diversity for more general scenarios, and summarizing all this work is outside the scope of this chapter. Instead, see [20, 26, 47] and the seminal work [58, 59]. Here, we consider two other candidates for transmit diversity in WiMAX systems and compare transmit and receive diversity.

5.3.2.1 2×2 STBC

The 2×2 STBC uses the same transmit-encoding scheme as for 2×1 transmit diversity. Now, the channel description—still flat fading and constant over two symbols—can be represented as a 2×2 matrix rather than a 2×1 vector:

$$\mathbf{H} = \begin{bmatrix} h_{11} & h_{12} \\ h_{21} & h_{22} \end{bmatrix}.$$

The resulting signals at times 0 and T on antennas 1 and 2 can be represented as

$$
\begin{aligned}
r_1(0) &= h_{11}s_1 + h_{21}s_2 + n_1(0), \\
r_1(T) &= -h_{11}s_2^* + h_{21}s_1^* + n_1(T), \\
r_2(0) &= h_{12}s_1 + h_{22}s_2 + n_2(0), \\
r_2(T) &= -h_{12}s_2^* + h_{22}s_1^* + n_2(T).
\end{aligned}
\tag{5.23}
$$

Using the following combining scheme

$$
\begin{aligned}
y_1 &= h_{11}^* r_1(0) + h_{21} r_1^*(T) + h_{12}^* r_2(0) + h_{22} r_2^*(T), \\
y_2 &= h_{21}^* r_1(0) - h_{11} r_1^*(T) + h_{22}^* r_2(0) - h_{21} r_2^*(T),
\end{aligned}
$$

yields the following decision statistics:

$$
\begin{aligned}
y_1 &= (|h_{11}|^2 + |h_{12}|^2 + |h_{21}|^2 + |h_{22}|^2)s_1 + 4 \text{ noise terms,} \\
y_2 &= (|h_{11}|^2 + |h_{12}|^2 + |h_{21}|^2 + |h_{22}|^2)s_2 + 4 \text{ noise terms,}
\end{aligned}
$$

and results in the following SNR:

$$
\gamma_\Sigma = \frac{\left(\sum_j \sum_i |h_{ij}|^2 \right)^2}{\sigma^2 \sum_j \sum_i |h_{ij}|^2} \frac{\mathcal{E}_x}{2} = \frac{\sum_{j=1}^2 \sum_{i=1}^2 |h_{ij}|^2}{\sigma^2} \frac{\mathcal{E}_x}{2}.
\tag{5.24}
$$

This SNR is like MRC with four receive antennas, where again there is a 3 dB penalty due to transmitting each symbol twice. An orthogonal, full-rate, full-diversity STBC over an $N_t \times N_r$ channel will provide a diversity gain equivalent to that of an MRC system with $N_t N_r$ antennas, with a $10\log_{10}N_t$ dB transmit power penalty owing to the N_t transmit antennas. In other words, in theory, it is generally beneficial to have somewhat evenly balanced antenna arrays, as this will maximize the diversity order for a fixed number of antenna elements. In practice, it is important to note that full-diversity, orthogonal STBCs exist only for certain combinations of N_t and N_r.

5.3.2.2 4 × 2 Stacked STBCs

The 2×2 Alamouti code achieves full diversity gain. In some cases, it may be possible to afford four transmit antennas at the base station. In this case, two data streams can be sent, using a double space/time transmit diversity (DSTTD) scheme that consists of operating two 2×1 Alamouti code systems in parallel [46, 61]. DSTTD, also called stacked STBCs, combines transmit diversity and MRC techniques, along with a form of spatial multiplexing, as shown in Figure 5.6.

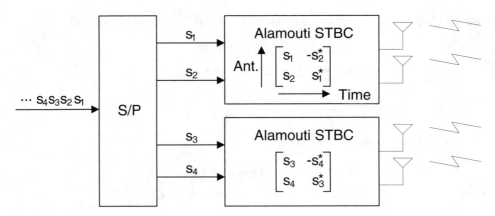

Figure 5.6 4×2 stacked STBC transmitter

The received signals at times 0 and T on antennas 1 and 2 can be represented with the equivalent channel model as

$$
\begin{bmatrix} r_1(0) \\ r_1^*(T) \\ \hline r_2(0) \\ r_2^*(T) \end{bmatrix} = \begin{bmatrix} h_{11} & h_{12} & h_{13} & h_{14} \\ h_{12}^* & -h_{11}^* & h_{14}^* & -h_{13}^* \\ \hline h_{21} & h_{22} & h_{23} & h_{24} \\ h_{22}^* & -h_{21}^* & h_{24}^* & -h_{23}^* \end{bmatrix} \begin{bmatrix} s_1 \\ s_2 \\ \hline s_3 \\ s_4 \end{bmatrix} + \begin{bmatrix} n_1(0) \\ (n_1^*) T \\ \hline n_2(0) \\ n_2^*(T) \end{bmatrix}. \tag{5.25}
$$

Then, the equivalent matrix channel model of DSTTD can be represented as

$$
\begin{bmatrix} \mathbf{r}_1 \\ \mathbf{r}_2 \end{bmatrix} = \begin{bmatrix} \mathbf{H}_{11} & \mathbf{H}_{12} \\ \mathbf{H}_{21} & \mathbf{H}_{22} \end{bmatrix} \begin{bmatrix} \mathbf{s}_1 \\ \mathbf{s}_2 \end{bmatrix} + \begin{bmatrix} \mathbf{n}_1 \\ \mathbf{n}_2 \end{bmatrix}. \tag{5.26}
$$

As shown in Equation (5.26), each $\mathbf{H}_{ij}$ channel matrix is the equivalent channel of the Alamouti code. Thus, DSTTD can achieve a diversity order of $N_d = 2N_r$ (ML, or maximum-likelihood, detection) or $N_d = 2$ (ZF, or zero forcing, detection) owing to the 2×1 Alamouti code while also transmitting two data streams (spatial multiplexing order of 2).

If the same linear combining scheme is used as in the 2×2 STBC case, the following decision statistics can be obtained:

$$
\begin{aligned}
y_1 &= (|h_{11}|^2 + |h_{12}|^2 + |h_{21}|^2 + |h_{22}|^2)s_1 + I_3 + I_4 + 4 \text{ noise terms,} \\
y_2 &= (|h_{11}|^2 + |h_{12}|^2 + |h_{21}|^2 + |h_{22}|^2)s_2 + I_3 + I_4 + 4 \text{ noise terms,} \\
y_3 &= (|h_{13}|^2 + |h_{14}|^2 + |h_{23}|^2 + |h_{24}|^2)s_3 + I_1 + I_2 + 4 \text{ noise terms,} \\
y_4 &= (|h_{13}|^2 + |h_{14}|^2 + |h_{23}|^2 + |h_{24}|^2)s_4 + I_1 + I_2 + 4 \text{ noise terms,}
\end{aligned} \tag{5.27}
$$

where I_i is the interference from the ith transmit antenna due to transmitting two simultaneous data streams. The detection process of DSTTD should attempt to suppress the interference between the two STBC encoders and for this purpose can turn to any of the spatial-multiplexing receivers (see Section 5.5.1). In contrast to OSTBCs (Alamouti codes), the ML receiver for stacked STBCs is *not* linear.

5.3.2.3 Transmit Diversity versus Receive Diversity

The three example space/time block codes showed that transmit and receive diversity are capable of providing an enhanced diversity that increases the robustness of communication over wireless fading channels. The manner in which this improvement is achieved is quite different, however.

Receive diversity: For MRC with N_r antennas and only one transmit antenna, the received SNR continuously grows as antennas are added, and the growth is linear:

$$\gamma_{MRC} = \frac{\mathcal{E}_x}{\sigma^2} \sum_{i=1}^{N_r} |h_i|^2 = \sum_{i=1}^{N_r} \gamma_i. \tag{5.28}$$

The expected value, or average combined SNR, can thus be found as

$$\overline{\gamma}_{MRC} = N_r \overline{\gamma}, \tag{5.29}$$

where $\overline{\gamma}$ is the average SNR on each branch. In other words, the SNR growth is *linear* with the number of receive antennas. Thus, from Shannon's capacity formula, it can be observed that since $C = B \log(1 + SNR)$, the throughput growth due to receive diversity is *logarithmic* with the number of receive antennas, since receive diversity serves to increase the SNR.

Transmit diversity: Due to the transmit-power penalty inherent to transmit diversity techniques, the received SNR does not always grow as transmit antennas are added. Instead, if there is a single receive antenna, the received combined SNR in an orthogonal STBC scheme is generally of the form

$$\gamma_\Sigma = \frac{\mathcal{E}_x}{N_t \sigma^2} \sum_{i=1}^{N_t} |h_i|^2. \tag{5.30}$$

As the number of transmit antennas grows large, this expression becomes

$$\gamma_\Sigma = \frac{\mathcal{E}_x}{\sigma^2} \frac{|h_1|^2 + |h_2|^2 + \ldots + |h_{N_t}|^2}{N_t} \to \frac{\mathcal{E}_x}{\sigma^2} E[|h_1|^2], \tag{5.31}$$

by the law of large numbers. Thus, open-loop transmit diversity causes the received SNR to "harden" to the average SNR. In other words, it eliminates the effects of fading but does not increase the average amount of useful received signal-to-noise ratio.

Example 5.1 Consider two possible antenna configurations that use a total of $N_a = 6$ antennas. In one system, we place two antennas at the transmitter and four at the receiver and implement the Alamouti STBC scheme. In the other system, we place one antenna at the transmitter and five at the receiver and perform MRC. Which configuration will achieve a lower BEP in a fading channel?

An exact calculation is not very simple and requires the BEP in AWGN to be integrated against a complex SNR expression. However, to get a feel, two things should be considered: the average output SNR (array gain) and the diversity order. The diversity order of the 2×4 STBC system is 8 but for the 1×5 MRC system is just 5. However, the average postcombining SNR is higher for the 1×5 MRC system, owing to array gain, since

$$\overline{\gamma}_{1 \times 5}^{MRC} = 5\overline{\gamma}, \tag{5.32}$$

whereas

$$\overline{\gamma}_{2 \times 4}^{STBC} = \frac{1}{2} 8\overline{\gamma} = 4\overline{\gamma}, \tag{5.33}$$

owing to the transmit-power penalty. Since the array gains of STBC and MRC over a single-input/single-output (SISO) system are both equal to the number of receive antennas Nr when the total number of transmit and receive antennas is fixed at $N_a = 6$, the diversity order at high SNR causes the occasional fades to be averaged out, and 2×4 STBC is therefore preferable to 5×1 MRC. On the other hand, at low SNR, a fixed-array gain is a more significant contribution than the SNR averaging provided by the diversity gain, and so pure MRC is generally preferable at low SNR.

Figure 5.7 compares the BEP performance of Alamouti STBC with MRC, using coherent BPSK with various N_a in a Rayleigh fading channel, and confirms this intuition. As expected, for a fixed $N_a > 3$, the Alamouti STBC outperforms MRC at high SNR owing to the diversity order, whereas MRC has better BEP performance than Alamouti STBC at low SNR owing to the array gain. In the case of $N_a = 6$, we observe that the BEP crossing point between 2×4 STBC and 1×5 MRC is at 2.03 dB average SNR on each branch.

5.3.3 Closed Loop-Transmit Diversity

If feedback is added to the system, the transmitter may be able to have knowledge of the channel between it and the receiver. Because the channel changes quickly in a highly mobile scenario, closed-loop transmission schemes tend to be feasible primarily in fixed or low-mobility scenarios. As we shall see, however, there is a substantial gain in many cases from possessing channel state information (CSI) at the transmitter, particularly in the spatial multiplexing setup discussed

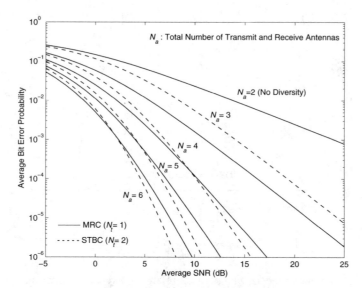

Figure 5.7 Comparison of the Alamouti STBC with MRC for coherent BPSK in a Rayleigh fading channel.

later in the chapter. This again has motivated intensive research on techniques for achieving low-rate prompt feedback, often specifically for the multiantenna channel [42].

The basic configuration for closed-loop transmit diversity is shown in Figure 5.8; in general, the receiver could also have multiple antennas, but we neglect that here for simplicity. An encoding algorithm is responsible for using the CSI to effectively use its N_t available channels. We will assume throughout this section that the transmitter has fully accurate CSI available to it, owing to the feedback channel. We now review two important types of closed-loop transmit diversity, focusing on how they affect the encoder design and on their achieved performance.

5.3.3.1 Transmit Selection Diversity

Transmit selection diversity (TSD) is the simplest form of transmit diversity, and also one of the most effective. In transmit selection diversity first suggested by Winters [65], only a subset $N^* < N_t$ of the available N_t antennas are used at a given time. The selected subset typically corresponds to the best channels between the transmitter and the receiver. TSD has the advantages of (1) significantly reduced hardware cost and complexity, (2) reduced spatial interference, since fewer transmit signals are sent, and (3) somewhat surprisingly, $N_t N_r$ diversity order, even though only N^* of the N_t antennas are used. Despite its optimal diversity *order*, TSD is not optimal in terms of diversity *gain*.

In the simplest case, a single transmit antenna is selected, where the chosen antenna results in the highest gain between the transmitter and the receive antenna. Mathematically, this is statistically identical to choosing the highest-gain receive antenna in a receive-diversity system, since they both result in an optimum antenna choice i^*:

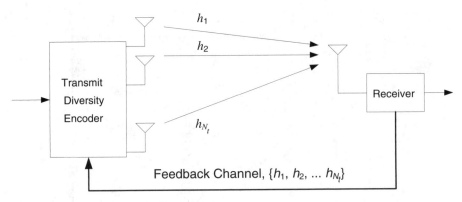

Figure 5.8 Closed-loop transmit diversity

$$i^* = \arg\max_{i \in (1, N_t)} |h_i|^2 . \tag{5.34}$$

Hence, TSD does not incur the power penalty relative to receive selection diversity that we observed in the case of STBCs versus MRC, while achieving the same diversity order. The average SNR with single-transmit antenna selection in a $N_t \times 1$ system with i.i.d. Rayleigh fading is thus

$$\gamma_{tsd} = \bar{\gamma} \sum_{i=1}^{N_t} \frac{1}{i}, \tag{5.35}$$

which is identical to Equation (5.13) for receiver selection combining. This is, however, a lower average SNR than can be achieved with beamforming techniques that use all the transmit antennas. In other words, transmit selection diversity captures the full diversity order—and so is robust against fading—but sacrifices some overall SNR performance relative to techniques that use or capture all the available energy at the transmitter and the receiver.

The feedback required for antenna transmit selection diversity is also quite low, since all that is needed is the index of the required antenna, not the full CSI. In the case of single-transmit antenna selection, only $\log_2 N_t$ bits of feedback are needed for each channel realization. For example, if there were $N_t = 4$ transmit antennas and the channel coherence time was $T_c = 10$ msec—corresponding to a Doppler of about 100Hz—only about 1kbps of channel feedback would be needed, assuming that the feedback rate was five times faster than the rate of channel decorrelation.

In the case of N^* active transmit antennas, choosing the best N^* out of the available N_t elements requires a potentially large search over

$$\binom{N_t}{N^*} \tag{5.36}$$

different possibilities, although for many practical configurations, the search is simple. For example, choosing the best two antennas out of four requires only six possible combinations to be checked. Even for very large antenna configurations, near-optimal results can be attained with much simpler searches. The required feedback for transmit antenna selection is about $N^* \log_2 N_t$ bits per channel coherence time. Because of its excellent performance versus complexity trade-offs, transmit selection diversity appears to be attractive as a technique for achieving spatial diversity, and has also been extended to other transmit diversity schemes such as space/time block codes [12, 31], spatial multiplexing [33], and multiuser MIMO systems [9]. An overview of transmit antenna selection can be found in [45].

In the context of WiMAX, a crucial drawback of transmit antenna selection is that its gain is often very limited in a frequency-selective fading channel. If the channel bandwidth is much wider than the channel coherence bandwidth, considerable frequency diversity exists, and the total received power in the entire bandwidth will be approximately equal regardless of which antenna is selected. If each OFDM subcarrier were able to independently choose the desired transmit antenna that maximized its subcarrier gain, TSD would be highly effective, but sending a different subset of subcarriers on each transmit antenna defeats the main purpose of transmit antenna selection: turning off (or not requiring) the RF chains for the $N_t - N^*$ antennas that were not selected. Additionally, in this case, the required feedback would increase in proportion to L (the number of subcarriers). Hence, despite its theoretical promise, *transmit selection diversity is likely to be useful only in deployments with small bandwidths and small delay spreads (low range)*, which is very limiting.

5.3.3.2 Linear Diversity Precoding

Linear precoding is a simple technique for improving the data rate, or the link reliability, by exploiting the CSI at the transmitter. In this section, we consider *diversity* precoding, a special case of linear precoding whereby the data rate is unchanged, and the linear precoder at the transmitter and a linear postcoder at the receiver are applied only to improve the link reliability. This will allow comparison with STBCs, and the advantage of transmit CSI will become apparent.

With linear precoding, the received data vector can be written as

$$\mathbf{y} = \mathbf{G}(\mathbf{HFx} + \mathbf{n}), \tag{5.37}$$

where the sizes of the transmitted ($\mathbf{x}$) and the received ($\mathbf{y}$) symbol vectors are $M \times 1$, the postcoder matrix $\mathbf{G}$ is $M \times N_r$, the channel matrix $\mathbf{H}$ is $N_r \times N_t$, the precoder matrix $\mathbf{F}$ is $N_t \times M$, and the noise vector $\mathbf{n}$ is $N_r \times 1$. For the case of diversity precoding (comparable to a rate 1 STBC), $M = 1$, and the SNR maximizing precoder $\mathbf{F}$ and postcoder $\mathbf{G}$ are the right- and left-singular vectors of $\mathbf{H}$ corresponding to its largest singular value, σ_{max}. In this $M = 1$ case, the equivalent channel model after precoding and postcoding for a given data symbol x is

$$y = \sigma_{max} \cdot x + n. \tag{5.38}$$

Sidebar 5.1 A Brief Primer on Matrix Theory

As this chapter indicates, linear algebra and matrix analysis are an inseparable part of MIMO theory. Matrix theory is also useful in understanding OFDM. In this book, we have tried to keep all the matrix notation as standard as possible, so that any appropriate reference will be capable of clarifying any of the presented equations.

In this sidebar, we simply define some of the more important notation for clarity. First, in this chapter, two types of transpose operations are used. The first is the *conventional transpose* $\mathbf{A}^T$, which is defined as

$$\mathbf{A}_{ij}^T = \mathbf{A}_{ji}$$

that is, only the rows and columns are reversed. The second type of transpose is the *conjugate transpose*, which is defined as

$$\mathbf{A}_{ij}^* = \left(\mathbf{A}_{ji}\right)^*.$$

That is, in addition to exchanging rows with columns, each term in the matrix is replaced with its complex conjugate. If all the terms in $\mathbf{A}$ are real, $\mathbf{A}^T = \mathbf{A}^*$. Sometimes, the conjugate transpose is called the *Hermitian transpose* and denoted as $\mathbf{A}^H$. They are equivalent.

Another recurring theme is matrix decomposition—specifically, the eigendecomposition and the singular-value decomposition, which are related to each other. If a matrix is square and diagonalizable ($M \times M$), it has the eigendecomposition

$$\mathbf{A} = \mathbf{T}\mathbf{\Lambda}\mathbf{T}^{-1},$$

where $\mathbf{T}$ contains the (right) eigenvectors of $\mathbf{A}$, and $\mathbf{\Lambda} = \text{diag}[\lambda_1 \ \lambda_2 \ ... \ \lambda_M]$ is a diagonal matrix containing the eigenvalues of $\mathbf{A}$. $\mathbf{T}$ is invertible as long as $\mathbf{A}$ is symmetric or has full rank (M nonzero eigenvalues).

When the eigendecomposition does not exist, either because $\mathbf{A}$ is not square or for the preceding reasons, a generalization of matrix diagonalization is the singular-value decomposition, which is defined as

$$\mathbf{A} = \mathbf{U}\mathbf{\Sigma}\mathbf{V}^*,$$

where $\mathbf{U}$ is $M \times r$, $\mathbf{V}$ is $N \times r$, and $\mathbf{\Sigma}$ is $r \times r$, and the rank of $\mathbf{A}$—the number of nonzero singular values—is r. Although $\mathbf{U}$ and $\mathbf{V}$ are no longer inverses of each other as in eigendecomposition, they are both unitary—$\mathbf{U}^*\mathbf{U} = \mathbf{V}^*\mathbf{V} = \mathbf{U}\mathbf{U}^* = \mathbf{V}\mathbf{V}^* = \mathbf{I}$—which means that they have orthonormal columns and rows. The singular values of $\mathbf{A}$ can be related to eigenvalues of $\mathbf{A}^*\mathbf{A}$ by

$$\sigma_i(\mathbf{A}) = \sqrt{\lambda_i(\mathbf{A}^*\mathbf{A})}.$$

Because $\mathbf{T}^{-1}$ is not unitary, it is not possible to find a more exact relation between the singular values and eigenvalues of a matrix, but these values generally are of the same order, since the eigenvalues of $\mathbf{A}^*\mathbf{A}$ are on the order of the square of those of $\mathbf{A}$.

Therefore, the received SNR is

$$\gamma = \frac{\mathcal{E}_x}{\sigma^2}\sigma_{\max}^2, \tag{5.39}$$

where σ^2 is the noise variance. Since the value or expected value of $\sigma_{\max}$ is not deterministic, the SNR can be bounded only as [47],

$$\frac{\|\mathbf{H}\|_{\mathbf{F}}^2}{N_t} \cdot \frac{\mathcal{E}_x}{\sigma^2} \le \gamma \le \|\mathbf{H}\|_{\mathbf{F}}^2 \cdot \frac{\mathcal{E}_x}{\sigma^2}, \tag{5.40}$$

where $\| \cdot \|_{\mathbf{F}}$ denotes the Frobenius norm and is defined as

$$\|\mathbf{H}\|_{\mathbf{F}} = \sqrt{\sum_{i=1}^{N_t}\sum_{j=1}^{N_r} h_{ij}^2}. \tag{5.41}$$

On the other hand, by generalizing the SNR expression for 2×2 STBCs—Equation (5.24)—the SNR for the case of STBC is given as

$$\gamma_{STBC} = \frac{\|\mathbf{H}\|_{\mathbf{F}}^2 \, \mathcal{E}_x}{N_t \sigma^2}. \tag{5.42}$$

By comparing Equation (5.40) and Equation (5.42), we see that linear precoding achieves a higher SNR than the open-loop STBCs, by up to a factor of N_t. When $N_r = 1$, the full SNR gain of $10\log_{10}N_t$ dB is achieved; that is, the upper bound on SNR in Equation (5.40) becomes an equality.

To use linear precoding, feeding back of CSI from the receiver to the transmitter is typically required. To keep the CSI feedback rate small, a codebook-based precoding method that requires only 3–6 bits of CSI feedback for each channel realization has been defined for WiMAX. More discussion on codebook-based precoding can be found in Section 5.8, with WiMAX implementations discussed in Chapter 8.

5.4 Beamforming

In contrast to the transmit diversity techniques of the previous section, the available antenna elements can instead be used to adjust the strength of the transmitted and received signals, based on their *direction,* which can be either the physical direction or the direction in a mathematical sense. This focusing of energy is achieved by choosing appropriate weights for each antenna element with a certain criterion. In this section, we look at the two principal classes of beamforming: direction of arrival (DOA)–based beamforming (physically directed) and eigenbeamforming (mathematically directed). It should be stressed that beamforming is an often misunderstood term, since these two classes of "beamforming" are radically different.

5.4.1 DOA-Based Beamforming

The incoming signals to a receiver may consist of desired energy and interference energy—for example, from other users or from multipath reflections. The various signals can be characterized in terms of the DOA or the angle of arrival (AOA) of each received signal. Each DOA can be estimated by using signal-processing techniques, such as the MUSIC, ESPRIT, and MLE algorithms (see [27, 38] and the references therein). From these acquired DOAs, a beamformer extracts a weighting vector for the antenna elements and uses it to transmit or receive the desired signal of a specific user while suppressing the undesired interference signals.

When the plane wave arrives at the d-spaced uniform linear array (ULA) with AOA θ, the wave at the first antenna element travels an additional distance of $d \sin \theta$ to arrive at the second element. This difference in propagation distance between the adjacent antenna elements can be formulated as an arrival-time delay, $\tau = d/c \sin \theta$. As a result, the signal arriving at the second antenna can be expressed in terms of signal at the first antenna element as

$$
\begin{aligned}
y_2(t) &= y_1(t)\exp(-j2\pi f_c \tau), \\
&= y_1(t)\exp(-j2\pi \frac{d \sin \theta}{\lambda}).
\end{aligned}
\tag{5.43}
$$

For an antenna array with N_r elements all spaced by d, the resulting received signal vector can therefore be expressed as

$$
\begin{aligned}
(t) &= [y_1(t)\, y_2(t)\, \dots\, y_{N_r}(t)]^T \\
&= y_1(t)\underbrace{[1\; \exp(-j2\pi \frac{d \sin \theta}{\lambda})\; \dots \exp(-j2\pi(N_r-1)\frac{d \sin \theta}{\lambda})]^T}_{\mathbf{a}(\theta)},
\end{aligned}
\tag{5.44}
$$

where $\mathbf{a}(\theta)$ is the *array response vector*.

In the following, we show an example to demonstrate the principle of DOA-based beamforming. Consider a three-element ULA with $d = \lambda/2$ spacing between the antenna elements. Assume that the desired user's signal is received with an AOA of $\theta_1 = 0$ —that is, the signal is coming from the broadside of the ULA—and two interfering signals are received with AOAs of $\theta_2 = \pi/3$ and $\theta_3 = -\pi/6$, respectively. The array response vectors are then given by

$$
\mathbf{a}(\theta_1) = \begin{bmatrix} 1 & 1 & 1 \end{bmatrix}^T, \quad \mathbf{a}(\theta_2) = \begin{bmatrix} 1 & e^{-j\frac{\sqrt{3}}{2}\pi} & e^{-j\sqrt{3}\pi} \end{bmatrix}^T, \quad and \quad \mathbf{a}(\theta_3) = \begin{bmatrix} 1 & e^{j\frac{\pi}{2}} & e^{j\pi} \end{bmatrix}^T.
\tag{5.45}
$$

The beamforming weight vector $\mathbf{w} = [w_1\, w_2\, w_3]^T$ should increase the antenna gain in the direction of the desired user while simultaneously minimizing the gain in the directions of interferers. Thus, the weight vector $\mathbf{w}$ should satisfy the following criterion:

$$
\mathbf{w}^* \begin{bmatrix} \mathbf{a}(\theta_1) & \mathbf{a}(\theta_2) & \mathbf{a}(\theta_3) \end{bmatrix} = \begin{bmatrix} 1 & 0 & 0 \end{bmatrix}^T,
\tag{5.46}
$$

and a unique solution for the weight vector is readily obtained as

$$\mathbf{w} = \begin{bmatrix} 0.3034 + j0.1966 & 0.3932 & 0.3034 - j0.1966 \end{bmatrix}^T. \tag{5.47}$$

Figure 5.9 shows the beam pattern using this weight vector. As expected, the beamformer has unity gain for the desired user and two nulls at the directions of two interferers. Since the beamformer can place nulls in the directions of interferers, the DOA-based beamformer in this example is often called the *null-steering beamformer* [27]. The null-steering beamformer can be designed to completely cancel out interfering signals only if the number of such signals is strictly less than the number of antenna elements. That is, if N_r is the number of receive antennas, $N_r - 1$ independent interferers can be canceled.[5] The disadvantage of this approach is that a null is placed in the direction of the interferers, so the antenna gain is not maximized at the direction of the desired user. Typically, there exists a trade-off between interference nulled and desired gain lost. A more detailed description on the DOA-beamformer with refined criterion can be found in [27, 38].

Thus far, we have assumed that the array response vectors of different users with corresponding AOAs are known. In practice, each resolvable multipath is likely to comprise several unresolved components coming from significantly different angles. In this case, it is not possible to associate a discrete AOA with a signal impinging the antenna array. Therefore, the DOA-based beamformer is viable only in LOS environments or in environments with limited local scattering around the transmitter.

5.4.2 Eigenbeamforming

Unlike DOA-based beamforming, eigenbeamforming does not have a similar physical interpretation. Instead of using the array-response vectors from AOAs of all different users, eigenbeamforming exploits the channel-impulse response of each antenna element to find array weights that satisfy a desired criterion, such as SNR maximization or MSE (mean squared error) minimization. By using channel knowledge at the transmitter, eigenbeamforming exploits the eigendecomposed channel response for focusing the transmit signal to the desired user even if there are cochannel interfering signals with numerous AOAs. Because eigenbeamforming is a *mathematical* technique rather than a *physical* technique for increasing the desired power and suppressing the interference signals, it is more viable in realistic wireless broadband environments, which are expected to have significant local scattering. When we refer to an *eigenchannel* in this section, we are referring to the complex channel corresponding to an eigenvalue in the channel matrix, which can be accessed by precoding with the (right) eigenvector of the channel matrix.

Consider a MIMO eigenbeamforming system using N_t antennas for transmission and N_r antennas for reception in a flat-fading channel. It is assumed that there are L effective cochannel

---

5. In some special cases, it may be possible to cancel more than $N_r - 1$ interferers, such as the special case in which a third interferer was at an angle of $2\pi/3$ or $7\pi/6$ as in Figure 5.9.

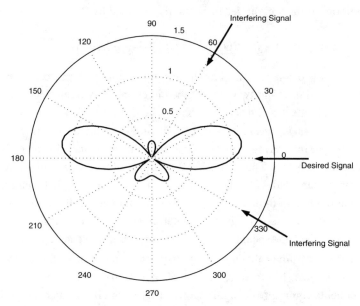

Figure 5.9 Null-steering beam pattern for the DOA-based beamforming using three-element ULA with $\lambda/2$ spacing at transmit antennas. The AOAs of the desired user and two interferers are 0, $\pi/3$, and $-\pi/6$, respectively.

interference signals, which is equivalent to L_I distinct cochannel interferers, each equipped with $N_{t,i}$ antenna elements, satisfying

$$L = \sum_{i=1}^{L_I} N_{t,i} \ .$$

Then, the N_r-dimensional received signal vector at the receiver is given by

$$\mathbf{y} = \mathbf{H}\mathbf{w}_t x + \mathbf{H}_I x_I + \mathbf{n} \ , \tag{5.48}$$

where $\mathbf{w}_t$ is the $N_t \times 1$ weighting vector at the desired user's transmitter, x is the desired symbol with energy $\mathcal{E}_x$, $\mathbf{x}_I = [x_1 \ x_2 \ \cdots \ x_L]^T$ is the interference vector, and n is the noise vector with covariance matrix $\sigma^2 \mathbf{I}$, $\mathbf{H}$ is the $N_r \times N_t$ channel gain matrix for the desired user, and $\mathbf{H}_I$ is the $N_r \times L$ channel gain matrix for the interferers. In order to maximize the output SINR at the receiver, joint optimal weighting vectors at both the transmitter and the receiver can be obtained as [67]

$$\mathbf{w}_t = \text{Eigenvector corresponding to the largest eigenvalue } \lambda_{max}\left(\mathbf{H}^*\mathbf{R}^{-1}\mathbf{H}\right), \tag{5.49}$$

and

$$\mathbf{w}_r = \alpha \mathbf{R}^{-1}\mathbf{H}\mathbf{w}_t, \tag{5.50}$$

where α is an arbitrary constant that does not affect the SNR, $\mathbf{R}$ is the interference-plus-noise covariance matrix, and $\lambda_{max}(\mathbf{A})$ is the largest eigenvalue of $\mathbf{A}$. We then have the maximum output SINR

$$\gamma = \lambda_{max}\left(\mathbf{H}^*\mathbf{R}^{-1}\mathbf{H}\right). \tag{5.51}$$

This result shows that the transmit power is focused on the largest eigenchannel among $\min(N_t, N_r)$ eigenchannels in order to maximize postbeamforming SINR. In this sense, this beamformer is often termed the *optimum eigenbeamformer, or optimum combiner* (OC).

It can be seen that conceptually, the eigenbeamformer is conceptually nearly identical to the linear diversity precoding scheme (Section 5.3.3), the only difference being that the eigenbeamformer takes interfering signals into account. If the interference terms are ignored, $\mathbf{R} \rightarrow \sigma^2\mathbf{I}$ and $\mathbf{w}_r \rightarrow \mathbf{G}$ and $\mathbf{w}_t \rightarrow \mathbf{F}$. This special case, which maximizes the received SNR, is also referred to as transmit MRC, or maximum ratio transmission (MRT) [41], which includes conventional MRC as a special case in which the transmitter has a single antenna element. In short, many of the proposed techniques going by different names have fundamental similarities or are special cases of general linear precoding/postcoding.

Figure 5.10 shows a performance comparison between the eigenbeamformer and other transmit/receive diversity schemes. The optimum beamformer cancels out a strong interferer by sacrificing a degree of freedom at the receiver. That is, the 2×2 optimum eigenbeamformer with one strong interferer is equivalent to the 2×1 MRT with no interference, which has also the same performance with 1×2 MRC. We also confirm that exploiting channel knowledge at the transmitter provides significant array gain and, especially in the case of a single receive antenna, the transmit diversity using MRT has the same array gain and diversity order of receive-diversity MRC.

To summarize, in the absence of interference, the output SNR of the optimum eigenbeam-former—that is, MRT—with $N_t > 1$ can be upper- and lower-bounded as follows:

$$\gamma^{STBC}_{N_t \times N_r} = \frac{\mathcal{E}_x}{N_t\sigma^2}\|\mathbf{H}\|^2_F < \gamma^{MRT}_{N_t \times N_r} = \frac{\mathcal{E}_x}{\sigma^2}\lambda_{max}\left(\mathbf{H}^H\mathbf{H}\right) \leq \frac{\mathcal{E}_x}{\sigma^2}\|\mathbf{H}\|^2_F = \gamma^{MRC}_{1 \times N_t N_r} \tag{5.52}$$

where the equality between MRT and MRC holds if and only if $N_r = 1$. The preceding inequality is a generalization of Equation (5.40). When L cochannel interferers exist, the average output SINR of the optimum eigenbeamformer with $N_r > L$ can be also bounded in terms of the average output SNR for several diversity schemes *without* interference, as follows:

$$\overline{\gamma}^{STBC}_{N_t \times (N_r - L)} < \overline{\gamma}^{MRT}_{N_t \times (N_r - L)} < \overline{\gamma}^{OC}_{N_t \times N_r} < \overline{\gamma}^{MRT}_{N_t \times N_r} < \overline{\gamma}^{MRC}_{1 \times N_t N_r}. \tag{5.53}$$

The eigenbeamformers of this section have been designed for transmission of a single data stream, using perfect channel state information at both the transmitter and the receiver. In order to further increase the system capacity using the acquired transmit CSI, up to rank $(\mathbf{H}) = \min(N_t, N_r)$ eigenchannels can be used for transmitting multiple data steams. This is known as spatial

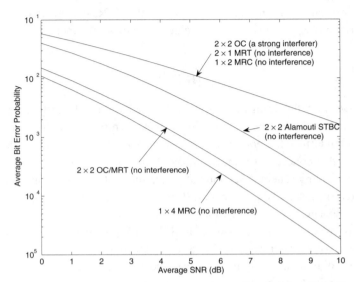

Figure 5.10 Performance comparison between eigen-beamforming and diversity. MRT and MRC have the same performance for the same number of antennas.

multiplexing and is discussed in the next section. In particular, Section 5.5.3 generalizes these results—in the absence of interference—to M data-bearing subchannels, where $1 \le M \le \min(N_r, N_t)$.

5.5 Spatial Multiplexing

From a data rate standpoint, the most exciting type of MIMO communication is spatial multiplexing, which refers to breaking the incoming high rate-data stream into N_t independent data streams, as shown in Figure 5.11. Assuming that the streams can be successfully decoded, the nominal spectral efficiency is thus increased by a factor of N_t . This is certainly exciting: It implies that adding antenna elements can greatly increase the viability of the high data rates desired for wireless broadband Internet access. However, this chapter adopts a critical view of spatial multiplexing and attempts to explain why many of the lauded recent results for MIMO will not prove directly applicable to WiMAX. Our goal is to help WiMAX designers understand the practical issues with MIMO and to separate viable design principles from the multitude of purely theoretical results that dominate much of the literature on the topic.

5.5.1 Introduction to Spatial Multiplexing

First, we summarize the classical results and widely used model for spatial multiplexing. The standard mathematical model for spatial multiplexing is similar to what was used for space/time coding:

$$\mathbf{y} = \mathbf{Hx} + \mathbf{n}, \tag{5.54}$$

where the size of the received vector $\mathbf{y}$ is $N_r \times 1$, the channel matrix $\mathbf{H}$ is $N_r \times N_t$, the transmit vector $\mathbf{x}$ is $N_t \times 1$, and the noise $\mathbf{n}$ is $N_r \times 1$. Typically, the transmit vector is normalized by N_t so that each symbol in $\mathbf{x}$ has average energy $\mathcal{E}_x/N_t$. This keeps the total transmit energy constant with the SISO case for comparison. The channel matrix in particular is of the form

$$\mathbf{H} = \begin{bmatrix} h_{11} & h_{12} & \cdots & h_{1N_t} \\ h_{21} & h_{22} & \cdots & h_{2N_t} \\ \vdots & \vdots & \ddots & \vdots \\ h_{N_{r1}} & h_{N_{r2}} & \cdots & h_{N_r N_t} \end{bmatrix}, \tag{5.55}$$

It is usually assumed that the entries in the channel matrix and the noise vector are complex Gaussian and i.i.d. with zero mean and covariance matrices that can be written as $\sigma_h^2 \mathbf{I}$ and $\sigma_z^2 \mathbf{I}$, respectively. Using basic linear algebra arguments, it is straightforward to confirm that decoding N_t streams is theoretically possible when there exist at least N_t nonzero eigenvalues in the channel matrix, that is rank($\mathbf{H}$) $\geq N_t$. This result has been generalized and made rigorous with information theory [23, 60].

This mathematical setup provides a rich framework for analysis based on random matrix theory [22, 63], information theory, and linear algebra. Using these tools, numerous insights on MIMO systems have been obtained; see [20, 30, 47, 60] for detailed summaries. Following are the key points regarding this single-link MIMO system model.

- The capacity, or maximum data rate, grows as $\min(N_t, N_r)\log(1 + SNR)$ when the SNR is large [60]. When the SNR is high, spatial multiplexing is optimal.
- When the SNR is low, the capacity-maximizing strategy is to send a single stream of data, using diversity precoding. Although much smaller than at high SNR, the capacity still grows approximately linearly with $\min(N_t, N_r)$, since capacity is linear with SNR in the low-SNR regime.
- Both of these cases are superior in terms of capacity to space/time coding, in which the data rate grows at best logarithmically with N_r.
- The average SNR of all N_t streams can be maintained without increasing the total transmit-power relative to a SISO system, since each transmitted stream is received at $N_r \geq N_t$ antennas and hence recovers the transmit power penalty of N_t due to the array gain. However, even a single low eigenvalue in the channel matrix can dominate the error performance.

5.5.2 Open-Loop MIMO: Spatial Multiplexing without Channel Feedback

As with multiantenna diversity techniques, spatial multiplexing can be performed with or without channel knowledge at the transmitter. We first consider the principal open-loop techniques; we always assume that the channel is known at the receiver, ostensibly through pilot symbols or other channel-estimation techniques. The open-loop techniques for spatial multiplexing attempt

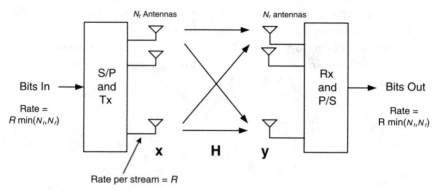

Figure 5.11 A spatial multiplexing MIMO system transmits multiple substreams to increase the data rate.

to suppress the interference that results from all N_t streams being received by each of the N_r antennas. The techniques discussed in this section are largely analogous to the interference-suppression techniques developed for equalization [48] and multiuser detection [64], as seen in Table 5.1.

5.5.2.1 Optimum Decoding: Maximum-Likelihood Detection

If the channel is unknown at the transmitter, the optimum decoder is the maximum-likelihood decoder, which finds the most likely input vector $\hat{\mathbf{x}}$ via a minimum-distance criterion

$$\hat{\mathbf{x}} = \arg\min \| \mathbf{y} - \mathbf{H}\hat{\mathbf{x}} \|^2 . \tag{5.56}$$

Unfortunately, there is no simple way to compute this, and an exhaustive search must be done over all M^{N_t} possible input vectors, where M is the order of the modulation (e.g., $M = 4$ for QPSK). The computational complexity is prohibitive for even a small number of antennas. Lower-complexity approximations of the ML detector, notably the sphere decoder, can be used to nearly achieve the performance of the ML detector in many cases [34], and these have some potential for high-performance open-loop MIMO systems. When optimum or near-optimum detection is achievable, the gain from transmitter channel knowledge is fairly small and is limited mainly to waterfilling over the channel eigenmodes, which provides significant gain only at low SNR.

5.5.2.2 Linear Detectors

As in other situations in which the optimum decoder is an intolerably complex maximum-likelihood detector, a sensible next step is to consider linear detectors that are capable of recovering the transmitted vector $\mathbf{x}$, as shown in Figure 5.12. The most obvious such detector is the *zero-forcing detector*, which sets the receiver equal to the inverse of the channel $\mathbf{G}_{zf} = \mathbf{H}^{-1}$ when $N_t = N_r$, or more generally to the pseudoinverse

$$\mathbf{G}_{zf} = (\mathbf{H}^*\mathbf{H})^{-1}\mathbf{H}^* . \tag{5.57}$$

Table 5.1 Similarity of Interference-Suppression Techniques for Various Applications, with Complexity Decreasing from Left to Right

	Optimum	Interference Cancellation	Linear
Equalization (ISI)	Maximum likelihood sequence detection (MLSD)	Decision feedback equalization (DFE)	Zero forcing minimum mean square error (MMSE)
Multiuser	Optimum multiuser detection (MUD)	Successive/parallel interference cancellation, iterative MUD	Decorrelating, MMSE
Spatial-multiplexing Receivers	ML detector sphere decoder (near optimum)	Bell Labs Layered Spaced Time (BLAST)	Zero forcing, MMSE

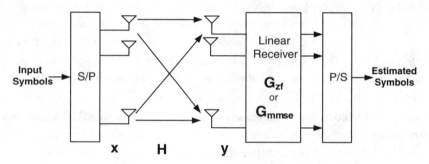

Figure 5.12 Spatial multiplexing with a linear receiver

As the name implies, the zero-forcing detector completely removes the spatial interference from the transmitted signal, giving an estimated received vector

$$\hat{\mathbf{x}} = \mathbf{G}_{zf}\mathbf{y} = \mathbf{G}_{zf}\mathbf{H}\mathbf{x} + \mathbf{G}_{zf}\mathbf{n} = \mathbf{x} + (\mathbf{H}^*\mathbf{H})^{-1}\mathbf{H}^*\mathbf{n}. \tag{5.58}$$

Because $\mathbf{G}_{fz}$ inverts the eigenvalues of $\mathbf{H}$, the bad spatial subchannels can severely amplify the noise in $\mathbf{n}$. This is particularly problematic in interference-limited MIMO systems and results in extremely poor performance. The zero-forcing detector is therefore not practical for WiMAX.

A logical alternative to the zero-forcing receiver is the MMSE receiver, which attempts to strike a balance between spatial-interference suppression and noise enhancement by simply minimizing the distortion. Therefore,

$$\mathbf{G}_{mmse} = \arg\min_{\mathbf{G}} \mathrm{E} \parallel \mathbf{G}\mathbf{y} - \mathbf{x} \parallel^2, \tag{5.59}$$

which can be derived using the well-known orthogonality principle as

$$\mathbf{G}_{mmse} = (\mathbf{H}^*\mathbf{H} + \frac{\sigma_z^2}{P_t}\mathbf{I})^{-1}\mathbf{H}^*, \tag{5.60}$$

where P_t is the transmitted power. In other words, as the SNR grows large, the MMSE detector converges to the ZF detector, but at low SNR, it prevents the worst eigenvalues from being inverted.

5.5.2.3 Interference cancellation: BLAST

The earliest known spatial-multiplexing receiver was invented and prototyped in Bell Labs and is called *Bell Labs layered space/time* (BLAST) [24]. Like other spatial-multiplexing MIMO systems, BLAST consists of parallel "layers" supporting multiple simultaneous data streams. The layers (substreams) in BLAST are separated by interference-cancellation techniques that decouple the overlapping data streams. The two most important techniques are the original *diagonal BLAST* (D-BLAST) [24] and its subsequent version, *vertical BLAST* (V-BLAST) [28].

D-BLAST groups the transmitted symbols into "layers" that are then coded in time independently of the other layers. These layers are then cycled to the various transmit antennas in a cyclical manner, resulting in each layer's being transmitted in a *diagonal* of space and time. In this way, each symbol stream achieves diversity in time via coding and in space by it rotating among all the antennas. Therefore, the N_t transmitted streams will equally share the good and bad spatial channels, as well as their priority in the decoding process now described.

The key to the BLAST techniques lies in the detection of the overlapping and mutually interfering spatial streams. The diagonal layered structure of D-BLAST can be detected by decoding one layer at a time. The decoding process for the second of four layers is shown in Figure 5.13a. Each layer is detected by *nulling* the layers that have not yet been detected and *canceling* the layers that have already been detected. In Figure 5.13, the layer to the left of the layer 2 block has already been detected and hence subtracted (canceled) from the received signal; those to the right remain as interference but can be nulled using knowledge of the channel. The time-domain coding helps compensate for errors or imperfections in the cancellation and nulling process. Two drawbacks of D-BLAST are that the decoding process is iterative and somewhat complex and that the diagonal-layering structure wastes space/time slots at the beginning and end of a D-BLAST block.

V-BLAST was subsequently addressed in order to reduce the inefficiency and complexity of D-BLAST. V-BLAST is conceptually somewhat simpler than D-BLAST. In V-BLAST, each antenna simply transmits an independent symbol stream—for example, QAM symbols. A variety of techniques can be used at the receiver to separate the various symbol stream from one another, including several of the techniques discussed elsewhere in this chapter. These techniques include linear receivers, such as the ZF and MMSE, which take the form at each receive antenna of a length N_r vector that can be used to null out the contributions from the $N_t - 1$ interfering data streams. In this case, the postdetection SNR for the ith stream is

$$\gamma_i = \frac{\mathcal{E}_x}{\sigma^2 \parallel \mathbf{w}_{r,i} \parallel^2} \quad i = 1, \cdots, N_t \tag{5.61}$$

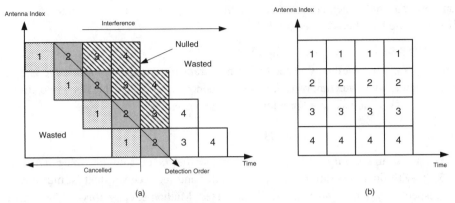

Figure 5.13 (a) D-BLAST detection of the layer 2 of four. (b) V-BLAST encoding. Detection is done dynamically; the layer (symbol stream) with the highest SNR is detected first and then canceled.

where $\mathbf{w}_{r,i}$ is the ith row of the zero-forcing or MMSE receiver $\mathbf{G}$ of Equation (5.57) and Equation (5.60), respectively.

Since this SNR is held hostage by the lower channel eigenvalues, the essence of V-BLAST is to combine a linear receiver with ordered successive interference cancellation. Instead of detecting all N_t streams in parallel, they are detected iteratively. First, the strongest symbol stream is detected, using a ZF or MMSE receiver, as before. After these symbols are detected, they can be subtracted out from the composite received signal. Then, the second-strongest signal is detected, which now sees effectively $N_t - 2$ interfering streams. In general, the ith detected stream experiences interference from only $N_t - i$ of the transmit antennas, so by the time the weakest symbol stream is detected, the vast majority of spatial interference has been removed. Using the ordered successive interference cancellation lowers the block error rate by about a factor of ten relative to a purely linear receiver, or equivalently, decreases the required SNR by about 4 dB [28]. Despite its apparent simplicity, V-BLAST prototypes have shown spectral efficiencies above 20 bps/Hz.

Despite demonstrating satisfactory performance in controlled laboratory environments, the BLAST techniques have not proved useful in cellular systems. One challenge is their dependence on high SNR for the joint decoding of the various streams, which is difficult to achieve in a multicell environment. In both BLAST schemes, these imperfections can quickly lead to catastrophic error propagation when the layers are detected incorrectly.

5.5.3 Closed-Loop MIMO: The Advantage of Channel Knowledge

The potential gain from transmitter channel knowledge is quite significant in spatial-multiplexing systems. First, we consider a simple theoretical example using *singular-value decomposition* that shows the potential gain of closed-loop spatial-multiplexing methods. Then we turn our attention to more practical linear-precoding techniques that could be considered in the near to medium term for multiantenna WiMAX systems as a means of raising the data rate relative to the diversity-based methods of Section 5.3.

5.5.3.1 SVD Precoding and Postcoding

A relatively straightforward way to see the gain of transmitter channel knowledge is by considering the singular-value decomposition (SVD, or generalized eigenvalue decomposition) of the channel matrix $\mathbf{H}$, which as noted previously can be written as

$$\mathbf{H} = \mathbf{U}\Sigma\mathbf{V}^*, \tag{5.62}$$

where $\mathbf{U}$ and $\mathbf{V}$ are *unitary* and Σ is a diagonal matrix of singular values. As shown in Figure 5.14, with linear operations at the transmitter and the receiver, that is, multiplying by $\mathbf{V}$ and $\mathbf{U}^*$, respectively, the channel can be diagonalized. Mathematically, this can be confirmed by considering a decision vector $\mathbf{d}$ that should be close to the input symbol vector $\mathbf{b}$. The decision vector can be written systematically as

$$
\begin{aligned}
\mathbf{d} &= \mathbf{U}^*\mathbf{y}, \\
&= \mathbf{U}^*(\mathbf{H}\mathbf{x} + \mathbf{n}), \\
&= \mathbf{U}^*(\mathbf{U}\Sigma\mathbf{V}^*\mathbf{V}\mathbf{b} + \mathbf{n}), \\
&= \mathbf{U}^*\mathbf{U}\Sigma\mathbf{V}^*\mathbf{V}\mathbf{b} + \mathbf{U}^*\mathbf{n}, \\
&= \Sigma\mathbf{b} + \mathbf{U}^*\mathbf{n},
\end{aligned}
\tag{5.63}
$$

which has diagonalized the channel and removed all the spatial interference without any matrix inversions or nonlinear processing. Because $\mathbf{U}$ is unitary, $\mathbf{U}^*\mathbf{n}$ still has the same variance as $\mathbf{n}$. Thus, the singular-value approach does not result in noise enhancement, as did the open-loop linear techniques. SVD-MIMO is not particularly practical, since the complexity of finding the SVD of an $N_t \times N_r$ matrix is on the order of $O(N_r N_t^2)$ if $N_r \geq N_t$ and requires a substantial amount of feedback. Nevertheless, it shows the promise of closed-loop MIMO as far as high performance at much lower complexity than the ML detector in open-loop MIMO.

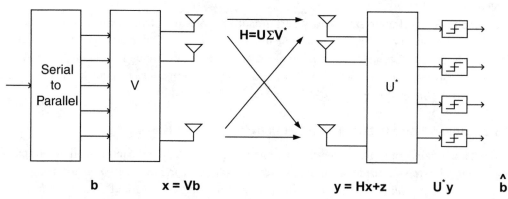

Figure 5.14 A MIMO system that has been diagonalized through SVD precoding.

5.5.3.2 Linear Precoding and Postcoding

The SVD illustratived how linear precoding and postcoding can diagonalize the MIMO channel matrix to provide up to $\min(N_r, N_t)$ dimensions to communicate data symbols through. More generally, the precoder and the postcoder can be jointly designed based on such criteria as the information capacity [50], the error probability [21], the detection MSE [52], or the received SNR [51]. From Section 5.3.3, recall that the general precoding formulation is

$$\mathbf{y} = \mathbf{G}(\mathbf{HFx} + \mathbf{n}), \qquad (5.64)$$

where $\mathbf{x}$ and $\mathbf{y}$ are $M\times1$, the postcoder matrix $\mathbf{G}$ is $M\times N_r$, the channel matrix $\mathbf{H}$ is $N_r \times N_t$, the precoder matrix $\mathbf{F}$ is $N_t \times M$, and $\mathbf{n}$ is $N_r \times 1$. For the SVD example, $M = \min(N_r, N_t)$, $\mathbf{G} = \mathbf{U}^*$, and $\mathbf{F} = \mathbf{V}$.

Regardless of the specific design criteria, the linear precoder and postcoder decompose the MIMO channel into a set of parallel subchannels as illustrated in Figure 5.15. Therefore, the received symbol for the ith subchannel can be expressed as

$$y_i = \alpha_i \sigma_i \beta_i x_i + \beta_i n_i, \quad i = 1, \cdots, M, \qquad (5.65)$$

where x_i and y_i are the transmitted and received symbols, respectively, with $E\,|\,x_i\,|^2 = \mathcal{E}_x$ as usual, σ_i are the singular values of $\mathbf{H}$, and α_i and β_i are the precoder and the postcoder weights, respectively. Through the precoder weights, the precoder can maximize the total capacity by distributing more transmission power to subchannels with larger gains and less to the others— referred to as waterfilling. The unequal power distribution based on the channel conditions is a principal reason for the capacity gain of linear precoding over the open-loop methods, such as BLAST. As in eigenbeamforming, the number of subchannels is bounded by

$$1 \leq M \leq \min(N_t, N_r), \qquad (5.66)$$

where $M = 1$ corresponds to the maximum diversity order, called *diversity* precoding in Section 5.3.3) and $M = \min(N_t, N_r)$ achieves the maximum number of parallel spatial streams. Intermediate values of M can be chosen to provide an attractive trade-off between raw throughput and link reliability or to suppress interfering signals, as shown in the eigenbeamforming discussion.

5.6 Shortcomings of Classical MIMO Theory

In order to realistically consider the gains that might be achieved by MIMO in a WiMAX systems, we emphasize that most of the well-known results for spatial multiplexing are based on the model in Equation (5.54) of the previous section, which makes the following critical assumptions.

- Because the entries of $\mathbf{H}$ are scalar random values, the multipath is assumed negligible, that is, the fading is frequency flat.
- Because the entries are i.i.d., the antennas are all uncorrelated.
- Usually, interference is ignored, and the background noise is assumed to be small.

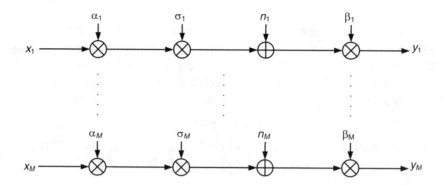

Figure 5.15 Spatial subchannels resulting from linear precoding and postcoding

Clearly, all these assumptions will be at least somewhat compromised in a cellular MIMO deployment. In many cases, they will be completely wrong. We now discuss how to address these important issues in a real system, such as WiMAX.

5.6.1 Multipath

Because WiMAX systems are expected to have moderate to high bandwidths over non-negligible transmission distances, the multipath in WiMAX is expected to be substantial, as discussed extensively in Chapters 3 and 4. Therefore, the flat-fading assumption appears to be unreasonable. However, OFDM can be introduced to convert a frequency-selective fading channel to L parallel flat-fading channels, as discussed in Chapter 4. If OFDM with sufficient subcarriers is combined with MIMO, the result is L parallel MIMO systems, and hence the model of Equation (5.54) is again reasonable. For this reason, OFDM and MIMO are a natural pair, and the first commercial MIMO system used OFDM in order to combat intersymbol interference [49]. MIMO-OFDM has been widely researched in recent years [5, 57]. Since WiMAX is based on OFDM, using the flat-fading model for MIMO is reasonable.

5.6.2 Uncorrelated Antennas

It is much more difficult to analyze MIMO systems with correlated antennas, so it is typically assumed that the spatial modes are uncorrelated and hence independent, assuming Gaussian and identically distributed. For a single user, identically distributed channels—that is, equal average power—are a reasonable assumption, since the antennas are colocated, but in general, the channels will be spatially correlated. On the other hand, if the antennas are considered to be at different MSs, the antennas will likely be uncorrelated, but the average power will be widely varying. Considering the case of a single-user MIMO channel, the main two causes of channel correlation are (1) insufficient spacing of the antenna elements, and (2) insufficient scattering in the channel. The first problem of insufficient spacing is prevalent when the platform is small, as is expected for MSs. Insufficient scattering is a frequent problem when the channel is approximately LOS, or when beamforming or directional antennas are used. In other words, MIMO's requirement for rich scattering directly conflicts with the desire for long-range transmission.

Encouragingly, research has shown that many of the MIMO results based on uncorrelated antennas are essentially accurate even with a modest degree of spatial correlation [13, 15, 39, 55, 72]. Due to the cost and difficultly in deploying more than two reasonably uncorrelated antennas in a MS, MIMO results should first be considered for an $N_t \times 2$ downlink and a $2 \times N_r$ uplink. Using polarization or other innovative methods, it may be possible in the future to have more uncorrelated antennas in the MS. Presently, WiMAX MSs are required to have two antennas. Having more than two is still considered impractical, but that may change soon owing to new antenna designs or other technology advancements.

5.6.3 Interference-Limited MIMO Systems

The third assumption—that the background noise is Gaussian and uncorrelated with the transmissions—is especially suspect in a cellular MIMO system. All well-designed cellular systems are by nature interference limited: If they were not, it would be possible to increase the spectral efficiency by lowering the frequency reuse or increasing the average loading per cell. In the downlink of a cellular system, where MIMO is expected to be the most profitable and viable, there will be an effective number of $N_t \cdot N_t$ interfering signals, whereas in Chapter 3, the number of non-negligible interfering neighboring base stations is N_I. Figure 5.16 illustrates the impact of other-cell interference in cellular MIMO systems. It is extremely difficult for a MIMO receiver at the MS to cope simultaneously with both the spatial interference, due to the N_t transmit antennas, and a high-level of other-cell interference. Although most researchers have neglected this problem, owing to its lack of tractability, it has been shown, using both information and communication theory, that the capacity of a MIMO cellular system can *decrease* as the number of transmit antennas increases if the spatial interference is not suitably addressed [2, 3, 4, 8, 14]. In summary, most theoretical MIMO results are for high-SNR environments with idealized (ML) decoding; in practice, MIMO must function in low-SINR environments with low-complexity receivers.

The other-cell interference problem is perhaps the most pressing problem confronting the use of spatial multiplexing in WiMAX systems. Various solutions for dealing with the other-cell interference have been suggested, including interference-aware receivers [19], multicell power control [10], distributed antennas [16], and multicell coordination [15, 73–76]. None of these techniques are explicitly supported by the WiMAX standard as of press time of this book, although the deployment of interference-aware receivers is certainly not precluded by the standard. We predict that creative approaches to the other-cell interference problem will be needed in order to make spatial multiplexing viable for users other than those very near the base station and hence experiencing a very low level of interference. Further, it should be noted that the sectorization methods detailed in Chapter 3 for increasing the SINR near the cell boundaries can result in less multipath diversity, and hence a more highly correlated spatial channel, as just discussed. Therefore, the requirement for rich scattering in MIMO systems may compete with the use of directional/sectorized antennas to reduce other-cell interference.

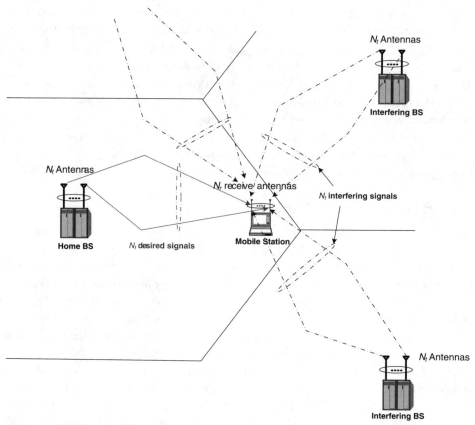

Figure 5.16   Other-cell interference in MIMO cellular systems

5.7 Channel Estimation for MIMO-OFDM

When OFDM is used with a MIMO transceiver, channel information is essential at the receiver in order to coherently detect the received signal and for diversity combining or spatial-interference suppression. Accurate channel information is also important at the transmitter for closed-loop MIMO. Channel estimation can be performed in two ways: training-based and blind. In training-based channel estimation, known symbols are transmitted specifically to aid the receiver's channel estimation-algorithms. In a blind channel-estimation method, the receiver must determine the channel without the aid of known symbols. Although higher-bandwidth efficiency can be obtained in blind techniques due to the lack of training overhead, the convergence speed and estimation accuracy are significantly compromised. For this reason, training-based channel-estimation techniques are more reliable, more prevalent, and supported by the WiMAX standard. This section considers the training-based techniques for MIMO-OFDM systems. Conventional OFDM channel estimation is the special case in which $N_r = N_t = 1$.

5.7.1 Preamble and Pilot

There are two ways to transmit training symbols: preamble or pilot tones. Preambles entail sending a certain number of training symbols prior to the user data symbols. In the case of OFDM, one or two preamble OFDM symbols are typical. Pilot tones involve inserting a few known pilot symbols among the subcarriers. Channel estimation in MIMO-OFDM systems can be performed in a variety of ways, but it is typical to use the preamble for synchronization[6] and initial channel estimation and the pilot tones for tracking the time-varying channel in order to maintain accurate channel estimates.

In MIMO-OFDM, the received signal at each antenna is a superposition of the signals transmitted from the N_t transmit antennas. Thus, the training signals for each transmit antenna need to be transmitted without interfering with one another in order to accurately estimate the channel. Figure 5.17 shows three MIMO-OFDM patterns that avoid interfering with one another: independent, scattered, and orthogonal patterns [37].

The independent pattern transmits training signals from one antenna at a time while the other antennas are silent, thus guaranteeing orthogonality between each training signal in the time domain. Clearly, an $N_t \times N_r$ channel can be estimated over N_t training signal times. The scattered-pilot pattern prevents overlap of training signals in the frequency domain by transmitting each antenna's pilot symbols on different subcarriers, while other antennas are silent on that subcarrier. Finally, the orthogonal pattern transmits training signals that are mathematically orthogonal, similar to CDMA. The independent pattern is often the most appropriate for MIMO-OFDM, since the preamble is usually generated the in time domain. For transmitting the pilot tones, any of these methods or some combination of them can be used.

In MIMO-OFDM, frequency-domain channel information is required in order to detect the data symbols on each subcarrier (recall the FEQ of Chapter 4). Since the preamble consists of pilot symbols on many of the subcarriers,[7] the channel-frequency response of each subcarrier can be reliably estimated from preamble with simple interpolation techniques. In normal data OFDM symbols, there are typically a very small number of pilot tones, so interpolation between these estimated subchannels is required [18, 35]. The training-symbol structure for the preamble and pilot tones is shown in Figure 5.18, with interpolation for pilot symbols. One-dimensional interpolation over either the time or frequency domain or two-dimensional interpolation over both the time and frequency domains can be performed with an assortment of well-known interpolation algorithms, such as linear and FFT. In the next section, we focus on channel estimation in the time and frequency domain, using the preamble and pilot symbols, and assume that interpolation can be performed by the receiver as necessary.

---

6. Synchronization for OFDM is discussed in detail in Chapter 4.
7. Each preamble uses only 1/3 or 1/6 of all the subcarriers in order to allow different sectors in the cell to be distinguished.

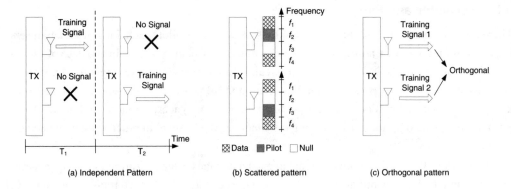

Figure 5.17 Three different patterns for transmitting training signals in MIMO-OFDM

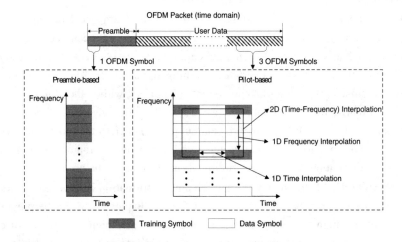

Figure 5.18 Training symbol structure of preamble-based and pilot-based channel estimation methods

5.7.2 Time versus Frequency-Domain Channel Estimation

MIMO-OFDM channels can be estimated in either the time or the frequency domain. The received time-domain signal can be directly used to estimate the channel impulse response; frequency-domain channel estimation is performed using the received signal after processing it with the FFT. Here, we review both the time- and the frequency-domain channel-estimation methods, assuming that each channel is clear of interference from the other transmit antennas, which can be ensured by using the pilot designs described previously. Thus, the antenna indices i and j are neglected in this section, and these techniques are directly applicable to single-antenna OFDM systems as well.

5.7.2.1 Time-Domain Channel Estimation

Channel-estimation methods based on the preamble and pilot tones are different due to the difference in the number of known symbols. For preamble-based channel estimation in the time domain with a cyclic prefix, the received OFDM symbol for a training signal can be expressed with a circulant matrix as

$$
\mathbf{y} =
\begin{bmatrix}
h(0) & \cdots & h(v) & 0 & \cdots & 0 \\
0 & h(0) & \cdots & h(v) & \cdots & 0 \\
\vdots & \vdots & \vdots & \vdots & \vdots & \vdots \\
h(1) & \cdots & h(v) & 0 & \cdots & h(0)
\end{bmatrix}
\begin{bmatrix}
x(L-1) \\
\vdots \\
x(0)
\end{bmatrix}
+ \mathbf{n}
$$

$$
=
\begin{bmatrix}
x(0) & x(L) & x(L-1) & \cdots & x(L-v+1) \\
x(1) & x(0) & x(L) & \cdots & x(L-v+2) \\
\vdots & \vdots & \vdots & \vdots & \vdots \\
x(L) & x(L-1) & \cdots & \cdots & x(L-v)
\end{bmatrix}
\begin{bmatrix}
h(0) \\
\vdots \\
h(v)
\end{bmatrix}
+ \mathbf{n}
$$

$$
= \mathbf{Xh} + \mathbf{n}, \tag{5.67}
$$

where $\mathbf{y}$ and $\mathbf{n}$ are the L samples of the received OFDM symbol and AWGN noise, $x(l)$ is the lth time sample of the transmitted OFDM symbol, and $h(i)$ is the ith time sample of the channel impulse response. Using this matrix description, the estimated channel $\hat{\mathbf{h}}$ can be readily obtained using the least-squares (LS) or MMSE method. For example, the LS—that is, zero forcing—estimate of the channel can be computed as

$$
\hat{\mathbf{h}} = (\mathbf{X}^*\mathbf{X})^{-1}\mathbf{X}^*\mathbf{y}, \tag{5.68}
$$

since $\mathbf{X}$ is deterministic and hence known *a priori* by the receiver. When pilot tones are used for time-domain channel estimation, the received signal can be expressed as

$$
\mathbf{y} = \overline{\mathbf{F}}^*\mathbf{X}_P\overline{\mathbf{F}}\mathbf{h} + \mathbf{n}, \tag{5.69}
$$

where $\mathbf{X}_P$ is a diagonal matrix whose diagonal elements are the pilot symbols in the frequency domain, $\overline{\mathbf{F}}$ is a $(P \times v)$ DFT matrix generated by selecting rows from $(L \times v)$ DFT matrix $\mathbf{F}$ according to the pilot subcarrier indices, and

$$
[\mathbf{F}]_{i,j} = \frac{1}{\sqrt{L}}\exp(-j2\pi(i-1)(j-1)/L) .
$$

Then, the LS pilot-based time-domain estimated channel is

$$
\hat{\mathbf{h}} = (\overline{\mathbf{F}}^*\mathbf{X}_P^*\mathbf{X}_P\overline{\mathbf{F}})^{-1}\overline{\mathbf{F}}^*\mathbf{X}_P^*\overline{\mathbf{F}}\mathbf{y}. \tag{5.70}
$$

5.7.2.2 Frequency-Domain Channel Estimation

Channel estimation is simpler in the frequency domain than in the time domain. For preamble-based frequency-domain channel estimation, the received symbol of the lth subcarrier in the frequency domain is

$$Y(l) = H(l)X(l) + N(l). \tag{5.71}$$

Since $X(l)$ is known *a priori* by the receiver, the channel frequency response of each subcarrier can easily be estimated. For example, lth frequency-domain estimated channel using LS is

$$\hat{H}(l) = X(l)^{-1}Y(l). \tag{5.72}$$

Similarly, for pilot-based channel estimation, the received symbols for the pilot tones are the same as Equation (5.71). To determine the complex channel gains for the data-bearing subcarriers, interpolation is required.

Least-squares channel estimation is often not very robust in high-interference or noisy environments, since these effects are ignored. This situation can be improved by averaging the LS estimates over numerous symbols or by using MMSE estimation. MMSE estimation is usually more reliable, since it forms a more conservative channel estimate based on the strength of the noise and statistics on the channel covariance matrix. The MMSE channel estimate in the frequency domain is

$$\hat{\mathbf{H}} = \mathbf{A}\mathbf{Y}, \tag{5.73}$$

where $\mathbf{H}$ and $\mathbf{Y}$ here are the L point DFT of $\mathbf{H}$ and the received signal on each output subcarrier, and the estimation matrix $\mathbf{A}$ is computed as

$$\mathbf{A} = \mathbf{R}_H(\mathbf{R}_H + \sigma^2(\mathbf{X}^*\mathbf{X})^{-1})^{-1}\mathbf{X}^{-1}, \tag{5.74}$$

and $\mathbf{R}_H = E[\mathbf{H}\mathbf{H}^*]$ is the channel covariance matrix, and it is assumed that the noise/interference on each subcarrier is uncorrelated and has variance σ^2. It can be seen by setting $\sigma^2 = 0$ that if noise is neglected, the MMSE and LS estimators are the same.

One of the drawbacks of conventional Linear MMSE frequency-domain channel estimation is that it requires knowledge of the channel covariance matrix in both the frequency and time domains. Since the receiver usually does not possess this information *a priori*, it also needs to be estimated, which can be performed based on past channel estimates. However, in mobile applications, the channel characteristics change rapidly, making it difficult to estimate and track the channel covariance matrix. In such cases, partial information about the channel covariance matrix may be the only possibility. For example, if only the maximum delay and the Doppler spread of the channel are known, bounds on the actual channel covariance matrix can be derived. Surprisingly, the LMMSE estimator with only partial information often results in performance that is comparable to the conventional LMMSE estimator with full channel covariance information. The performance of these channel-estimation and tracking schemes for WiMAX are provided in Chapter 11.

5.8 Channel Feedback

As shown in previous sections, closed-loop techniques, such as linear precoding and transmit beamforming, yield better throughput and performance than do open-loop techniques, such as STBC. The key requirement for closed-loop techniques is knowledge of the channel at the transmitter, referred to as transmit CSI. Two possible methods exist for obtaining transmit CSI. First, CSI is sent back by the receiver to the transmitter over a feedback channel. Second, in TDD systems, CSI can be acquired at the transmitter by exploiting *channel reciprocity*, or inferring the downlink channel from the uplink channel, and can be directly measured. Our discussion focuses on the feedback channel, namely on an efficient technique based on quantized feedback [43]. Quantized feedback will be discussed for linear precoding, but it is applicable for other types of closed-loop communication, such as beamforming [44], adaptive modulation [69], or adaptive STBC [42].

The development of quantized precoding is motivated by the need for reducing the channel feedback rate in a MIMO linear precoding system. Ideally, the transmit precoder would be informed by the instantaneous and exact value of the matrix channel between the transmit and receive antenna arrays. But accurate quantization and feedback of this matrix channel can require a large number of bits, especially for a MIMO-OFDM system with numerous antennas, subcarriers, and a rapidly varying channel. Quantized precoding techniques provide a solution for this problem by quantizing the optimal precoder at the receiver. Specifically, the precoder is constrained to be one of N distinct matrices, which as a group is called a *precoding codebook*. If the precoding codebook of N matrices is known to both the receiver and the transmitter, only $\log_2 N$ bits of feedback are required for indicating the index of the appropriate precoder matrix. The number of required feedback bits for acceptable distortion is usually small, typically 3–8 bits. Figure 5.19 illustrates a quantized precoding system.

Typically, the precoding codebook is designed to minimize the difference between the quantized precoder and the optimal one, which is referred to as the *distortion*. The MMSE is a typical distortion measure; another is the Fubini-Study distance

$$d(\mathbf{A}, \mathbf{B}) = \arccos |\det(\mathbf{AB}^*)|, \tag{5.75}$$

where $\mathbf{A}$ and $\mathbf{B}$ are two different matrices. Other possible distortion measures include chordal distance and the projection 2-norm, but these distortion measures do not easily allow for optimal precoding codebooks to be derived analytically and so are usually computed using numerical methods, such as the Lloyd algorithm [25]. These techniques have been shown to provide near-optimal performance even with only a few bits of channel feedback [43].

The effectiveness and efficiency of quantized precoding has led to its inclusion in the WiMAX standard, which has defined precoding codebooks for various channel configurations. It also should be noted that in the WiMAX standard, *channel sounding*, a method for obtaining transmit CSI through reciprocity, has been defined for TDD systems.

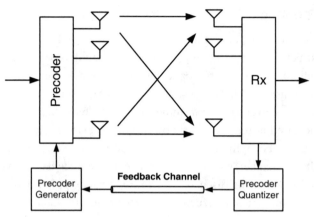

Figure 5.19 Linear precoding with quantized feedback

5.9 Advanced Techniques for MIMO

In addition to the single-user MIMO systems that use diversity, beamforming, or spatial multiplexing, these techniques can be combined and also used to service multiple users—mobile stations—simultaneously. In this section, we briefly look at some of these advanced concepts for increasing the capacity, reliability, and flexibility of MIMO systems.

5.9.1 Switching Between Diversity and Multiplexing

In order to achieve the reliability of diversity and the high raw data rate of spatial multiplexing, these two MIMO techniques can be used simultaneously or alternately, based on the channel conditions. There is a fundamental trade-off between diversity and multiplexing: One cannot have full diversity gain and also attempt spatial multiplexing. Essentially, the choice comes down to the following question: Would you rather have a thin but very reliable pipe or a wide but not very reliable pipe? Naturally, a compromise on each would often be the preference.

The notion of switching between diversity and multiplexing was first introduced by Heath [32], and then developed into an elegant theory [71]. In practice, the most likely approach is simply to switch between a few preferred modes—for example, STBCs, stacked STBCs, and closed-loop spatial multiplexing—with error-correction coding, frequency interleaving, and adaptive modulation used to provide diversity. As seen in Figure 5.20, simple diversity is likely to give better performance for moderate numbers of antennas, so spatial multiplexing is not likely to be desirable unless there are more than two antennas at the transmitter or the receiver. Our simulation results in Chapters 11 and 12 cast further light on which schemes are most promising under various configurations.

5.9.2 Multiuser MIMO Systems

The MIMO schemes developed in this chapter have implicitly assumed that only a single user is active on all the antennas at each instant in time and on each frequency channel. In fact, multiple

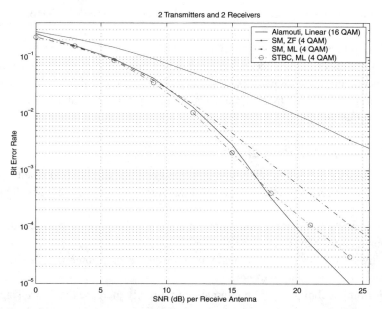

Figure 5.20 BER versus SNR for configurations of 2×2 MIMO. The simple Alamouti code outperforms spatial multiplexing at the same data rate owing to its superior diversity. Figure from [26], courtesy of IEEE.

users can use the spatial channels simultaneously, which can be advantageous versus users having to take turns sharing the channel. For example, imagine a downlink scenario with $N_t = 4$ transmit antennas at the BS and $N_r = 1$ antennas at each MS. Using the techniques presented thus far in this chapter, only a single stream could be sent to single user. Using multiuser encoding techniques, though, four streams could be sent to the four users. Since each MS has only a single receive antenna, it received signal-processing capabilities are quite limited: In this example, it is not possible for the MS to cancel out the interfering three streams and successfully receive its own stream. Therefore, in a multiuser MIMO system, the base station must proactively cancel the spatial interference so that the mobile stations can receive their desired data streams.

Multiuser MIMO has generated a large amount of recent interest; see [29] for a summary. The main idea emerging from this research is that multiple users can be simultaneously multiplexed to take simultaneous advantage of multiuser and spatial diversity. The optimal BS interference-cancellation strategy is the so-called dirty paper coding [7, 36], but it is not directly practical. More realistic linear multiuser precoding techniques have been developed [11, 56, 68], but these too require accurate channel knowledge for all the candidate MSs. In addition to these downlink strategies, it is possible for multiple users to transmit in the uplink at the same time, using a subset of their antennas, to create a *virtual MIMO* system. For example, if three spatially distributed users each transmitted on a single antenna, as long as there were $N_r \geq 3$ antennas at the base station, this could be treated as a $3 \times N_r$ virtual MIMO channel, the significant difference from conventional

MIMO being that the geographically separated transmitters can directly collaborate on their transmissions.

In addition to these implementation challenges, it is debatable whether there is much potential gain from multiuser MIMO techniques in a wideband MIMO-OFDMA system, due to the substantial spatial and multiuser diversity already present. Hence, it is likely that WiMAX systems will continue to use TDMA (time division multiple access) and OFDMA for multiple access in the foreseeable future.

5.10 Summary and Conclusions

This chapter has presented the wide variety of techniques that can be used when multiple antennas are present at the receiver and/or the transmitter. Table 5.2 summarizes MIMO techniques.

- Spatial diversity offers incredible improvements in reliability, comparable to increasing the transmit power by a factor of 10–100.
- These diversity gains can be attained with multiple receive antennas, multiple transmit antennas, or a combination of both.
- Beamforming techniques are an alternative to directly increase the desired signal energy while suppressing, or nulling, interfering signals.
- In contrast to diversity and beamforming, spatial multiplexing allows multiple data streams to be simultaneously transmitted using sophisticated signal processing.
- Since multiple-antenna techniques require channel knowledge, the MIMO-OFDM channel can be estimated, and this channel knowledge can be relayed to the transmitter for even larger gains.
- Throughout the chapter, we adopted a critical view of MIMO systems and explained the practical issues and shortcomings of the various techniques in the context of a cellular broadband system like WiMAX.
- It is possible to switch between diversity and multiplexing modes to find a desirable reliability-throughput operating point; multiuser MIMO strategies can be harnessed to transmit to multiple users simultaneously over parallel spatial channels.

Table 5.2 Summary of MIMO Techniques

Technique	(N_r, N_t)	Feedback?	Rate r[a]	Comments
Reliability-Enchancement Techniques ($r \leq 1$)				
Selection combining	$N_r \geq 1$ $N_t = 1$	Open loop	$r = 1$	Increases average SNR by $(1 + 1/2 + 1/3 + \ldots 1/N_r)$
Maximal ratio combining	$N_r \geq 1$ $N_t = 1$	Open loop	$r = 1$	Increases SNR to $\gamma_\Sigma = \gamma_1 + \gamma_2 + \ldots + \gamma_{Nr}$
Space/time block codes	$N_r \geq 1$ $N_t > 1$	Open loop	$r \leq 1$	Increases SNR to $\gamma_\Sigma = \gamma \|\mathbf{H}\|_F / N_t$
Transmit selection diversity	$N_r \geq 1$ $N_t > 1$	Closed loop: Feedback desired antenna index	$r = 1$ usually $(r < N_t)$	Same SNR as selection combining
DOA beamforming	$N_r \geq 1$ $N_t \geq 1$ $N_r + N_t > 2$	Open loop if $N_t = 1$ Closed loop if $N_t > 1$ or used for interference suppression	$r = 1$	Can suppress up to $(N_r - 1) + (N_t - 1)$ interference signals and increase gain in desired direction. Ineffective in multipath channels
Precoding Techniques				
Linear diversity precoding	$N_r \geq 1$ $N_t > 1$	Closed loop: Feedback channel matrix	$r = 1$	Special case of linear beamforming; only one data stream is sent. Increases SNR to $\gamma_\Sigma = \gamma \|\mathbf{H}\|_F$
Eigenbeam-forming	$N_r \geq 1$ $N_t > 1$	Closed loop: Feedback channel matrix	$1 \leq r \leq \min(N_r - L, N_t - L)$	Can be used to both increase desired signal gain and suppress L interfering users
General linear precoding	$N_r > 1$ $N_t > 1$	Closed loop: Feedback channel matrix	$1 \leq r \leq \min(N_r, N_t)$	Similar to eigenbeamforming, but interfering signals generally not suppressed; goal is to send multiple data streams
Spatial Multiplexing				
Open-loop spatial multiplexing	$N_r > 1$ $N_t > 1$	Open loop	$r = \min(N_r, N_t)$	Can receive in a variety of ways: linear receiver (MMSE), ML receiver, sphere decoder. If $N_r > N_t$, select best N_r antennas to send streams
BLAST	$N_r > 1$ $N_t > 1$	Open loop	$r = \min(N_r, N_t)$	Successively decode transmitted streams
General linear precoding	$N_r > 1$ $N_t > 1$	Closed loop: Feedback channel matrix	$1 \leq r \leq \min(N_r, N_t)$	Same as preceding; both a precoding technique and a spatial-multiplexing technique

a. r is similar to the number of streams M but slightly more general, since $r < 1$ is possible for some of the transmit-diversity techniques.

5.11 Bibliography

[1]   S. M. Alamouti. A simple transmit diversity technique for wireless communications. *IEEE Journal on Selected Areas in Communications*, 16(8):1451–1458, October 1998.

[2]   J. G. Andrews, W. Choi, and R. W. Heath. Overcoming interference in multi-antenna cellular networks. *IEEE Wireless Communications Magazine*, forthcoming (available at: www.ece.utexas.edu/jandrews).

[3]   R. Blum. MIMO capacity with interference. *IEEE Journal on Selected Areas in Communications*, 21(5):793–801, June 2003.

[4]   R. Blum, J. Winters, and N. Sollenberger. On the capacity of cellular systems with MIMO. *IEEE Communications Letters*, 6:242–244, June 2002.

[5]   H. Bolcskei, R. W. Heath, and A. J. Paulraj. Blind channel identification and equalization in OFDM-based multiantenna systems. *IEEE Transactions on Signal Processing*, 50(1):96–109, January 2002.

[6]   H. Bolcskei and A. J. Paulraj. Space-frequency coded broadband OFDM systems. In *Proceedings, IEEE Wireless Networking and Communications Conference*, pp. 1–6, Chicago, September 2000.

[7]   G. Caire and S. Shamai. On the achievable throughput of a multi-antenna gaussian broadcast channel. *IEEE Transactions on Information Theory*, 49(7):1691–1706, July 2003.

[8]   S. Catreux, P. Driessen, and L. Greenstein. Attainable throughput of an interference-limited multiple-input multiple-output (MIMO) cellular system. *IEEE Transactions on Communications*, 49(8):1307–1311, August 2001.

[9]   R. Chen, J. Andrews, and R. W. Heath. Transmit selection diversity for multiuser spatial multiplexing systems. In *Proceedings, IEEE Globecom*, pp. 2625–2629, Dallas, TX, December 2004.

[10] R. Chen, J. G. Andrews, R. W. Heath, and A. Ghosh. Uplink power control in multi-cell spatial multiplexing wireless systems. *IEEE Transactions on Wireless Communications*, forthcoming.

[11] R. Chen, R. W. Heath, and J. G. Andrews. Transmit selection diversity for multiuser spatial division multiplexing wireless systems. *IEEE Transactions on Signal Processing*, March 2007.

[12] Z. Chen, J. Yuan, B. Vucetic, and Z. Zhou. Performance of Alamouti scheme with transmit antenna selection. *Electronics Letters*, pp. 1666–1667, November 2003.

[13] M. Chiani, M. Z. Win, and A. Zanella. On the capacity of spatially correlated MIMO Rayleigh-fading channels. *IEEE Transactions on Information Theory*, 49(10):2363–2371, October 2003.

[14] W. Choi and J. G. Andrews. Spatial multiplexing in cellular MIMO CDMA systems with linear receivers. In *Proceedings, IEEE International Conference on Communications*, Seoul, Korea, May 2005.

[15] W. Choi and J. G. Andrews. Base station cooperatively scheduled transmission in a cellular MIMO TDMA system. In *Proceedings, Conference on Information Sciences and Systems (CISS)*, March 2006.

[16] W. Choi and J. G. Andrews, Downlink Performance and Capacity of Distributed Antenna Systems in a Multicell Environment, *IEEE Transactions on Wireless Communications*, 6(1), January 2007.

[17] C.-N. Chuah, D. N. C. Tse, J. M. Kahn, and R. A. Valenzuela. Capacity scaling in MIMO wireless systems under correlated fading. *IEEE Transactions on Information Theory*, 48:637–651, March 2002.

[18] L. J. Cimini. Analysis and simulation of a digital mobile channel using orthogonal frequency division multiplexing. *IEEE Transactions on Communications*, 33(7):665–675, July 1985.

[19] H. Dai, A. Molisch, and H. V. Poor. Downlink capacity of interference-limited MIMO systems with joint detection. *IEEE Transactions on Wireless Communications*, 3(2):442–453, March 2004.

[20] S. Diggavi, N. Al-Dhahir, A. Stamoulis, and A. Calderbank. Great expectations: The value of spatial diversity in wireless networks. *Proceedings of the IEEE*, pp. 219–270, February 2004.

[21] Y. Ding, N. Davidson, Z. Q. Luo, and K. M. Wong. Minimum BER block precoders for zero-forcing equalization. *IEEE Transactions on Signal Processing*, 51:2410–2423, September 2003.

[22] A. Edelman. Eigenvalues and condition number of random matrices. PhD thesis, MIT, 1989.

[23] G. Foschini and M. Gans. On limits of wireless communications in a fading environment when using multiple antennas. *Wireless Personal Communications*, 6:311–335, March 1998.

[24] G. J. Foschini. Layered space-time architecture for wireless communication in a fading environment when using multiple antennas. *Bell Labs Technical Journal*, 1(2):41–59, 1996.

[25] A. Gersho and R. M. Gray. *Vector Quantization and Signal Compression*. Kluwer, 1992.

[26] D. Gesbert, M. Sha, D. Shiu, P. J. Smith, and A. Naguib. From theory to practice: An overview of MIMO spacetime coded wireless systems. *IEEE Journal on Selected Areas in Communications*, 21(3):281–302, April 2003.

[27] L. C. Godara. Application of antenna arrays to mobile communications, part II: Beam-forming and direction-of-arrival considerations. *Proceedings of the IEEE*, 85(8):1195–1245, August 1997.

[28] G. D. Golden, G. J. Foschini, R. A. Valenzuela, and P. W. Wolniansky. Detection algorithm and initial laboratory results using V-BLAST space-time communication architecture. *IEE Electronics Letters*, 35:14–16, January 1999.

[29] A. Goldsmith, S. Jafar, N. Jindal, and S. Vishwanath. Capacity limits of MIMO channels. *IEEE Journal on Selected Areas in Communications*, 21(5):684–702, June 2003.

[30] A. J. Goldsmith. *Wireless Communications*. Cambridge University Press, 2005.

[31] D. Gore and A. Paulraj. Space-time block coding with optimal antenna selection. In *Proceedings, IEEE International Conference on Acoustics, Speech, and Signal Processing (ICASSP)*, pp. 2441–2444, May 2001.

[32] R. W. Heath and A. J. Paulraj. Switching between multiplexing and diversity based on constellation distance. In *Proceedings, Allerton Conference on Communications, Control, and Computing*, September 2000.

[33] R. W. Heath, S. Sandhu, and A. Paulraj. Antenna selection for spatial multiplexing systems with linear receivers. *IEEE Communications Letters*, 5(4):142–144, April 2001.

[34] B. M. Hochwald and S. ten Brink. Achieving near-capacity on a multiple-antenna channel. *IEEE Transactions on Communications*, 51(3):389–399, March 2003.

[35] M. Hsieh and C. Wei. Channel estimation for OFDM systems based on comb-type pilot arrangement in frequency selective fading channels. *IEEE Transactions Consumer Electronics*, 44(1):217–225, February 1998.

[36] N. Jindal, S. Vishwanath, and A. Goldsmith. On the duality of Gaussian multiple-access and broadcast channels. *IEEE Transactions on Information Theory*, 50(5):768–783, May 2004.

[37] T. Kim and J. G. Andrews. Optimal pilot-to-data power ratio for MIMO-OFDM. In *Proceedings, IEEE Globecom*, 3:1481–1485, St. Louis, MO, December 2005.

[38] J. C. Liberti and T. S. Rappaport. *Smart Antennas for Wireless Communications: IS-95 and Third Generation CDMA Applications*. Prentice Hall, 1999.

[39] K. Liu, V. Raghavan, and A. M. Sayeed. Capacity scaling and spectral efficiency in wide-band correlated MIMO channels. *IEEE Transactions on Information Theory*, 49(10):2504–2527, October 2003.

[40] Z. Liu, G. Giannakis, and B. Hughes. Double differential space-time block coding for time-selective fading channels. *IEEE Transactions on Communications*, 49(9):1529–1539, September 2001.

[41] T. Lo. Maximum ratio transmission. *IEEE Transactions on Communications*, pp. 1458–1461, October 1999.

[42] D. J. Love and R. W. Heath. Limited feedback unitary precoding for orthogonal space-time block codes. *IEEE Transactions on Signal Processing*, pp. 64–73, January 2005.

[43] D. J. Love and R. W. Heath. Limited feedback unitary precoding for spatial multiplexing systems. *IEEE Transactions on Information Theory*, 51(8):1967–1976, August 2005.

[44] D. J. Love, R. W. Heath, and T. Strohmer. Grassmannian beamforming for MIMO wireless systems. *IEEE Transactions on Information Theory*, 49(10):2735–2747, October 2003.

[45] A. F. Molisch and M. Z. Win. MIMO systems with antenna selection. *IEEE Microwave Magazine*, 5(1):46–56, March 2004.

[46] E. N. Onggosanusi, A. G. Dabak, and T. A. Schmidl. High rate space-time block coded scheme: Performance and improvement in correlated fading channels. In *Proceedings, IEEE Wireless Communications and Networking Conference*, 1:194–199, Orlando, FL, March 2002.

[47] A. Paulraj, D. Gore, and R. Nabar. *Introduction to Space-Time Wireless Communications*. Cambridge University Press, Cambridge, 2003.

[48] J. G. Proakis. *Digital Communications*. 3rd ed., McGraw-Hill, 1995.

[49] H. Sampath, S. Talwar, J. Tellado, V. Erceg, and A. Paulraj. A fourth-generation MIMO-OFDM broadband wireless system: Design, performance, and field trial results. *IEEE Communications Magazine*, 40(9):143–149, September 2002.

[50] A. Scaglione, S. Barbarossa, and G. B. Giannakis. Filterbank transceivers optimizing information rate in block transmissions over dispersive channels. *IEEE Transactions on Information Theory*, 45:1988–2006, April 1999.

[51] A. Scaglione, G. B. Giannakis, and S. Barbarossa. Redundant filterbank precoders and equalizers, part I and II. *IEEE Transactions on Signal Processing*, 47:1988–2022, July 1999.

[52] A. Scaglione, P. Stoica, S. Barbarossa, G. B. Giannakis, and H. Sampath. Optimal designs for space-time linear precoders and decoders. *IEEE Transactions on Signal Processing*, 50(5):1051–1064, May 2002.

[53] P. Schniter. Low-complexity equalization of OFDM in doubly-selective channels. *IEEE Transactions on Signal Processing*, 52(4):1002–1011, April 2004.

[54] N. Seshadri and J. H. Winters. Two schemes for improving the performance of frequency-duplex (FDD) transmission systems using transmitter antenna diversity. *International Journal Wireless Information Networks*, 1:49–60, January 1994.

[55] D. Shiu, G. J. Foschini, M. J. Gans, and J. M. Kahn. Fading correlation and its effect on the capacity of multi-element antenna systems. *IEEE Transactions on Communications*, 48:502–513, March 2000.

[56] Q. Spencer, A. Swindlehurst, and M. Haardt. Zero-forcing methods for downlink spatial multiplexing in multiuser MIMO channels. *IEEE Transactions on Signal Processing*, 52:461–471, February 2004.

[57] G. L. Stuber, J. R. Barry, S. W. McLaughlin, Y. Li, M. Ingram, and T. G. Pratt. Broadband MIMO-OFDM wireless communications. *Proceedings of the IEEE*, pp. 271–294, February 2004.

[58] V. Tarokh, H. Jafarkhani, and A. R. Calderbank. Space-time block codes from orthogonal designs. *IEEE Transactions on Information Theory*, 45(5):1456–1467, July 1999.

[59] V. Tarokh, N. Seshadri, and A. R. Calderbank. Space-time codes for high data rate wireless communication: Performance criterion and code construction. *IEEE Transactions on Information Theory*, 44(2):744–765, March 1998.

[60] E. Teletar. Capacity of multi-antenna gaussian channels. *European Transactions Telecommunications,* 6:585–595, November–December 1999.

[61] Texas Instruments. Double-STTD scheme for HSDPA systems with four transmit antennas: Link level simulation results. TSG-R WG1 document, TSGR1#20(01)0458, May 2001.

[62] D. Tse and P. Viswanath. *Fundamentals of Wireless Communication.* Cambridge University Press, 2005.

[63] A. Tulino and S. Verdu. Random matrix theory and wireless communications. *Foundations and Trends in Communications and Information Theory,* 1(1):1–186, 2004.

[64] S. Verdu. *Multiuser Detection.* Cambridge University Press, 1998.

[65] J. H. Winters. Switched diversity with feedback for DPSK mobile radio systems. *IEEE Transactions on Vehicular Technology,* 32:134–150, February 1983.

[66] A. Wittneben. A new bandwidth efficient transmit antenna modulation diversity scheme for linear digital modulation. In *Proceedings, IEEE International Conference on Communications,* pp. 1630–1634, Geneva, Switzerland, May 1993.

[67] K.-K. Wong, R. D. Murch, and K. B. Letaief. Optimizing time and space MIMO antenna system for frequency selective fading channels. *IEEE Journal on Selected Areas in Communications,* 19(7):1395–1407, July 2001.

[68] K. K. Wong, R. D. Murch, and K. B. Letaief. A joint-channel diagonalization for multiuser MIMO antenna systems. *IEEE Transactions on Wireless Communications,* 2(4):773–786, July 2003.

[69] P. Xia, S. Zhou, and G. B. Giannakis. Multiantenna adaptive modulation with beamforming based on bandwidth-constrained feedback. *IEEE Transactions on Communications,* 53(3):526–536, March 2005.

[70] H. Yang. A road to future broadband wireless access: MIMO-OFDM-based air interface. *IEEE Communications Magazine,* 43(1):53–60, January 2005.

[71] L. Zheng and D. Tse. Diversity and multiplexing: A fundamental trade-off in multiple antenna channels. *IEEE Transactions on Information Theory,* 49(5), May 2003.

[72] A. M. Tulino, A. Lozano, and S. Verdu, Impact of antenna correlation on the capacity of multiantenna channels. *IEEE Transactions on Information Theory,* 51(7):2491–2509, July 2005.

[73] S. A. Jarar, G. Foschini, and A. J. Goldsmith. Phantomnet: Exploring optimal multicellular multiple antenna systems. *EURASIP Journal on Applied Signal Processing, Special issue on MIMO Communication and Signal Processing,* pp. 591–605, May 2004.

[74] W. Yu and T. Lan. Transmitter optimization for the multiantenna downlink with per-antenna power constraints. Submitted to *IEEE Transactions on Signal Processing,* December 2005.

[75] H. Zhang, H. Dai, and Q. Zhou. Base station cooperation for multiuser MIMO: Joint transmission and BS selection. In *Proceedings, Conference on Information Sciences and Systems (CISS),* March 2004.

[76] O. Somekh, B. M. Zaidel, and S. Shamai. Sum rate characterization of joint multiple cell-site processing. Submitted to *IEEE Transactions on Information Theory,* August 2005.

CHAPTER 6

Orthogonal Frequency Division Multiple Access

WiMAX presents a very challenging *multiuser* communication problem: Many users in the same geographic area requiring high on-demand data rates in a finite bandwidth with low latency. Multiple-access techniques allow users to share the available bandwidth by allotting each user some fraction of the total system resources. Experience has shown that dramatic performance differences are possible between various multiple-access strategies. For example, the lively CDMA versus TDMA debate for cellular voice systems went on for some time in the 1990s. The diverse nature of anticipated WiMAX traffic—VoIP, data transfer, and video streaming—and the challenging aspects of the system deployment—mobility, neighboring cells, high required bandwidth efficiency—make the multiple-access problem quite complicated in WiMAX. The implementation of an efficient and flexible multiple-access strategy is critical to WiMAX system performance.

OFDM is not a multiple-access strategy but rather a modulation technique that creates many independent streams of data that can be used by different users. Previous OFDM systems, such as DSL, 802.11a/g, and the earlier versions of 802.16/WiMAX, use single-user OFDM: All the subcarriers are used by a single user at a time. For example, in 802.11a/g, colocated users share the 20MHz bandwidth by transmitting at different times after contending for the channel. WiMAX (802.16e-2005) takes a different approach, known as orthogonal frequency division multiple access (OFDMA), whereby users share subcarriers *and* time slots. As this chapter will describe, this additional flexibility allows for increased multiuser diversity, increased freedom in scheduling the users, and several other subtle but important implementation advantages. OFDMA does come with a few costs, such as overhead in both directions: The transmitter needs channel information for its users, and the receiver needs to know which subcarriers it has been assigned.

This chapter explains OFDMA in the following four steps.

1. Multiple-access techniques are summarized, with special attention to their interaction with OFDM modulation.
2. The two key sources of capacity gain in OFDMA are overviewed: multiuser diversity and adaptive modulation.
3. Algorithms that harness the multiuser diversity and adaptive modulation gains are described and compared.
4. OFDMA's implementation in WiMAX is briefly discussed, along with challenges and opportunities to improve OFDMA performance.

6.1 Multiple-Access Strategies for OFDM

Multiple-access strategies typically attempt to provide *orthogonal*, or noninterfering, communication channels for each active link. The most common way to divide the available dimensions among the multiple users is through the use of frequency, time, or code division multiplexing. In frequency division multiple access (FDMA), each user receives a unique carrier frequency and bandwidth. In time division multiple access (TDMA), each user is given a unique time slot, either on demand or in a fixed rotation. Wireless TDMA systems almost invariably also use FDMA in some form, since using the entire electromagnetic spectrum is not allowable. Orthogonal code division multiple access (CDMA) systems allow each user to share the bandwidth and time slots with many other users and rely on orthogonal binary codes to separate out the users. More generally, all CDMA system, including the popular nonorthogonal ones, share in common that many users share time and frequency.

It can be easily proved that TDMA, FDMA, and orthogonal CDMA all have the same capacity in an additive noise channel [12, 19], since they all can be designed to have the same number of orthogonal dimensions in a given bandwidth and amount of time.[1] For example, assume that it takes one unit of bandwidth to send a user's signal and that eight units of bandwidth are available. Eight users can be accommodated with each technique. In FDMA, eight orthogonal frequency slots would be created, one for each user. In TDMA, each user would use all eight frequency slots but would transmit only one eighth of the time. In CDMA, each user would transmit all the time over all the frequencies but would use one of eight available orthogonal codes to ensure that there was no interference with the other seven users.

So why all the debate over multiple access? One reason is that orthogonality is not possible in dense wireless systems. The techniques guarantee orthogonality only between users in the *same* cell, whereas users in *different*, potentially neighboring, cells will likely be given the same time or frequency slot. Further, the orthogonality is additionally compromised owing imperfect bandpass filtering (FDMA) and multipath channels and imperfect synchronization (TDMA and

---

1. It may be complicated to find orthogonal codes for divisions that are not factors of 2. Nevertheless, they are provably the same in their efficiency.

especially CDMA). In practice, each multiple-access technique (FDMA, TDMA, CDMA) entails its own list of pros and cons. One of the principal merits of OFDMA is that many of the best features of each technique can be achieved.

6.1.1 Random Access versus Multiple Access

Before describing in more detail how TDMA, FDMA, and CDMA can be applied to OFDM, it is useful to consider an alternative *random-access* technique: carrier sense multiple access (CSMA), commonly used in packet-based communication systems, notably Ethernet and wireless LANs, such as 802.11. In random access, users *contend* for the channel rather than being allocated a reserved time, frequency, or code resource. Well-known random-access techniques include ALOHA and slotted ALOHA, as well as CSMA. In ALOHA, users simply transmit packets at will without regard to other users. A packet not acknowledged by the receiver after some period is assumed lost and is retransmitted. Naturally, this scheme is very inefficient and delay prone as the intensity of the traffic increases, as most transmissions result in collisions. Slotted ALOHA improves on this by about a factor of 2, since users transmit on specified time boundaries, and hence collisions are about half as likely.

CSMA improves on ALOHA and slotted ALOHA through carrier sensing; users "listen" to the channel before transmitting, in order to not cause avoidable collisions. Numerous contention algorithms have been developed for CSMA systems; one of the most well known is the distributed coordination function (DCF) of 802.11, whereby users wait for a random amount of time after the channel is clear before transmitting, in order to reduce the probability of two stations transmitting immediately after the channel becomes available. Although the theoretical efficiency of CSMA is often around 60 percent to 70 percent, in wireless LANs, the efficiency is often empirically observed to be less than 50 percent, even when there is only a single user [30].

Although random access is almost always pursued in the time dimension, there is no reason that frequency and code slots couldn't be contended for in an identical fashion. However, because random access tends to be inefficient, systems sophisticated enough to have frequency and especially code slots generally opt for multiple access rather than random access. Hence, CSMA systems can generally be viewed as a type of TDMA, where some inefficiency due to contention and collisions is tolerated in order to have a very simple distributed channel-acquisition procedure in which users acquire resources only when they have packets to send. It should be noted that although FDMA and TDMA are certainly more efficient than CSMA when all users have packets to send, wasted (unused) frequency and time slots in FDMA and TDMA can also bring down the efficiency considerably. In fact, around half the bandwidth is typically wasted in TDMA and FDMA voice systems, which is one major reason that CDMA has proved so successful for voice. Assuming full queues, the efficiency of a connection-oriented MAC can approach 90 percent, compared to at best 50 percent or less in most CSMA wireless systems, such as 802.11. The need for extremely high spectral efficiency in WiMAX thus precludes the use of CSMA, and the burden of resource assignment is placed on the base stations.

6.1.2 Frequency Division Multiple Access

FDMA can be readily implemented in OFDM systems by assigning each user a set of subcarriers. This allocation can be performed in a number of ways. The simplest method is a static allocation of subcarriers to each user, as shown in Figure 6.1a. For example, in a 64-subcarrier OFDM system, user 1 could take subcarriers 1–16, with users 2, 3, and 4 using subcarriers 17–32, 33–48, and 49–64, respectively. The allocations are enforced with a multiplexer for the various users before the IFFT operation. Naturally, there could also be uneven allocations with high-data-rate users being allocated, more subcarriers than to lower-rate users.

An improvement upon static allocation is *dynamic subcarrier allocation,* based on channel-state conditions. For example, owing to frequency-selective fading, user 1 may have relatively good channels on subcarriers 33–48, whereas user 3 might have good channels on subcarriers 1–16. Obviously, it would be mutually beneficial for these users to swap the static allocations given previously. In the next section, we discuss well-developed theories for how the dynamic allocation of subcarriers should be performed.

6.1.3 Time Division Multiple Access—"Round Robin"

In addition to or instead of FDMA, TDMA can accommodate multiple users. In reality, WiMAX systems use both FDMA and TDMA, since there will generally be more users in the system than can be carried simultaneously on a single OFDM symbol. Furthermore, users often will not have data to send, so it is crucial for efficiency's sake that subcarriers be dynamically allocated in order to avoid waste.

Static TDMA is shown Figure 6.1a. Such a static TDMA methodology is appropriate for constant data rate—circuit-switched—applications such as voice and streaming video. But in general, a packet-based system such as WiMAX, can use more sophisticated scheduling algorithms based on queue lengths, channel conditions, and delay constraints to achieve much better performance than static TDMA. In the context of a packet-based system, static TDMA is often called *round-robin* scheduling: Each user simply waits for a turn to transmit.

6.1.4 Code Division Multiple Access

CDMA is the dominant multiple-access technique for present cellular systems but is not particularly appropriate for high-speed data, since the entire premise of CDMA is that a bandwidth much larger than the data rate is used to suppress the interference, as shown in Figure 6.2. In wireless broadband networks, the data rates already are very large, so spreading the spectrum farther is not viable. Even the nominally CDMA broadband standards, such as HSDPA and 1xEV-DO, have very small spreading factors and are dynamic TDMA systems, since users' transmitting turns are based on scheduling objectives, such as channel conditions and latency.

OFDM and CDMA are not fundamentally incompatible; they can be combined to create a multicarrier CDMA (MC-CDMA) waveform [15]. It is possible to use spread-spectrum signaling and to separate users by codes in OFDM by spreading in either the time or the frequency domain. Time-domain spreading entails each subcarrier transmitting the same data symbol on several con-

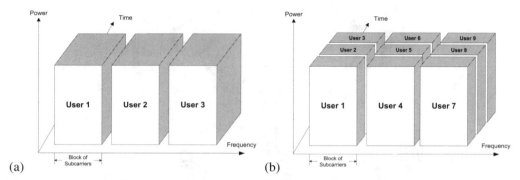

Figure 6.1 (a) FDMA and (b) a combination of FDMA with TDMA

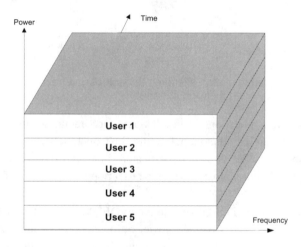

Figure 6.2 CDMA's users share time and frequency slots but use codes that allow the users to be separated by the receiver

secutive OFDM symbols; that is, the data symbol is multiplied by a length N code sequence and then sent on a specific subcarrier for the next N OFDM symbols. Frequency-domain spreading, which generally has slightly better performance than time-domain spreading [13], entails each data symbol being sent simultaneously on N different subcarriers. MC-CDMA is not part of the WiMAX standards but could be deemed appropriate in the future, especially for the uplink.

6.1.5 Advantages of OFDMA

OFDMA is essentially a hybrid of FDMA and TDMA: Users are dynamically assigned subcarriers (FDMA) in different time slots (TDMA) as shown in Figure 6.3. The advantages of OFDMA start with the advantages of single-user OFDM in terms of robust multipath suppression and frequency diversity. In addition, OFDMA is a flexible multiple-access technique that can accommodate many

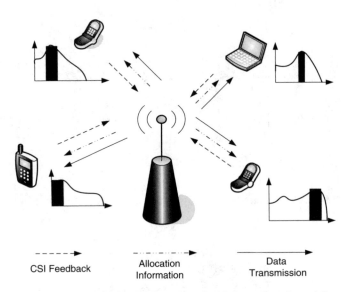

CSI Feedback   Allocation   Data
                     Information   Transmission

Figure 6.3 In OFDMA, the base station allocates to each user a fraction of the subcarriers, preferably in a range where they have a strong channel.

users with widely varying applications, data rates, and QoS requirements. Because the multiple access is performed in the digital domain, before the IFFT operation, dynamic and efficient bandwidth allocation is possible. This allows sophisticated time- and frequency- domain scheduling algorithms to be integrated in order to best serve the user population. Some of these algorithms are discussed in the next section.

One significant advantage of OFDMA relative to OFDM is its potential to reduce the transmit power and to relax the peak-to-average-power ratio (PAPR) problem, which was discussed in detail in Chapter 4. The PAPR problem is particularly acute in the uplink, where power efficiency and cost of the power amplifier are extremely sensitive quantities. By splitting the entire bandwidth among many MSs in the cell, each MS uses only a small subset of subcarriers. Therefore, each MS transmits with a lower PAPR—recall that PAPR increases with the number of subcarriers—and with far lower total power than if it had to transmit over the entire bandwidth. Figure 6.4 illustrates this. Lower data rates and bursty data are handled much more efficiently in OFDMA than in single-user OFDM or with TDMA or CSMA, since rather than having to blast at high power over the entire bandwidth, OFDMA allows the same data rate to be sent over a longer period of time using the same total power.

6.2 Multiuser Diversity and Adaptive Modulation

In OFDMA, the subcarrier and the power allocation should be based on the channel conditions in order to maximize the throughput. In this section, we provide necessary background discussion on the key two principles that enable high performance in OFDMA: multiuser diversity and adaptive modulation. Multiuser diversity describes the gains available by selecting a user or sub-

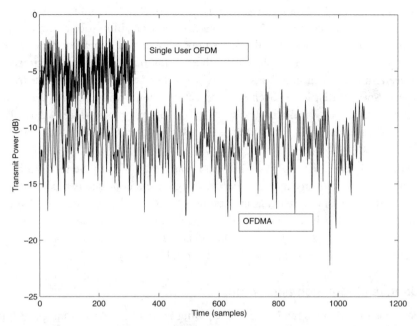

Figure 6.4 OFDM with 256 subcarriers and OFDMA with only 64 of the 256 subcarriers used. The total power used is the same, but OFDMA allows much lower peak power.

set of users having "good" conditions. Adaptive modulation is the means by which good channels can be exploited to achieve higher data rates.

6.2.1 Multiuser Diversity

The main motivation for adaptive subcarrier allocation in OFDMA systems is to exploit multiuser diversity. Although OFDMA systems have a number of subcarriers, we focus temporarily on the allocation for a single subcarrier among multiple users.

Consider a K-user system in which the subcarrier of interest experiences i.i.d. Rayleigh fading—that is, each user's channel gain is independent of the others—and is denoted by h_k. The probability density function (PDF) of user k's channel gain $p(h_k)$ is given by

$$p(h_k) = \begin{cases} 2h_k e^{-h_k^2} & \text{if } h_k \geq 0 \\ 0 & \text{if } h_k < 0 \end{cases} \tag{6.1}$$

Now suppose that the base station transmits only to the user with the highest channel gain, denoted as $h_{max} = \max\{h_1, h_2, \cdots, h_K\}$. It is easy to verify that the PDF of h_{max} is

$$p(h_{max}) = 2Kh_{max}\left(1 - e^{-h_{max}^2}\right)^{K-1} e^{-h_{max}^2}. \tag{6.2}$$

Figure 6.5 shows the PDF of h_{max} for various values of K. As the number of users increases, the PDF of h_{max} shifts to the right, which means that the probability of getting a large channel gain improves. Figure 6.6 shows how this increased channel gain improves the capacity and bit error rate for uncoded QPSK. Both plots show that the multiuser diversity gain improves as the number of users in the system increases, but the majority of the gain is achieved from only the first few users. Specifically, it has been proved, using extreme-value theory, that in a K-user system, the average capacity scales as $\log\log K$[31], assuming just Rayleigh fading. If i.i.d. lognormal shadowing is present for each of the users, which is a reasonable assumption, the scaling improves to $\sqrt{\log K}$ [5].

In a WiMAX system, the multiuser diversity gain will generally be reduced by averaging effects, such as spatial diversity and the need to assign users contiguous blocks of subcarriers. This conflict is discussed in more detail in Section 6.4.3. Nevertheless, the gains from multiuser diversity are considerable in practical systems. Although we focus on the gains in terms of throughput (capacity) in this chapter, it should be noted that in some cases, the largest impact from multiuser diversity is on link reliability and overall coverage area.

6.2.2 Adaptive Modulation and Coding

WiMAX systems use adaptive modulation and coding in order to take advantage of fluctuations in the channel. The basic idea is quite simple: Transmit as high a data rate as possible when the channel is good, and transmit at a lower rate when the channel is poor, in order to avoid excessive dropped packets. Lower data rates are achieved by using a small constellation, such as QPSK, and low-rate error-correcting codes, such as rate 1/2 convolutional or turbo codes. The higher data rates are achieved with large constellations, such as 64 QAM, and less robust error-correcting codes; for example, rate 3/4 convolutional, turbo, or LDPC codes. In all, 52 configurations of modulation order and coding types and rates are possible, although most implementations of WiMAX offer only a fraction of these. These configurations are referred to as *burst profiles* and are enumerated in Table 8.4.

A block diagram of an AMC system is given in Figure 6.7. For simplicity, we first consider a single-user system attempting to transmit as quickly as possible through a channel with a variable SINR—for example, due to fading. The goal of the transmitter is to transmit data from its queue as rapidly as possible, subject to the data being demodulated and decoded reliably at the receiver. Feedback is critical for adaptive modulation and coding: The transmitter needs to know the "channel SINR" γ, which is defined as the received SINR γ_r divided by the transmit power P_t, which itself is usually a function of γ. The received SINR is thus $\gamma_r = P_t\gamma$.

Figure 6.8 shows that by using six of the common WiMAX burst profiles, it is possible to achieve a large range of spectral efficiencies. This allows the throughput to increase as the SINR increases following the trend promised by Shannon's formula $C = \log_2(1 + SNR)$. In this case, the lowest offered data rate is QPSK and rate 1/2 turbo codes; the highest data-rate burst profile is with 64 QAM and rate 3/4 turbo codes. The achieved throughput normalized by the bandwidth is defined as

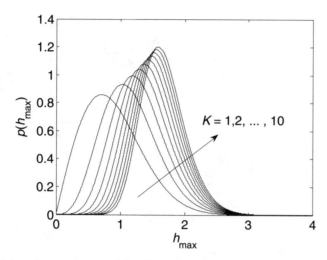

Figure 6.5 PDF of h_{max}, the maximum of the K users' channel gains

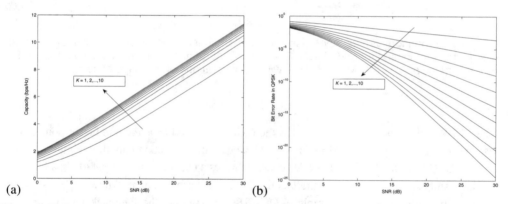

Figure 6.6 For various numbers of users K, (a) average capacity and (b) QPSK bit error rate

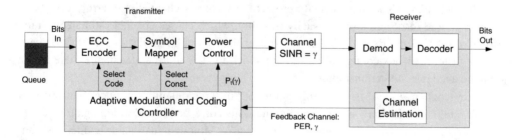

Figure 6.7 Adaptive modulation and coding block diagram

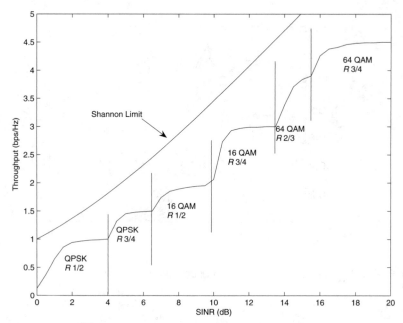

Figure 6.8 Throughput versus SINR, assuming that the best available constellation and coding configuration are chosen for each SINR. Only six configurations are used in this figure, and the turbo decoder is a max log MAP decoder with eight iterations of message passing.

$$T = (1 - \text{BLER})r\log_2(M)\text{ bps/Hz,} \tag{6.3}$$

where BLER is the block error rate, $r \leq 1$ is the coding rate, and M is the number of points in the constellation. For example, 64 QAM with rate 3/4 codes achieves a maximum throughput of 4.5bps/Hz, when BLER $\rightarrow$ 0; QPSK with rate 1/2 codes achieves a best-case throughput of 1bps/Hz.

The results shown here are for the idealized case of perfect channel knowledge and do not consider retransmissions—for example, with ARQ. In practice, the feedback will incur some delay and perhaps also be degraded owing to imperfect channel estimation or errors in the feedback channel. WiMAX systems heavily protect the feedback channel with error correction, so the main source of degradation is usually mobility, which causes channel estimates to rapidly become obsolete. Empirically, with speeds greater than about 30 km/hr on a 2,100MHz carrier, even the faster feedback configurations do not allow timely and accurate channel state information to be available at the transmitter.

A key challenge in AMC is to efficiently control three quantities at once: transmit power, transmit rate (constellation), and the coding rate. This corresponds to developing an appropriate policy for the AMC controller shown in Figure 6.7. Although reasonable guidelines can be developed from a theoretical study of adaptive modulation, in practice, the system engineer needs to develop and fine-tune the algorithm, based on extensive simulations, since performance depends on many factors. Some of these considerations are

- **BLER and received SINR:** In adaptive-modulation theory, the transmitter needs to know only the statistics and instantaneous channel SINR. From the channel SINR, the transmitter can determine the optimum coding/modulation strategy and transmit power [8]. In practice, however, the BLER should be carefully monitored as the final word on whether the data rate should be increased (if the BLER is low) or decreased to a more robust setting.

- **Automatic repeat request (ARQ):** ARQ allows rapid retransmissions, and hybrid-ARQ generally increases the ideal BLER operating point by about a factor of 10: for example, from 1 percent to 10 percent. For delay-tolerant applications, it may be possible to accept a BLER approaching even 70 percent, if Chase combining is used in conjunction with H-ARQ to make use of unsuccessful packets.

- **Power control versus waterfilling:** In theory, the best power-control policy from a capacity standpoint is the so-called waterfilling strategy, in which more power is allocated to strong channels and less power allocated to weak channels [11, 12]. In practice, the opposite may be true in some cases. For example, in Figure 6.8, almost nothing is gained with a 13dB SINR versus an 11dB SINR: In both cases, the throughput is 3bps/Hz. Therefore, as the SINR improved from 11dB to 13dB, the transmitter would be well advised to lower the transmit power, in order to save power and generate less interference to neighboring cells [3].

- **Adaptive modulation in OFDMA:** In an OFDMA system, each user is allocated a block of subcarriers, each having a different set of SINRs. Therefore, care needs to be paid to which constellation/coding set is chosen, based on the varying SINRs across the subcarriers.

6.3 Resource-Allocation Techniques for OFDMA

There are a number of ways to take advantage of multiuser diversity and adaptive modulation in OFDMA systems. Algorithms that take advantage of these gains are not specified by the WiMAX standard, and all WiMAX developer are free to develop their own innovative procedures. The idea is to develop algorithms for determining which users to schedule, how to allocate subcarriers to them, and how to determine the appropriate power levels for each user on each subcarrier. In this section, we will consider some of the possible approaches to resource allocation. We focus on the class of techniques that attempt to balance the desire for high throughput with fairness among the users in the system. We generally assume that the outgoing queues for each user are full, but in practice, the algorithms discussed here can be modified to adjust for queue length or delay constraints, which in many applications may be as, if not more, important than raw throughput.[2]

Referring to the downlink OFDMA system shown in Figure 6.3, users estimate and feedback the channel state information (CSI) to a centralized base station, where subcarrier and power allocation are determined according to users' CSI and the resource-allocation procedure. Once the subcarriers for each user have been determined, the base station must inform each user

---

2. Queueing theory and delay-constrained scheduling is a rich topic in its own right, and doing it justice here is outside the scope of this chapter.

which subcarriers have been allocated to it. This subcarrier mapping must be broadcast to all users whenever the resource allocation changes: The format of these messages is discussed in Chapter 8. Typically, the resource allocation must be performed on the order of the channel coherence time, although it may be performed more frequently if a lot of users are competing for resources.

The resource allocation is usually formulated as a constrained optimization problem, to either (1) minimize the total transmit power with a constraint on the user data rate [21, 39] or (2) maximize the total data rate with a constraint on total transmit power [18, 24, 25, 43]. The first objective is appropriate for fixed-rate applications, such as voice, whereas the second is more appropriate for bursty applications, such as data and other IP applications. Therefore, in this section, we focus on the rate-adaptive algorithms (category 2), which are more relevant to WiMAX systems. We also note that considerable related work on resource allocation has been done for multicarrier DSL systems [2, 6, 7, 41]; the coverage and references in this section are by no means comprehensive. Unless otherwise stated, we assume in this section that the base station has obtained perfect instantaneous channe-station information for all users. Table 6.1 summarizes the notation that will be used throughout this section.

6.3.1 Maximum Sum Rate Algorithm

As the name indicates, the objective of the maximum sum rate (MSR) algorithm, is to maximize the sum rate of all users, given a total transmit power constraint [43]. This algorithm is optimal if the goal is to get as much data as possible through the system. The drawback of the MSR algorithm is that it is likely that a few users close to the base station, and hence having excellent channels, will be allocated all the system resources. We now briefly characterize the SINR, data rate, and power and subcarrier allocation that the MSR algorithm achieves.

Table 6.1 Notations

Notation	Meaning
K	Number of users
L	Number of subcarriers
$h_{k,l}$	Envelope of channel gain for user k in subcarrier l
$P_{k,l}$	Transmit power allocated for user k in subcarrier l
σ^2	AWGN power spectrum density
P_{tot}	Total transmit power available at the base station
B	Total transmission bandwidth

Let $P_{k,l}$ denote user k's transmit power in subcarrier l. The signal-to-interference-plus-noise ratio for user k in subcarrier l, denoted as $\text{SINR}_{k,l}$, can be expressed as

$$\text{SINR}_{k,l} = \frac{P_{k,l} h_{k,l}^2}{\sum_{j=1, j \neq k}^{K} P_{j,l} h_{k,l}^2 + \sigma^2 \frac{B}{L}}. \tag{6.4}$$

Using the Shannon capacity formula as the throughput measure,[3] the MSR algorithm maximizes the following quantity:

$$\max_{P_{k,l}} \sum_{k=1}^{K} \sum_{l=1}^{L} \frac{B}{L} \log\left(1 + \text{SINR}_{k,l}\right), \tag{6.5}$$

with the total power constraint $\sum_{k=1}^{K} \sum_{l=1}^{L} P_{k,l} \leq P_{tot}$.

The sum capacity is maximized if the total throughput in each subcarrier is maximized. Hence, the maximum sum capacity optimization problem can be decoupled into L simpler problems, one for each subcarrier. Further, the sum capacity in subcarrier l, denoted as C_l, can be written as

$$C_l = \sum_{k=1}^{K} \log\left(1 + \frac{P_{k,l}}{P_{tot,l} - P_{k,l} + \frac{\sigma^2}{h_{k,l}^2} \frac{B}{L}}\right), \tag{6.6}$$

where $P_{tot,l} - P_{k,l}$ denotes other users' interference to user k in subcarrier l. It is easy to show that C_l is maximized when all available power $P_{tot,l}$ is assigned to the single user with the largest channel gain in subcarrier l. This result agrees with intuition: Give each channel to the user with the best gain in that channel. This is sometimes referred to as a "greedy" optimization. The optimal power allocation proceeds by the waterfilling algorithm discussed previously, and the total sum capacity is readily determined by adding up the rate on each of the subcarriers.

6.3.2 Maximum Fairness Algorithm

Although the total throughput is maximized by the MSR algorithm, in a cellular system such as WiMAX, in which the pathloss attenuation varies by several orders of magnitude between users, some users will be extremely underserved by an MSR-based scheduling procedure. At the alternative

---

3. Throughout this section, we use the Shannon capacity formula as the throughput measure. In practice, there is a gap between the achieved data rate and the maximum (Shannon) rate, which can be simply characterized with a SINR gap of a few dB. Therefore, this approach to resource allocation is valid, but the exact numbers given here are optimistic.

extreme, the maximum fairness algorithm [29] aims to allocate the subcarriers and power such that the *minimum* user's data rate is maximized. This essentially corresponds to equalizing the data rates of all users; hence the name "maximum fairness."

The maximum fairness algorithm can be referred to as a *max-min* problem, since the goal is to maximize the minimum data rate. The optimum subcarrier and power allocation is considerably more difficult to determine than in the MSR case, because the objective function is not concave. It is particularly difficult to simultaneously find the optimum subcarrier and power allocation. Therefore, low-complexity suboptimal algorithms are necessary, in which the subcarrier and power allocation are done separately.

A common approach is to assume initially that equal power is allocated to each subcarrier and then to iteratively assign each available subcarrier to a low-rate user with the best channel on it [29, 40]. Once this generally suboptimal subcarrier allocation is completed, an optimum (waterfilling) power allocation can be performed. It is typical for this suboptimal approximation to be very close to the performance obtained with an exhaustive search for the best joint subcarrier-power allocation, in terms of both the fairness achieved and the total throughput.

6.3.3 Proportional Rate Constraints Algorithm

A weakness of the maximum fairness algorithm is that the rate distribution among users is not flexible. Further, the total throughput is limited largely by the user with the worst SINR, as most of the resources are allocated to that user, which is clearly suboptimal. In a wireless broadband network, it is likely that different users require application-specific data rates that vary substantially. A generalization of the maximum fairness algorithm is a the proportional rate constraints (PRC) algorithm, whose objective is to maximize the sum throughput, with the additional constraint that each user's data rate is proportional to a set of predetermined system parameters $\{\beta_k\}_{k=1}^K$. Mathematically, the proportional data rates constraint can be expressed as

$$\frac{R_1}{\beta_1} = \frac{R_2}{\beta_2} = \cdots = \frac{R_K}{\beta_K} \, , \tag{6.7}$$

where each user's achieved data rate R_k is

$$R_k = \sum_{l=1}^L \frac{\rho_{k,n} B}{L} \log_2 \left(1 + \frac{P_{k,l} h_{k,l}^2}{\sigma^2 \frac{B}{L}} \right), \tag{6.8}$$

and $\rho_{k,l}$ can be the value only of either 1 or 0, indicating whether subcarrier l is used by user k. Clearly, this is the same setup as the maximum fairness algorithm if $\beta_k = 1 \, \forall k$. The advantage is that any arbitrary data rates can be achieved by varying the β_k values.

The PRC optimization problem is also generally very difficult to solve directly, since it involves both continuous variables $p_{k,l}$ and binary variables $\rho_{k,l}$, and the feasible set is not convex. As for the maximum fairness case, the prudent approach is to separate the subcarrier and power allocation and to settle for a near-optimal subcarrier and power allocation that can be achieved with manageable complexity. The near-optimal approach is derived and outlined in [32, 33] and a low-complexity implementation developed in [40].

6.3.4 Proportional Fairness Scheduling

The three algorithms discussed thus far attempt to *instantaneously* achieve an objective such as the total sum throughput (MSR algorithm), maximum fairness (equal data rates among all users), or preset proportional rates for each user. Alternatively, one could attempt to achieve such objectives over time, which provides significant additional flexibility to the scheduling algorithms. In this case, in addition to throughput and fairness, a third element enters the trade-off: *latency*. In an extreme case of latency tolerance, the scheduler could simply wait for the user to get close to the base station before transmitting. In fact, the MSR algorithm achieves both fairness *and* maximum throughput if the users are assumed to have the same average channels in the long term—on the order of minutes, hours, or more—and there is no constraint with regard to latency. Since latencies, even on the order of seconds, are generally unacceptable, scheduling algorithms that balance latency and throughput and achieve some degree of fairness are needed. The most popular framework for this type of scheduling is proportional fairness (PF) scheduling [36, 38].

The PF scheduler is designed to take advantage of multiuser diversity while maintaining comparable long-term throughput for all users. Let $R_k(t)$ denote the instantaneous data rate that user k can achieve at time t, and let $T_k(t)$ be the average throughput for user k up to time slot t. The PF scheduler selects the user, denoted as k^*, with the highest $R_k(t)/T_k(t)$ for transmission. In the long term, this is equivalent to selecting the user with the highest instantaneous rate relative to its mean rate. The average throughput $T_k(t)$ for all users is then updated according to

$$T_k(t+1) = \begin{cases} \left(1 - \dfrac{1}{t_c}\right)T_k(t) + \dfrac{1}{t_c}R_k(t) & k = k^* \\ \left(1 - \dfrac{1}{t_c}\right)T_k(t) & k \neq k^* \end{cases}. \tag{6.9}$$

Since the PF scheduler selects the user with the largest instantaneous data rate relative to its average throughput, "bad" channels for each user are unlikely to be selected. On the other hand, consistently underserved users receive scheduling priority, which promotes fairness. The parameter t_c controls the latency of the system. If t_c is large, the latency increases, with the benefit of higher sum throughput. If t_c is small, the latency decreases, since the average throughput values change more quickly, at the expense of some throughput.

The PF scheduler has been widely adopted in packet date systems, such as HSDPA and 1xEV-DO, where t_c is commonly set between 10 and 20. One interesting property of PF scheduling is that as $t_c \to \infty$, the sum of the logs of the user data rates is maximized. That is, PF scheduling maximizes

$$\sum\nolimits_{k=1}^{K} \log T_k \,.$$

Although originally designed for a single-channel time-slotted system, the PF scheduler can be adapted to an OFDMA system. In an OFDMA system, due to the multiple parallel subcarriers in the frequency domain, multiple users can transmit on different subcarriers simultaneously. The original PF algorithm can be extended to OFDMA by treating each subcarrier independently. Let $R_k(t,n)$ be the supportable data rate for user k in subcarrier n at time slot t. Then for each subcarrier, the user with the largest $R_k(t,n)/T_k(t)$ is selected for transmission. Let $\Omega_k(t)$ denote the set of subcarriers in which user k is scheduled for transmission at time slot t, then the average user throughput is updated as

$$T_k(t+1) = \left(1 - \frac{1}{t_c}\right) T_k(t) + \frac{1}{t_c} \sum_{n \in \Omega_k(t)} R_k(t,n) \tag{6.10}$$

for $k = 1, 2, \cdots, K$. Other weighted adaptations and evolutions of PF scheduling of OFDMA are certainly possible.

6.3.5 Performance Comparison

In this section, we briefly compare the performance of the various scheduling algorithms for OFDMA that we have discussed, in order to gain intuition on their relative performance and merits. In these results, an exponentially decaying multipath profile with six multipath components was used to generate the frequency diversity. All users have the same average SNR. The absolute-capacity numbers are not especially important, what is important are the trends between the curves.

6.3.5.1 Throughput
First, we consider the multiuser diversity gains of the various types of algorithms. Figure 6.9 shows the capacity, normalized by the total bandwidths for static TDMA (round-robin), proportional fairness, and the MSR algorithm. As expected, the MSR algorithm achieves the best total throughput, and the gain increases as the number of users increases, on the order of $\log \log K$. Static TDMA achieves no multiuser gain, since the users transmit independent of their channel realizations. It can be seen that the PF algorithm approaches the throughput of the MSR algorithm, with an expected penalty owing to its support for underserved users.

6.3.5.2 Fairness
Now, let's consider how the worst user in the system does (Figure 6.10). As expected, the MF algorithm achieves the best performance for the most underserved user, with a slight gain for optimal power allocation over its allocated subcarriers (waterfilling) relative to an equal-power allocation. Also as expected, the MSR algorithm results in a starved worst-case user; in fact, it is typical for several users to receive no resources at all for substantial periods of time. Static TDMA performs in between the two, with the percentage loss relative to the MF algorithm

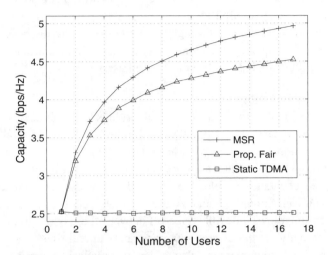

Figure 6.9 Sum capacity versus number of users, for a single-carrier system with scheduling in the time domain only

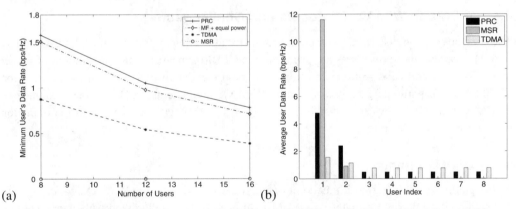

(a)   (b)

Figure 6.10 (a) Minimum user's capacity in multiuser OFDM versus the number of users; (b) normalized average throughput per user in a heterogeneous environment

increasing as the number of users increases, since TDMA does not take advantage of multiuser diversity.

Next, we consider a heterogeneous environment with eight users. The first user has an average SINR of 20dB, the second user has an average SINR of 10dB, and the other users 3–8 have average SINRs of 0dB. This is a reasonable scenario in which user 1 is the closest to the base station, users 3–8 are near the cell edge, and user 2 is in between. Clearly, the bulk of the resources will be allocated to users 1 and 2 by the MSR algorithm, and this can be readily observed in Figure 6.10b. The downside of this approach, of course, is that users 3–8 have a throughput of approximately zero.

A more balanced approach would be to use the PRC algorithm and adopt proportional rate constraints equal to the relative SINRs: $\beta_1 = 100, \beta_2 = 10, \beta_3 = 1, \ldots \beta_8 = 1$. This allows the underserved users to get at least some throughput, while preserving the bulk of the multiuser diversity gains. Naturally, a more equal assignment of the β_is will increase the fairness, with the extreme case $\beta_i = 1 \, \forall \, i$ equalizing the data rates for all users.

6.3.5.3   Summary of Comparison

Table 6.2 compares the four resource-allocation algorithms that this chapter introduced for OFDMA systems. In summary, the MSR allocation is best in terms of total throughput and achieves a low computational complexity but has a terribly unfair distribution of rates. Hence, the MSR algorithm is viable only when all users have nearly identical channel conditions and a relatively large degree of latency is tolerable. The MF algorithm achieves complete fairness while sacrificing significant throughput and so is appropriate only for fixed, equal-rate applications. The PRC algorithm allows a flexible trade-off between these two extremes, but it may not always be possible to aptly set the desired rate constraints in real time. The popular PF algorithm, which is fairly simple to implement, also achieves a practical balance between throughput and fairness.

6.4 OFDMA in WiMAX: Protocols and Challenges

The previous section discussed several algorithms for allocating system resources to users. In an OFDMA system, those resources are primarily the OFDM subcarriers and the amount of power given to each user. In this section, we summarize important details of a practical implementation of OFDMA. In particular, we consider how WiMAX implements OFDMA, the challenge of OFDMA in a cellular system, and how diversity in OFDMA can be exploited in conjunction with other types of diversity.

6.4.1 OFDMA Protocols

Although the scheduling algorithms do not need to be specified by the WiMAX standard—and so are not—several key attributes of OFDMA do need to be standardized: subchannelization, mapping messages, and ranging. The details of which are elaborated on in Chapters 8 and 9.

Table 6.2 Comparison of OFDMA Rate-Adaptive Resource-Allocation Schemes

Algorithm	Sum Capacity	Fairness	Complexity	Simple Algorithm?
Maximum sum rate (MSR)	Best	Poor and inflexible	Low	Not necessary [18]
Maximum fairness (MF)	Poor	Best but inflexible	Medium	Available [29]
Proportional rate constraints (PRC)	Good	Most flexible	High	Available [33]
Proportional fairness (PF)	Good	Flexible	Low	Available [38]

6.4.1.1 Subchannelization

In WiMAX, users are allocated blocks of subcarriers rather than individual subcarriers, in order to lower the complexity of the subcarrier-allocation algorithm and simplify the mapping messages. Assuming that a user k is allocated a block of L_k subcarriers, these L_k subcarriers can be either spread out over the entire bandwidth—*distributed subcarrier permutation*—or all in the same frequency range—*adjacent subcarrier permutation*. The primary benefit of a distributed permutation is improved frequency diversity and robustness; the benefit of adjacent permutation is increased multiuser diversity. More details on these allocations are given in Section 8.6.

6.4.1.2 Mapping Messages

In order for each MS to know which subcarriers are intended for it, the BS must broadcast this information in DL MAP messages. Similarly, the BS tells each MS which subcarriers to transmit on in a UL MAP message. In addition to communicating the DL and UL subcarrier allocations to the MS, the MS must also be informed of the *burst profile* used in the DL and the UL. The burst profile is based on the measured SINR and BLER in both links and identifies the appropriate level of modulation and coding. These burst profiles, identified in Table 8.4, are how adaptive modulation and coding are implemented in WiMAX. Details on the DL MAP and UL MAP messages are given in Section 8.7.

6.4.1.3 Ranging

Since each MS has a unique distance from the base station, it is critical in the uplink to synchronize the symbols and equalize the received power levels among the various active MSs. This process is known as *ranging;* when initiated, ranging requires the BS to estimate the channel strength and the time of arrival for the MS in question. Downlink synchronization is not needed, since this link is already synchronous, but in the uplink, the active users need to be synchronized to at least within a cyclic prefix guard time of one another. Otherwise, significant intercarrier and intersymbol interference can result. Similarly, although downlink power control is recommended in order to reduce spurious other-cell interference, it is not strictly required. Uplink power control is needed to (1) improve battery life, (2) reduce spurious other-cell interference, and (3) avoid drowning out faraway users in the same cell who are sharing an OFDM symbol with them. The third point arises from degraded orthogonality between cocell uplink users, owing to such practical issues as analog-to-digital dynamic range, carrier offset from residual Doppler and oscillators mismatching that is not corrected by ranging, and imperfect synchronization. The uplink power-control problem in WiMAX is similar to the near/far problem in CDMA, although considerably less strict; in uplink CDMA, the power control must be extremely accurate.

   In WiMAX, four types of ranging procedures exist: initial ranging, periodic ranging, bandwidth request, and handover ranging. Ranging is performed during two or four consecutive symbols with no phase discontinuity, which allows the BS to listen to a misaligned MS that has a timing mismatch larger than the cyclic prefix. If the ranging procedure is successful, the BS sends a ranging response (RNG-RES) message that instructs the MS on the appropriate timing-offset adjustment, frequency-offset correction, and power setting. If ranging was unsuccessful, the MS increases its power level and sends a new ranging message, continuing this process until success. Sections 8.10 and 9.5 have more details on the ranging procedure for WiMAX.

6.4.2 Cellular OFDMA

Note that since the scheduling algorithms discussed thus far in this chapter are all very dependent on the perceived SINR for each user, the scheduling choices of each base station affect the users in the adjacent cells. For example, if a certain MS near the cell edge, presumably with a low SINR, is selected to transmit in the uplink at high power, the effective SINRs of all the users in the cell next to it will be lowered, hence perhaps changing the ideal subcarrier allocation and burst profile for that cell. Therefore, a cellular OFDMA system greatly benefits from methods for suppressing or avoiding the interference from adjacent cells.

A simple approach is to use a unique frequency-hopping pattern for each base station to randomize to the other-cell interference [27], an approach popularized by the Flarion (now QUALCOMM) scheme called Flash-OFDM. Although this scheme reduces the probability of a worst-case interference scenario, under a high-system load, the interference levels, can still rapidly approach untenable levels and the probability of collision can grow large [35, 37]. A more sophisticated approach is to develop advanced receivers that are capable of canceling the interference from a few dominant interference sources. This is a challenging proposition even in a single-carrier system [1], and its viability in a cellular OFDMA system is open to debate.

An appealing approach is to revisit the resource-allocation algorithms discussed in Section 6.3 in the context of a multicell system. If each base station is unaware of the exact conditions in the other cells, and if no cooperation among neighboring base stations is allowed, the subcarrier and power allocation follows the theory of noncooperative games [9, 14, 41] and results in a so-called Nash equilibrium. Simply put, this scenario is the equivalent of gridlock: The users reach a point at which neither increasing nor decreasing their power autonomously improves their capacity.

Naturally, better performance can be obtained if the base stations cooperate with one another. For example, a master scheduler for all the base stations could know the channels in each base station and make multicell resource-allocation schedules accordingly. This would be prohibitively complex, though, owing to (1) transferring large amounts of real-time information to and from this centralized scheduler, and (2) the computational difficulties involved in processing this quantity of information to determine a globally optimal or near-optimal resource allocation. Simpler approaches are possible: For example, neighboring base stations could share simple information to make sure they don't assign the same subcarriers to vulnerable users. Research on cellular cooperation and encoding has been very active recently, including fundamental work from an information theory perspective [4, 10, 16, 17, 26, 34, 42], as well as more heuristic techniques specifically for cellular OFDMA [20, 23, 28]. As of press time, it appears promising that in the next few years, WiMAX systems will begin to adopt some of these techniques to improve their coverage and spectral efficiency.

6.4.3 Limited Diversity Gains

Diversity is a key source of performance gain in OFDMA systems. In particular, OFDMA exploits multiuser diversity among the various MSs, frequency diversity across the subcarriers, and time diversity by allowing latency. Spatial diversity is also a key aspect of WiMAX systems. One important observation is that these sources of diversity generally compete with one another. For example, imagine that the receiver has two sufficiently spaced antennas. If two-branch selection diversity is used for each subcarrier, the amount of variation between each subcarrier will decrease significantly, since most of the deep fades will be eliminated by the selection process. Now, if ten users were to execute an OFDMA scheduling algorithm, although the overall performance would increase further, the multiuser diversity gain would be less than without the selection diversity, since each user has already eliminated their worst channels with the selection combining. The intuition of this simple example can be extended to other diversity-exploiting techniques, such as coding and interleaving, space/time and space/frequency codes, and so on. In short, *the total diversity gain will be less than the sum of the diversity gains from the individual techniques.*

Figure 6.11 shows the combined effect of multiuser and spatial diversity for five configurations of 2×1 MIMO systems: single antenna (SISO), opportunistic beamforming (BF) [38], Alamouti STBCs, and transmit beamforming with limited feedback (1-bit CSI) and perfect CSI. For a single user, the SISO and opportunistic BF are least effective, since opportunistic BF requires multiuser diversity to get a performance gain over SISO. Alamouti codes increase performance, in particular reducing the probability of a very low SINR from occurring. The CSI-endowed techniques do the best; notably, the perfect-CSI case is always 3dB better than Alamouti codes regardless of the number of users. When the system does have 50 users, however, some of the conclusions change considerably. Now, Alamouti codes perform worse than single-antenna transmission! The reason is that Alamouti codes harden the received SINR toward the average, and so the SINR difference between the users is attenuated, but this is exactly what is exploited by a multiuser scheduler that picks the best of the 50 users. The advantage of perfect CSI is also narrowed relative to SISO and opportunistic beamforming. The key point here is that the diversity gains from various techniques may interfere with one another; only a complete system characterization can reliably predict the overall system performance.

6.5 Summary and Conclusions

Although the main idea of OFDMA is quite simple in concept—share an OFDM symbol among several users at the same time—efficiently assigning subcarriers, data rates, and power levels to each user in the downlink and the uplink is a challenging task. In particular, this chapter emphasized the following points.

- Traditional multiple-access techniques—FDMA, TDMA, CDMA, CSMA—can all be applied to OFDM. The recommended approach is an FDMA-TDMA hybrid called OFDMA.

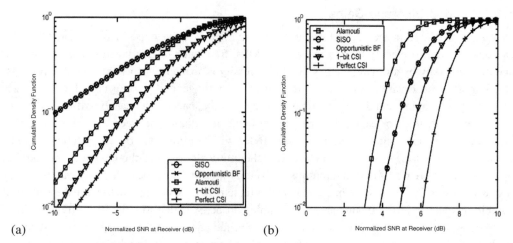

(a)   (b)

Figure 6.11 The SINR of multiuser diversity combined with antenna-diversity techniques for (a) $K = 1$ users and (b) $K = 50$ users. Figure from [22], courtesy of IEEE.

- OFDMA achieves its high performance and flexible accommodation of many users through *multiuser diversity and adaptive modulation.*

- A number of resource-allocation procedures are possible for OFDMA. We introduced and compared four such algorithms that achieve various trade-offs in terms of sum throughput, fairness to underserved users, and complexity.

- To implement OFDMA, some overhead messaging is required. We summarized how this is done in WiMAX.

- Challenges posed by OFDMA include (1) interfering neighboring cells and (2) limited total diversity gains.

6.6 Bibliography

[1] J. G. Andrews. Interference cancellation for cellular systems: A contemporary overview. *IEEE Wireless Communications Magazine*, 12(2):19–29, April 2005.

[2] R. Cendrillon, M. Moonen, J. Verliden, T. Bostoen, and W. Yu. Optimal multiuser spectrum management for digital subscriber lines. In *Proceedings IEEE International Conference on Communications*, 1:1–5, June 2004.

[3] R. Chen, J. G. Andrews, R. W. Heath, and A. Ghosh. Uplink power control in multi-cell spatial multiplexing wireless systems. *IEEE Transactions on Wireless Communications*. Forthcoming.

[4] W. Choi and J. G. Andrews. Base station cooperatively scheduled transmission in a cellular MIMO TDMA system. In *Proceedings, Conference on Information Sciences and Systems (CISS)*, March 2006.

[5] W. Choi and J. G. Andrews. The capacity gain from base station cooperative scheduling in a MIMO DPC cellular system. In *Proceedings, IEEE International Symposium on Information Theory*, Seattle, WA, June 2006.

[6] J. Chow, J. Tu, and J. Cioffi. A discrete multitone transceiver system for HDSL applications. *IEEE Journal on Selected Areas in Communications*, 9(6):895–908, August 1991.

[7] P. Chow, J. Cioffi, and J. Bingham. A practical discrete multitone transceiver loading algorithm for data transmission over spectrally shaped channels. *IEEE Transactions on Communications*, 43(2/3/4):773–775, February–April 1995.

[8] S. G. Chua and A. Goldsmith. Adaptive coded modulation for fading channels. *IEEE Transactions on Communications*, 46(5):595–602, May 1998.

[9] S. T. Chung, S. J. Kim, J. Lee, and J. M. Cioffi. A game-theoretic approach to power allocation in frequency selective Gaussian interference channels. In *Proceedings, IEEE International Symposium on Information Theory*, p. 316, July 2003.

[10] G. J. Foschini, H. Huang, K. Karakayali, R. A. Valenzuela, and S. Venkatesan. The value of coherent base station coordination. In *Proceedings, Conference on Information Sciences and Systems (CISS)*, Johns Hopkins University, March 2005.

[11] A. Goldsmith and P. Varaiya. Capacity of fading channels with channel side information. *IEEE Transactions on Information Theory*, pp. 1986–1992, November 1997.

[12] A. J. Goldsmith. Wireless Communications. Cambridge University Press, 2005.

[13] X. Gui and T. S. Ng. Performance of asynchronous orthogonal multicarrier system in a frequency selective fading channel. *IEEE Transactions on Communications*, 47(7):1084–1091, July 1999.

[14] Z. Han, Z. Ji, and K. R. Liu. Power minimization for multi-cell OFDM networks using distributed noncooperative game approach. In *Proceedings, IEEE Globecom*, pp. 3742–3747, December 2004.

[15] S. Hara and R. Prasad. Overview of multicarrier CDMA. *IEEE Communications Magazine*, 35(12):126–133, December 1997.

[16] H. Huang and S. Venkatesan. Asymptotic downlink capacity of coordinated cellular network. In *Proceedings, IEEE Asilomar*, March 2004.

[17] S. A. Jafar, G. Foschini, and A. J. Goldsmith. Phantomnet: Exploring optimal multicellular multiple antenna systems. *EURASIP Journal on Applied Signal Processing, Special Issue on MIMO Communications and Signal Processing*, pp. 591–605, May 2004.

[18] J. Jang and K. Lee. Transmit power adaptation for multiuser OFDM systems. *IEEE Journal on Selected Areas in Communications*, 21(2):171–178, February 2003.

[19] P. Jung, P. Baier, and A. Steil. Advantages of CDMA and spread spectrum techniques over FDMA and TDMA in cellular mobile radio applications. *IEEE Transactions on Vehicular Technology*, pp. 357–364, August 1993.

[20] H. Kim, Y. Han, and J. Koo. Optimal subchannel allocation scheme in multicell OFDMA systems. In *Proceedings, IEEE Vehicular Technology Conference*, pp. 1821–1825, May 2004.

[21] D. Kivanc, G. Li, and H. Liu. Computationally efficient bandwidth allocation and power control for OFDMA. *IEEE Transactions on Wireless Communications*, 2(6):1150–1158, November 2003.

[22] E. G. Larsson. On the combination of spatial diversity and multiuser diversity. *IEEE Communications Letters*, 8(8):517–519, August 2004.

[23] G. Li and H. Liu. Downlink dynamic resource allocation for multi-cell OFDMA system. In *Proceedings, IEEE Vehicular Technology Conference*, pp. 1698–1702, October 2003.

[24] G. Li and H. Liu. On the optimality of the OFDMA network. *IEEE Communications Letters*, 9(5):438–440, May 2005.

[25] C. Mohanram and S. Bhashyam. A sub-optimal joint subcarrier and power allocation algorithm for multiuser OFDM. *IEEE Communications Letters*, 9(8):685–687, August 2005.

[26] T. C. Ng and W. Yu. Joint optimization of relay strategies and power allocation in a cooperative cellular network. Submitted to *IEEE Journal on Selected Areas in Communications.*Forthcoming.

[27] H. Olofsson, J. Naslund, and J. Skold. Interference diversity gain in frequency hopping GSM. In *Proceedings, IEEE Vehicular Technology Conference*, 1:102–106, August 1995.

[28] S. Pietrzyk and G. J. Janssen. Subcarrier allocation and power control for QoS provision in the presence of CCI for the downlink of cellular OFDMA systems. In *Proceedings, IEEE Vehicular Technology Conference*, pp. 2221–2225, April 2003.

[29] W. Rhee and J. M. Cioffi. Increase in capacity of multiuser OFDM system using dynamic subchannel allocation. In *Proceedings, IEEE Vehicular Technology Conference*, pp. 1085–1089, Tokyo, May 2000.

[30] S. Shakkottai, T. S. Rappaport, and P. Karlson. Cross-layer design for wireless networks. *IEEE Communications Magazine*, pp. 74–80, October 2003.

[31] M. Sharif and B. Hassibi. Scaling laws of sum rate using time-sharing, DPC, and beamforming for mimo broadcast channels. In *Proceedings, IEEE International Symposium on Information Theory*, p. 175, July 2004.

[32] Z. Shen, J. G. Andrews, and B. Evans. Optimal power allocation for multiuser OFDM. In *Proceedings, IEEE Globecom*, pp. 337–341, San Francisco, December 2003.

[33] Z. Shen, J. G. Andrews, and B. L. Evans. Adaptive resource allocation for multiuser OFDM with constrained fairness. *IEEE Transactions on Wireless Communications*, 4(6):2726–2737, November 2005.

[34] S. Shamai, O. Somekh, and B. M. Zaidel. Multi-cell communications: An information theoretic perspective. In *Joint Workshop on Communications and Coding (JWCC)*, Florence, Italy, October 2004.

[35] K. Stamatiou and J. Proakis. A performance analysis of coded frequency-hopped OFDMA [cellular system]. In *Proceedings, IEEE Wireless Communications and Networking Conference*, 2:1132–1137, March 2005.

[36] D. Tse. Multiuser diversity in wireless networks. In *Stanford Wireless Communications Seminar*, www.stanford.edu/group/wcs/. April 2001.

[37] S. Tsumura, R. Mino, S. Hara, and Y. Hara. Performance comparison of OFDM-FH and MC-CDMA in single and multi-cell environments. In *Proceedings, IEEE Vehicular Technology Conference*, 3: 1730–1734, May 2005.

[38] P. Viswanath, D. Tse, and R. Laroia. Opportunistic beamforming using dumb antennas. *IEEE Transactions on Information Theory*, 48(6):1277–1294, June 2002.

[39] C. Wong, R. Cheng, K. Letaief, and R. Murch. Multiuser OFDM with adaptive subcarrier, bit, and power allocation. *IEEE Journal on Selected Areas in Communications*, 17(10):1747–1758, October 1999.

[40] I. Wong, Z. Shen, B. Evans, and J. Andrews. A low complexity algorithm for proportional resource allocation in OFDMA systems. In *Proceedings, IEEE Signal Processing Workshop*, pp. 1–6, Austin, TX, October 2004.

[41] W. Yu, G. Ginis, and J. Cioffi. Distributed multiuser power control for digital subscriber lines. *IEEE Journal on Selected Areas in Communications*, 20(5):1105–1115, June 2002.

[42] H. Zhang, H. Dai, and Q. Zhou. Base station cooperation for multiuser MIMO: Joint transmission and BS selection. In *Proceedings, Conference on Information Sciences and Systems (CISS)*, March 2004.

[43] Y. J. Zhang and K. B. Letaief. Multiuser adaptive subcarrier-and-bit allocation with adaptive cell selection for OFDM systems. *IEEE Transactions on Wireless Communications*, 3(4):1566–1575, September 2004.

Networking and Services Aspects of Broadband Wireless

S o far in Part II of this book, we have discussed only the air-interface aspects of broadband wireless networks. In particular, we have discussed physical (PHY)-layer techniques to transport bits over the air at high rates, and media access (MAC)-layer techniques for sharing the available radio resources among multiple users and services. Those aspects are definitely among the most critical and challenging ones for broadband wireless system design, and, in fact, most of IEEE 802.16e and WiMAX specifications deal with those aspects. But from a standpoint of delivering broadband wireless *services* to end users, there are several other aspects and challenges that require consideration. Some of the additional challenges that need to be addressed include

1. How do we provide end-to-end quality of service (QoS)? After all, quality as perceived by the customer is what is provided by the overall network, not only the wireless air interface.
2. How do we provide call/session control services, particularly for multimedia sessions, including voice telephony? How are these sessions set up, managed, and terminated?
3. How do we provide security services in the network? How can subscribers be assured that their communications are safe, and how can the network be protected from unauthorized use?
4. How do we locate a mobile user and how do we maintain an ongoing session while a user moves from the coverage area of one base station to another?

To answer these questions, we need to go beyond the wireless air-interface and look at broadband wireless systems from an end-to-end network perspective. We need to look at the overall network architecture, higher-layer protocols, and the interaction among several network elements beyond the mobile station and the base station. To be sure, the air interface also plays a role in the answers to these questions.

The purpose of this chapter is to provide an end-to-end network and services perspective to broadband wireless. The first four sections attempt to answer the four questions listed earlier, which pertain to *quality of service (QoS), multimedia session management, security,* and *mobility management.*

Since WiMAX is designed primarily to provide IP-based services—be it data, voice, video, messaging or multimedia—a good part of the discussion in this chapter is around IP-based protocols and architecture and how they are used to meet the end-to-end service requirements. As pointed out in Chapter 1, IP was designed primarily for survivability and not so much for efficiency. IP was also designed for best-effort data and not for supporting services that require QoS. The need to support multimedia and other services with stringent QoS needs has led to new developments in IP protocols and architecture. Developments have also occurred for optimizing IP over a capacity-constrained and unreliable wireless medium. Although significant progress has been made over the past several years, adapting IP to the special challenges of wireless and multimedia services continues to be an area of active research and development. This chapter reviews some of these developments.

The topics covered in this chapter have a very broad scope, and our intent is to provide only a brief overview. More detailed exposition can be found in [5, 34, 48, 65].

7.1 Quality of Service

In this section, we discuss QoS from an end-to-end network perspective. How is QoS provided for communication between the two end points of a broadband wireless packet network, which in addition to the wireless link may include several other links interconnected via routers, switches, and other network nodes? The links between intermediate nodes may use a variety of layer 2 technologies, such as ATM, frame relay, and Ethernet, each of which may have its own methods to provide QoS. It is not our intent to cover how QoS is handled in each of these layer 2 technologies. Instead, we provide a brief overview of the general requirements and methods for providing QoS in packet networks and focus on how this is done end to end using emerging layer 3 IP QoS technologies. Since WiMAX is envisioned to provide end-to-end IP services and will likely be deployed using an IP core network, IP QoS and its interaction with the wireless link layer are what is most relevant to WiMAX network performance.

First, what is QoS? This rather elusive term denotes some form of assurance that a service will perform to a certain level. The performance level is typically specified in terms of throughput, packet loss, delay, and jitter, and the requirements vary, based on the application and service. The form of assurance can also vary from a hard quantitative measure, such as a guarantee that all voice packets will be delivered with less than 100ms delay 99 percent of the time—to a soft qualitative guarantee that certain applications and users will be given priority over others.

Resource limitations in the network is what makes providing assurances a challenge. Although typically, the most-constrained resource is the wireless link, the other intermediate nodes and links that have to be traversed for an end-to-end service also have resource limitations.[1] Each link has its own bandwidth-capacity limits, and each node has limited memory for buffering packets before forwarding. Overbuilding the network to provide higher bandwidth capacity and larger buffers is an expensive and inefficient way to provide quality, particularly when the quality requirements are very high. Therefore, more clever methods for providing QoS must be devised and these methods must take into account the particular needs of the application or service and optimize the resources used. Different applications require a different mix of resources. For example, latency-intolerant applications require faster access to bandwidth resources and not memory, whereas latency-tolerant applications can use memory resources to avoid packets being dropped, while waiting for access to bandwidth resources. This fact may be exploited to deliver QoS efficiently. In short, a QoS-enabled network should provide guarantees appropriate for various application and service types while making efficient use of network resources.

7.1.1 QoS Mechanisms in Packet Networks

Providing end-to-end QoS requires mechanisms in both the control plane and the data plane. Control plane mechanisms are needed to allow the users and the network to negotiate and agree on the required QoS specifications, identify which users and applications are entitled to what type of QoS, and let the network appropriately allocate resources to each service. Data plane mechanisms are required to enforce the agreed-on QoS requirements by controlling the amount of network resources that each application/user can consume.

7.1.1.1 Control Plane Mechanisms

Such mechanisms include *QoS policy management, signaling,* and *admission control.* QoS policy management is about defining and provisioning the various levels and types of QoS services, as well as managing which user and application gets what QoS. Figure 7.1 shows a generalized policy-management system as described by IETF that may be used for managing QoS policies.[2] The components of the system include (1) a policy repository, which typically is a directory containing the policy data, such as username, applications, and the network resources to which these are entitled; (2) policy decision points (PDP), which translate the higher-level policy data into specific configuration information for individual network nodes; (3) policy enforcement points (PEP), which are the data path nodes that act on the decisions made by the PDP; and (4) protocols for communication among the data store, PDP, and PEP. Examples of these protocols include LDAP (lightweight directory access protocol) [30] for communication between data source and PDP, and COPS (common open protocol services) [21] for communication between PDP to PEP.

Signaling is about how a user communicates QoS requirements to a network. Signaling mechanisms may be either static or dynamic. In the static case, the PDP takes the high-level

---

1. Note, though, that unlike wireless, other links are generally considered reliable. Therefore, packets losses there stem mostly from buffer overflow caused by congestion, not from channel-induced bit errors.
2. A similar model is often used for security policies as well.

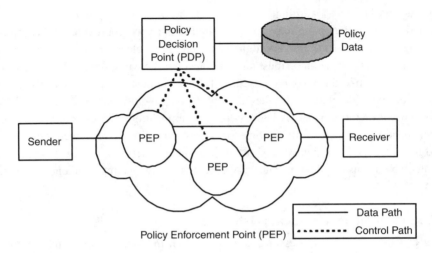

Figure 7.1 A QoS policy-management system

policy information in the policy data and creates configuration information that is pushed down to each PEP that enforces the policies. Policy data is usually created based on service-level agreements (SLA) between the user and the network provider. In the dynamic case, QoS requirements are signaled by the user or application as needed just prior to the data flow. RSVP (resource reservation protocol) is a protocol used for such signaling and is covered in Section 7.1.2. When a request for a certain QoS arrives at the PEP, it checks with the PDP for approval, and, if accepted, allocates the necessary resources for delivering the requested QoS.

Admission control, the other important control plane function, is the ability of a network to control admission to new traffic, based on resource availability. Admission control is necessary to ensure that new traffic is admitted into the network only if such admission will not compromise the performance of existing traffic. Admission control may be done either at each node on a per-hop basis, just at the ingress-edge node, or by a centralized system that has knowledge of the end-to-end network conditions.

7.1.1.2 Data Plane Mechanisms

These methods enforce the agreed-on QoS by classifying the incoming packets into several queues and allocating appropriate resources to each queue. Classification is done by inspecting the headers of incoming packets; resource allocation is done by using appropriate scheduling algorithms and buffer-management techniques for storing and forwarding packets in each queue.

There are fundamentally two different approaches to how these queues are defined. The first approach called *per-flow* handling, is to have a separate queue for each individual session or flow. In this case, packets belonging to a given session or flow need to be uniquely identified. For IP traffic, this is typically the five fields in the IP header: source and destination IP addresses, source and destination port addresses, and transport-layer protocol fields. The IntServ methods defined by the IETF use per-flow handling of IP packets. From an end user perspective, per-flow handling tends to enhance the experienced quality, since a given session is granted resources independent of other sessions. Per-flow handling, however, requires that each network node keep state of individual sessions and apply independent processing, which becomes very difficult or impractical when the number of flows becomes very large, particularly in the core of the network.

The second approach is to classify packets into a few different generic classes and put each class in a different queue. This approach is called *aggregate* handling, since queues here will consist of packets from multiple sessions or flows. Here again, some form of identification in the packet header is used to determine which aggregate class the packet belongs to. DiffServ and 802.1p are examples of aggregate traffic-handling mechanisms for IP and Ethernet packets, respectively. Aggregate handling reduces the state maintenance and processing burden on network nodes and is much more scalable than per-flow methods. The user-experienced quality, however, may be somewhat compromised, since it is affected by traffic from others.

7.1.1.3 Tradeoffs

Both control plane and data plane mechanisms involve trade-offs. Higher complexity in both cases can provide better QoS guarantees. In the control plane, for example, admission-control decisions and resource-allocation efficiency can be improved if the user signals the requirements in greater detail to the network. This, however, increases the signaling load. Enforcing fine-grained QoS requirements increases the complexity of the data plane mechanisms, such as scheduling and buffer management. Network designers need to strive for reducing unnecessary complexity while delivering meaningful QoS.

7.1.2 IP QoS Technologies

So far, we have covered general QoS principles as applied to a packet network. We now describe some of the emerging protocols and architecture for delivering QoS in an IP network. As already mentioned, traditional IP networks were designed for best-effort data and did not include any provision for QoS. Some form of QoS can be provided by relying on different end to end transport-layer protocols that run over IP. For example, TCP (transport control protocol) ensures that data is transferred end-to-end reliably without errors.[3] Similarly, RTP (real time transport protocol) ensures that packets are delivered in sequence and in a manner that allows for continuous playout of media streams. These transport-layer protocols, however, do not have any mechanism for controlling the end-to-end delay or throughput that is provided by the network. For ensuring end-to-end latency and throughput, QoS mechanisms need to be in place in the network layer, and traditional IP did not have any.

Recognizing this deficiency, the IETF developed a number of new architectures and protocols for delivering end-to-end QoS in an IP network. Three of the more important developments are (1) *integrated services (IntServ)*, (2) *differentiated services (DiffServ)*, and (3) *multiprotocol*

---

3. TCP has performance issues when operating in a wireless link. We cover these in Section 7.5.

label switching (MPLS). We briefly cover each of these now. These three developments together will likely transform the traditional best-effort, free-for-all IP network into a QoS-capable and more manageable network.

7.1.2.1 Integrated Services Architecture

The IntServ architecture is designed to provide hard QoS guarantees on a *per-flow* basis with significant granularity by using end-to-end dynamic signaling and resource reservation throughout the IP network. The architecture supports three types of QoS.

1. *Guaranteed services* provide hard guarantees on quality, including quantified upper bounds on end-to-end delay and jitter and zero packet loss owing to buffer overflows. This service aims to emulate a dedicated rate circuit-switched service in an IP network.
2. *Controlled load services* provide qualitative guarantees aimed at approximating the service a user would experience from a lightly loaded best-effort network. This service provides a guaranteed sustained rate but no assurance on delay or packet loss.
3. *Best-effort services* provide no guarantees and require no reservation.

IntServ uses the *resource reservation protocol* (RSVP) for signaling end-to-end QoS requirements and making end-to-end resource reservations. RSVP messages carry information on how the network can identify a particular flow, quantitative parameters describing the flow, the service type required for the flow, and policy information, such as user identity and application.

The quantitative parameters describing the flow are specified using the *TSpec* (traffic specifications) [59] standard. Service guarantees are provided if and only if the packets in the traffic flow conform to the parameters in the TSpec. TSpec characterizes traffic by using a token-bucket model with the following parameters: peak rate (p), minimum policed unit[4] (m), maximum packet size (M), bucket depth (b), and bucket rate (r). A flow is considered conforming to the TSpec as long as the amount of traffic generated in any time interval t is less than $min[(M + pt), (b + rt))]$. In essence, the user can send up to b bytes of data at its full peak rate of p but must lower its rate down to r after that. TSpec is used by the sender to characterize its traffic; a similar specification, called *FlowSpec*, is used by the receiver to describe the profile of the traffic it would like to receive. For controlled load services, FlowSpec parameters are the same as that of TSpec, though the values may be different. For guaranteed service, it is TSpec parameters plus a rate (R) and slack (S) parameter, where $R \geq r$ is the rate required and S is the difference between the desired delay and the delay that would be achieved if the rate R were used. A node may use S to reduce the amount of resources reserved for the flow, if necessary.

Here is how RSVP works [11]. The transmitting application sends a PATH message toward the receivers. PATH messages include a TSpec description of the data the transmitter wishes to send and follows the path that the data will take. All RSVP-aware nodes in the data path establish state for the flow and forward it to the next router if it can support the request. RSVP states are

---

4.   This implies that packets smaller than m bytes are treated as if they are m bytes long.

soft and need to be periodically refreshed. Each router may also advertise its capabilities, such as link delay and throughput, through another object, called *ADSpec* (advertised specifications). Receivers respond to the PATH message by sending an RESV message with a QoS request back to the sender via the same path. The RESV message contains a FlowSpec that indicates back to the network and the sender application the profile of the traffic the receiver would like to receive. The receiver may use ADSpec to ensure that it does not make a request that exceeds the advertised capabilities of the network. All RSVP-aware nodes that receive an RESV message verify that they have the resources necessary to meet the QoS requested. If resources are available, they are allocated, and the RESV message is sent forward to the next node toward the sender. If a node cannot accommodate the resource request, a rejection is sent back to the receiver. Each node makes its own admission-control decisions. RSVP also facilitates admission control based on network policies: Nodes may also extract policy information from PATH/RESV messages and verify them against network policies. This verification may be done using the COPS protocol [21] under the model described in Figure 7.1.

Note that it is the receiver that specifies the required QoS. This is done not only because the receiver usually pays for the service but also to accommodate multicast reception, where different users may receive different portions or versions of the service. Multicast is further facilitated by allowing RESV messages from multiple receivers to be "merged" as they make their way from multiple receivers back to the sender. It should also be noted that RSVP makes reservations only in one direction and therefore requires two separate reservation for two-way QoS.

Although the IntServ architecture with RSVP provides the highest level of IP QoS guarantee, it does have some major limitations. First, it uses per flow traffic handling and therefore suffers from the attendant scalability issues. Imagine having to control flows associated with millions of individual sessions in the core of the network. Second, the need for periodic refreshing of the soft state information can be an intolerable overhead in large networks. Third, since RSVP does not run over a reliable transport protocol, such as TCP, signaling messages may be lost. Fourth, IntServ and RSVP are relatively complex to provision and implement. Fifth, RSVP also requires an authentication infrastructure to ensure the validity of reservation requests. Although some of these issues are being addressed, they have rendered IntServ unusable in large IP core networks. They are, however, quite effective in smaller networks. An alternative architecture, DiffServ, overcomes some of the issues with IntServ.

7.1.2.2 Differentiated Services

Recognizing the scalability problems that prevented the widespread deployment of IntServ, IETF started developing a new model in 1997 to provide QoS without the overhead of signaling and state maintenance. Called differentiated services, or DiffServ, the new model relies on aggregate traffic handling, not the per flow traffic handling used in IntServ. DiffServ divides the traffic into a small number of classes and treats each class differently. DiffServ uses the previously ignored Type of Service (TOS) field in the IP header for marking the packets to a particular class. The marking is a 6-bit label called DiffServ code point (DSCP), as shown in Figure 7.2.

Figure 7.2 also shows a collection of routers that make up a DiffServ network domain. Typically, a user or an application sending traffic into a Diffserv network marks each transmitted packet with the appropriate DSCP. The ingress-edge router classifies the packets into queues, based on the DSCP. The router then measures the submitted traffic for conformance to the agreed-on profiles[5] and, if packets are found nonconforming, changes the DSCP of the offending packets. The ingress-edge router may also do traffic shaping by delaying or dropping the packets as necessary. In a DiffServ network, the edge router does admission control and ensures that only acceptable traffic is injected into the network. All other routers within the DiffServ network simply use the DSCP to apply specific queuing or scheduling behavior—known as a *per hop behavior* (PHB)—appropriate for the particular class.

A number of PHBs may be defined and enforced throughout a DiffServ network. For example, a PHB may guarantee a minimum fraction of available bandwidth to a particular class. The IETF has standardized two PHBs.

1. *The expedited forwarding* (EF) PHB is defined in RFC (request for comments) 2598 [32]. Packets marked for expedited forwarding are given the highest priority. Each router is required to allocate a fixed minimum bandwidth on each interface for EF traffic and forward the packets with minimal delay. EF is typically used to emulate a virtual circuit for delay-sensitive applications. To avoid EF traffic being dropped or delayed, the edge router should ensure that sufficient resources are available before admitting inside the DiffServ network. Packets may be dropped if a user exceeds the agreed-on peak rate.

2. *Assured forwarding* (AF) is a group of PHBs defined in RFC 2597 [29]. AF has four independent classes, each having three levels of drop precedence. Each class is allocated bandwidth separately, but none are guaranteed. If buffers allocated for a given class get filled up, packets will be discarded from that class, based on the level of drop precedence. Here again, it is the job of the ingress router to mark the traffic with the appropriate class and precedence levels. For example, the ingress router may mark packets with different levels of drop precedence, based on how well they conform to the service-level agreements (SLAs).

It should be noted that PHBs are individual forwarding rules applied at each router and by themselves do not make any guarantees of end-to-end QoS. It is, however, possible to ensure end-to-end QoS within the DiffServ domain by concatenating routers with the same PHBs and limiting the rate at which packets are submitted for any PHB. For example, a concatenation of EF PHBs along a prespecified route, with careful admission control, can yield a service similar to a constant bit rate virtual circuit suitable for voice telephony. Other concatenations of PHBs may yield a service suitable for streaming video, and so forth.

---

5. Traffic profiles are typically agreed-on a priori using service-level agreements (SLAs) between the network provider and the user. QoS signaling, such as RSVP, is not typically used with DiffServ.

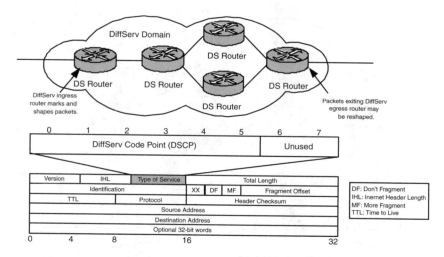

Figure 7.2 Differentiated services network and DSCP

Although it lacks the degree of service assurance and granularity offered by IntServ, DiffServ does offer good QoS that is scalable and easy to deploy. DiffServ mechanisms will likely be deployed for achieving QoS, particularly in the core of large IP networks. It is also possible to build a network that is made up of IntServ regions on the edges and DiffServ in the core, so as to get the best of both architectures. In this case, IntServ signaling, service definition, and admission control are maintained, with all flows mapped onto a few DiffServ classes at the boundary between the edge and the core. As far as IntServ is concerned, the entire DiffServ core is treated like a single logical link, which is realized by tunneling all IntServ traffic through the DiffServ core.

7.1.2.3 Multiprotocol Label Switching

MPLS is another recent development aimed at improving the performance of IP networks [29]. Originally developed as a method for improving the forwarding speed of routers, MPLS is now being used as a traffic engineering tool and as a mechanism to offer differentiated services. MPLS also allows for tighter integration between IP and ATM, improving the performance of IP traffic over ATM networks.

The basic idea behind MPLS is to insert between the layer 2 and IP headers of a packet a new fixed-length "label" that can be used as shorthand for how the packet should be treated within the MPLS network (see Figure 7.3). Within an MPLS network, packets are not routed using IP headers but instead are switched using the information in the label.

Figure 7.3 shows the components of an MPLS network. The router at the ingress edge of an MPLS network is called the ingress label-edge router (LER) and is responsible for inserting the label into each incoming packet and mapping the packet to an appropriate forward equivalence class (FEC). All packets belonging to an FEC are routed along the same path, called the *label switched path* (LSP) and given the same QoS treatment. The LSP is fixed prior to the data transmission via manual configuration or using signaling protocols. The intermediate routers, called

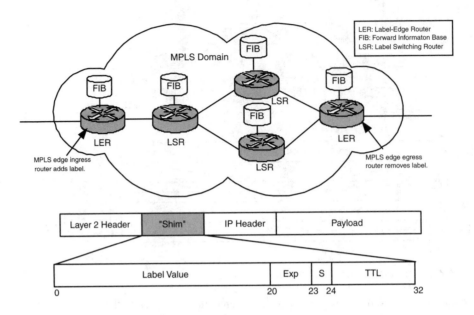

Figure 7.3 MPLS network and components

label switching routers (LSR), maintain a forward information base (FIB) and forward MPLS packets by looking up the next hop in the FIB. It should be noted that the "label" has only local significance and that it is replaced by the LSR with a new label as packets are forwarded from one node to another along the LSP. This is similar to the virtual path identifier/virtual circuit identifier (VPI/VCI) concept used in ATM. In fact, an ATM switch could be used as an LSR, where the label is simply the VPI/VCI. The label swapping and forwarding continue until the packet reaches the egress-edge router, where the label is deleted before forwarding to a non-MPLS node.

By having predetermined paths, MPLS speeds up the forwarding process, albeit at the cost of additional processing at the edge router that converts IP packets to MPLS packets. MPLS can also be used to alleviate congestion through traffic engineering (TE). Unlike traditional IP networks that route traffic automatically through the shortest path, MPLS can route traffic through engineered paths that balance the load of various links, routers, and nodes in the network. Along with such signaling protocols as RSVP-TE or LDP-CR,[6] it is possible in an MPLS network to compute paths with a variety of constraints and to reserve resources accordingly. Dynamic traffic management allows the network to operate closer to its peak efficiency. It is also possible to engineer paths for specific applications—for example, to set up dedicated circuits by configuring permanent LSPs for voice traffic or for virtual private network (VPN) applications.

6. RSVP-TE is an extension of RSVP for use in traffic engineering applications. LDP-CR stands for Label Distribution Protocol-constraint-based routing.

Although not by itself an end-to-end IP QoS mechanism, MPLS does provide a good infrastructure over which IP QoS may be implemented. Both IntServ and DiffServ mechanisms may be implemented on an MPLS infrastructure, though MPLS-DiffServ is a more common choice. MPLS, however, breaks the end-to-end principle of IP protocols and puts control in the hands of the network operator.

7.2 Multimedia Session Management

A *session* may loosely be defined as a set of meaningful communications between two or more users or devices over a limited time duration. In the context of multimedia communications, the term *session* includes voice telephony, audio and video streaming, chat and instant messaging, interactive games, virtual reality sessions, and so on. A session may also have multiple connections associated with it; for example, a video conference, in which the audio and video parts are separate connections.

Session management encompasses more than transfer of bits from a transmitter to a receiver. It includes support for locating and getting consent from the parties involved in the communication, negotiating the parameters and characteristics of the communication, modifying it midstream as necessary, and terminating it. For traditional IP data applications, such as Web browsing and e-mail, session management is rather simple. For example, for a Web download, a DNS (domain name server) is used to identify the appropriate Web site, TCP is used to reliably transfer the content, and the application itself—hypertext transfer protocol (HTTP)—is used to provide basic session management. Session management follows a "one size fits all" policy, with everyone pretty much getting to view the same Web pages without being able to specify preferences in any meaningful way.[7] IP multimedia communications, however, need a more robust session-management scheme, primarily because of the need to support a large variety of applications and terminals. Such session management tasks as capabilities negotiation become very important when different terminals support different encoding schemes, for example. Or say, if one party wants to listen to the audio while others receive both video and audio of a multicast stream.

Clearly, there is a need for a session-control protocol to support multimedia services, including telephony using IP. The ITU standard H.323 was the protocol that traditionally served this purpose in most IP telephony and multimedia systems. Recently, a much more simple and lightweight protocol, called the *session initiation protocol* (SIP), has emerged as the leading contender for this task, and it will likely become the standard session control protocol used in WiMAX networks. SIP has already been chosen as the session control protocol for third generation (3G) cellular networks (see Sidebar 7.1). Also needed is a transport-layer protocol that meets the requirements of multimedia communications. Real-time transport protocol (RTP) was designed for this purpose. SIP and RTP work well together to provide the session-control and media-transport functions required for IP multimedia sessions. We provide a brief overview of these two protocols.

---

7. Strictly speaking, MIME types are used to tell the browser the type of information it receives. But a browser couldn't choose if it wanted a .gif or .jpg file, for instance.

7.2.1 Session Initiation Protocol

SIP is a transaction-oriented text-based application-layer protocol that runs over IP [55]. When compared to H.323, SIP is designed as a flexible, lightweight protocol that is extensible, easy to implement, and quite powerful. Its design philosophy was to decouple the signaling protocol from the service itself and thereby make it useful for a range of unknown future services as well. A partial list of currently supported services includes multimedia call establishment, user mobility, conference call, multicast, call redirection and other supplementary services, unified messaging, presence detection, and instant messaging. SIP integrates well with other IP-based protocols to provide full multimedia session capabilities. For example, it may use RTP for media exchange, transport-layer security (TLS) for security, session description protocol (SDP) for session description, and DNS for discovery. SIP can run over a variety of transport protocols: TCP, user datagram protocol (UDP), stream control transport protocol (SCTP), and TLS over TCP. Obviously, media streams, such as voice and video for real-time communications, use UDP rather than TCP owing to delay constraints.

An important feature of SIP is its programmability. SIP follows the HTTP programming model, which allows users and third-party providers to develop SIP-based customized services rather easily. Many arbitrary services can be built on top of SIP. For example, one could build a service to redirect calls from unknown callers during office hours to a secretary or reply with a Web page if unavailable. Call-control services, such as third-party call control, that are very difficult to implement in traditional intelligent network (IN)–based circuit-switched networks are very easy to set up using SIP. SIP programming may be done using call-processing language (CPL), common gateway interface (CGI), and application programming interfaces (APIs), such as JAIN[8] and Parlay/OSA.[9] The call/session processing logic in SIP may live in either the network or the end devices, depending on the particular application. For example, call distribution may be implemented in the network, distinctive ringing may be implemented in the end device, and forward-on-busy may be implemented in both places.

7.2.1.1 SIP Components and Architecture

The basic components of a SIP architecture are illustrated in Figure 7.4.

- SIP end points are called *user agents (UA)* and are responsible for making or responding to calls on behalf of a SIP user or application. Every SIP user or application is given a SIP URL (universal resource locator) that resembles an e-mail address: sip: user-name@domainname. The UA acts as either a client or a server, depending on whether it is generating requests or responding to requests on behalf of the user. The UA is typically implemented in the subscriber terminal but may also be on an application server located elsewhere—for example, a video server in the network.

---

8. JAIN is Java for Advanced Intelligent Network.
9. OSA is Open Systems Access.

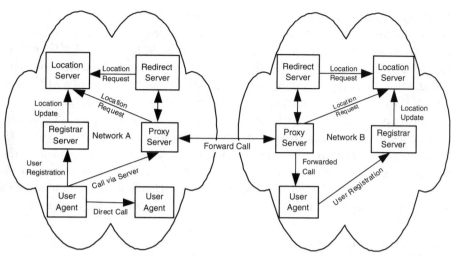

Figure 7.4 Basic SIP architecture

- A SIP *proxy server* relays session signaling and acts as both client and server. The proxy server typically operates in a transactional manner and does not keep the session-state information. This makes it extremely scalable and robust. It may, however, be required to keep state information for certain applications. Proxy servers may rewrite parts of a SIP message before relaying, if required. For example, if a user has moved to a new location, the proxy server may need to change the destination address. The proxy server determines the current location of the user by querying the location server. The proxy server may also provide authentication and security services as needed and interact with other proxies belonging to different SIP domains.

- A SIP *redirect server* responds to a UA request with a redirection response indicting the current location of the called party. The UA has to then establish a new session to the indicated location. This function is analogous to that of a DNS server, which provides the current IP address for a given URL.

- A *registrar server* is where a SIP UA registers its current location information and preferences. The location information typically includes the current IP address of the SIP UA but may also have additional link-layer-specific details, such as base station identity or access router identity. The registration message also includes the transport protocol to be used—such as TCP or UDP—port number, and optional fields, such as timestamp, validity period, and contact preferences.

- The redirect or proxy server contacts the *location server* to determine the current location of the user. The location server may be colocated with other servers, such as the registrar server. As shown in Figure 7.4, SIP users may initiate a session by directly contacting one another or via the proxy server.

7.2.1.2   SIP Transactions and Session-Control Examples

SIP has only six basic message types, called *methods*, defined in RFC 3261 but allows a large number of extensions [55]. The basic messages are INVITE, ACK, CANCEL, OPTIONS, REGISTER, and BYE. Table 7.1 lists some of the main methods used in SIP and describes their functions. SIP also defines a range of extensible response codes that a recipient of a SIP message can use to respond. The response-code classes and examples are listed in Table 7.2.

The transactions involved in a simple SIP call are illustrated in Figure 7.5. The call setup involves a simple three-way handshake. First, the initiator user agent A sends an INVITE message to the recipient. The recipient can respond to the INVITE message with a range of responses: provisional or final, such as BUSY, DECLINED, QUEUED. In the example shown, user agent B responds with a provisional RINGING information message, followed by a call-accepted 200 OK message. Finally, user agent A sends an ACK back to the recipient to complete the call setup.

Using this simple mechanism, SIP can establish calls within 1.5 times round-trip time, whereas H.323 protocol typically would take 3–7 round-trip times for call setup. Once the session is established, multimedia packets can begin to flow between the two end points, using RTP. It should be noted that the media flow and session control are decoupled and may travel by different routes. To terminate the call, a BYE message is sent, and a final 200 OK response is received to complete the call termination.

The structure of the INVITE message is also provided in Figure 7.5 to show how a typical SIP message looks. The message body here is made of a session description protocol (SDP), which contains information about the session, such as origin (o), subject (s), media type (m), and attributes (a), which can be negotiated during setup. SIP and SDP go together in most applications.

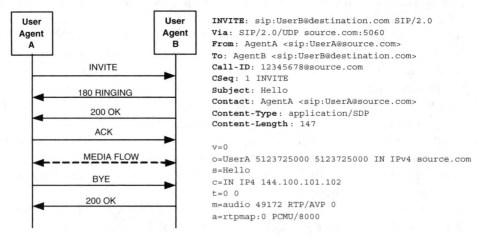

Figure 7.5 A simple call setup using SIP

Table 7.1 The Main SIP Methods

Method	Functional Description	Reference
ACK	Acknowledge receiving an INVITE	RFC 3261
BYE	Terminate sessions	RFC 3261
CANCEL	Cancel sessions and pending transactions	RFC 3261
INFO	Signaling during a session	RFC 2976
MESSAGE	Short instant messages without having to establish a session	RFC 3428
NOTIFY	Notifying user of event	RFC 3265
OPTIONS	Check for capabilities	RFC 3261
PRACK	Provisional acknowledgment	RFC 3262
REFER	Instruct user to establish session with a third party	RFC 3515
REGISTER	Register URL with a SIP server	RFC 3261
SUBSCRIBE	Allow a user to request notification of events	RFC 3265

Table 7.2 SIP Response-Code Classes with Examples

Class	Type	Functional Description	Examples
1XX	Provisional	Request was received and is being processed	100 Trying 180 Ringing, 181 Forwarding, 182 Queueing, 183 In Progress
2XX	Success	Request was successful	200 OK
3XX	Redirection	Sender redirected to try another location	301 Moved Permanently 302 Moved Temporarily
4XX	Client error	Syntax error in message	401 Unauthorized, 404 Not Found, 420 Bad Extension, 486 Busy
5XX	Server error	Problem with server	503 Service unavailable
6XX	Global error	Called party information error	600 Busy, 603 Decline

Figure 7.5 shows a direct call set up between two SIP user agents, but the same can be done using similar messages via a proxy server. Figure 7.6 shows a session setup using a proxy server, with the additional function of *forking*, whereby a SIP server attempts to set up a session with multiple devices or user agents associated with a particular user. In Figure 7.6, it is used for simultaneous ringing and establishing the session with the user agent that accepts the session first, while canceling the other.

7.2.1.3   Other SIP Uses

So far, we have used simple call session setups to illustrate SIP transactions. It should, however, be emphasized that SIP can be used to build much more complex signaling for a wide range of multimedia session-control applications.

SIP can also be used for QoS signaling and mobility management. For example, invitations may indicate in SDP that QoS assurance is mandatory. In this case, call setup may proceed only if the QoS preconditions are met. A SIP extended method, COMET, can be used to indicate the success or failure of the precondition. In SIP, users are identified by a SIP URL but are located via an IP address, phone number, e-mail address, or a variety of other locators. This separation of user identity from contact location address allows SIP to be used to provide *personal mobility*—a person moving to a different terminal, but still remaining in contact; *session mobility*—maintaining an ongoing media session while changing terminals; and *service mobility*—getting access to user services while moving or changing devices and/or networks. SIP may be also used, albeit suboptimally, for *terminal mobility* (maintaining session while the terminal changes its point of attachment to the network). For example, in the middle of a session, if a terminal moves from one network to another, a SIP client could send another INVITE request to the correspondent host with the same session ID but with an updated session description that includes the new IP address. SIP does not by itself ensure seamlessness in the sense that packets are not lost or that the transfer of connection is quick. That depends on lower-layer network detection, selection, and hand-off functions.

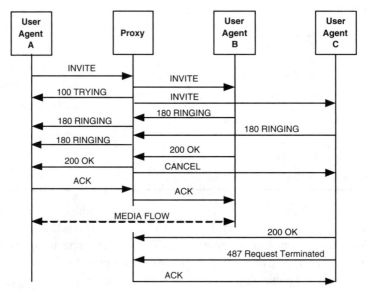

Figure 7.6 A simple call setup and termination using SIP

Sidebar 7.1 The IPMS (IP Multimedia Subsystem) Architecture

IMS is the first standards-based next-generation architecture that fully exploits the flexibility offered by IP and SIP [48]. Developed for 3G wireless networks by the Third-Generation Partnership Project (3GPP), IMS is access network independent and is likely to be deployed by fixed-line providers and WiMAX operators as well. The IMS architecture divides the network into three layers: a *media and end-point layer* that transports the IP bearer traffic, a SIP-based *session-control layer*, and an *application layer* that supports open interfaces. By decoupling session control, transport, and applications, IMS provides several advantages.

1. It can serve as a common core network for a variety of access networks, including fixed-line and wireless networks.
2. It facilitates sharing of common resources, such as subscriber data bases, authentication, and billing, across all services.
3. It provides an open interface for rapid application development by third parties using standards-based interfaces, such as Open Services Gateway (OSA) and Parley.
4. It provides a very flexible platform for delivering a variety of IP services, including integrated converged services—services that innovatively combine voice, video, data, conferencing, messaging, and push-to-talk—that can be delivered on a variety of devices and networks, using presence and location information.

Figure 7.7 shows a layered representation of the various elements in a network with IMS. The media and end-point-layer network elements are user equipment (UE), which contains a SIP user agent; media gateway (MG), which supports media conversion and processing (codecs), and interworks between legacy circuit streams and packet streams; media gateway control function (MGCF), which maintains call states and controls multiple media gateways; and media resource function, which mixes and processes multiple media streams.
The session-control-layer network elements are

- The *call session control functions* (CSCF), which are essentially SIP servers used for controlling the sessions, applying policies, and coordinating with other network elements. The proxy CSCF (P-CSCF) is the entry point to IMS for devices, interrogating CSCF (I-CSCF) is the entry point to IMS from other networks, and the serving CSCF (S-CSCF) is the ultimate session-control entity for end points.
- The *breakout gateway control function* (BGCF) selects the network to which a PSTN breakout occurs.
- The *home subscriber server* (HSS) is the master database with subscriber profiles used for authentication and authorization.
- The *domain name system* (DNS) translates between SIP URI, IP addresses and telephone numbers.
- Charging and billing functions.

The application-layer elements could include SIP application servers; OSA gateway, which can interface to parley application servers and Web-based services.

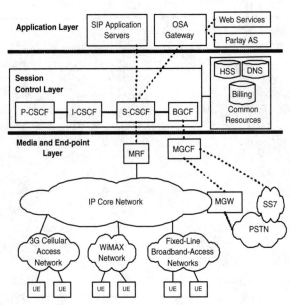

Figure 7.7 The 3GPP IMS architecture

7.2.2 Real-Time Transport Protocol

SIP provides the necessary session-control functions but is not used for transporting the media stream. RTP, defined in RFC 1889 [57], is the most popular transport protocol used for transferring data in multimedia sessions. RTP was developed because traditional transport protocols, such as TCP and UDP, are not suitable for multimedia sessions: TCP offers no delay bounds, and UDP does not guarantee delay or packet loss. RTP typically runs over UDP and provides ordering and timing information suitable for real-time applications, such as voice and video, over both unicast and multicast network services. The RTP header contains content identification, the audio/video encoding method, sequence numbers, and timing information to ensure that packets are played out in the right order and at a constant rate. The timing information facilitates jitter calculation that allows receivers to adopt appropriate buffering strategies for smooth play-out. RTP is implemented along with RTCP (real-time control protocol), which manages the traffic flow. RTCP provides feedback on the quality of the link, which can be used to modify encoding schemes, if necessary. By using timing information, RTCP also facilitates synchronization of multiple streams, such as audio and video streams associated with a session. Synchronization across multiple sources, however, requires use of the network timing protocol (NTP). RTCP also provides support for real-time conferencing of groups. This support includes source identification and support for audio and video bridges, as well as multicast-to-unicast translators. RTP and RTCP do not reduce the overall delay of the real-time information or make any guarantees concerning quality of service.

7.3 Security

Security is a broad and complex subject, and this section provides only a brief introduction to it. We cover the basic security issues, introduce some terminology, and provide a brief overview of some of the security mechanisms, using examples that are relevant to broadband wireless services, especially WiMAX.

A well-designed security architecture for a wireless communication system should support the following basic requirements:

- **Privacy:** Provide protection from eavesdropping as the user data traverses the network from source to destination.
- **Data integrity:** Ensure that user data and control/management messages are protected from being tampered with while in transit.
- **Authentication:** Have a mechanism to ensure that a given user/device is the one it claims to be. Conversely, the user/device should also be able to verify the authenticity of the network that it is connecting to. Together, the two are referred to as mutual authentication.
- **Authorization:** Have a mechanism in place to verify that a given user is authorized to receive a particular service.
- **Access control:** Ensure that only authorized users are allowed to get access to the offered services.

Security is typically handled at multiple layers within a system. Each layer handles different aspects of security, though in some cases, there may be redundant mechanisms. As a general principle of security, it is considered good to have more than one mechanism providing protection so that security is not compromised in case one of the mechanisms is broken. Table 7.3 shows how security is handled at various layers of the IP stack. At the link layer, strong encryption should be used for wireless systems to prevent over-the-air eavesdropping. Also needed at the link layer is access control to prevent unauthorized users from using network resources: precious over-the-air resources.

Table 7.3 Examples of Security Mechanisms at Various Layers of the IP Stack

Layer	Security Mechanism	Notes
Link	AES encryption, device authentication, port authentication (802.1X)	Typically done only on wireless links
Network	Firewall, IPsec, AAA infrastructure (RADIUS, DIAMETER)	Protects the network and the information going across it
Transport	Transport-layer security (TLS)	Provides secure transport-layer services, using certificate architecture
Application	Digital signatures, certificates, secure electronic transactions (SET), digital rights management (DRM)	Can provide both privacy and authentication; relies mostly on public key infrastructure

Link-layer encryptions are not often used in wired links, where eavesdropping is considered more difficult to do. In those cases, privacy is ensured by the end-to-end security mechanisms used at the higher layers. At the network layer, a number of methods provide security. For example, IPsec could be used to provide authentication and encryption services. The network itself may be protected from malicious attack through the use of firewalls. Authentication and authorization services are typically done through the use of AAA (authentication, authorization, and accounting) protocols, such as RADIUS (Remote Access Dial-In User Service) [50] and DIAMETER[10] [13]. At the transport layer, TLS—its precedent was called SSL secure sockets layer—may be used to add security to transport-layer protocols and packets [20]. At the application layer, digital signatures, certificates, digital rights management, and so on are implemented, depending on the sensitivity of the application.

In the following subsections, we review a few of the security mechanisms that are relevant to WiMAX. Our focus here is mostly on the concepts involved rather than on the specified implementation detail described in WiMAX and relevant IETF standards.

7.3.1 Encryption and AES

Encryption is the method used to protect the confidentiality of data flowing between a transmitter and a receiver. Encryption involves taking a stream or block of data to be protected, called *plaintext,* and using another stream or block of data, called the *encryption key,* to perform a reversible mathematical operation to generate a *ciphertext.* The ciphertext is unintelligible and hence can be sent across the network without fear of being eavesdropped. The receiver does an operation called *decryption* to extract the plaintext from the ciphertext, using the same or different key. When the same key is used for encryption and decryption, the process is called *symmetric key* encryption. This key is typically derived from a shared secret between the transmitter and the receiver and for strong encryption typically should be at least 64 bytes long. When different keys are used for encryption and decryption, the process is called *asymmetric key* encryption. Both symmetric and asymmetric key encryptions are typically used in broadband wireless communication systems, each serving different needs. In this section, we describe a symmetric key encryption system called AES (*advanced encryption standard*); the next section covers asymmetric key encryption system.

AES is the new data encryption standard adopted by the National Institute of Standards as part of FIPS 197 [41] and is specified as a link-layer encryption method to be used in WiMAX. AES is based on the Rijndael algorithm [17], which is a block-ciphering method believed to have strong cryptographic properties. Besides offering strong encryption, AES is fast, easy to implement in hardware or software, and requires less memory than do other comparable encryption schemes. The computational efficiency of AES has been a key reason for its rapid widespread adoption.

---

10. DIAMETER is not an acronym but a pun on the name RADIUS, implying that it is twice as good.

The AES algorithm operates on a 128-bit block size of data, organized in a 4×4 array of bytes called a *state*. The encryption key sizes could be 128, 192, or 256 bits long; WiMAX specifies the use of 128-bit keys. The ciphering process can be summarized using the following pseudocode:

```
Cipher(input, output, roundkey)
begin
    state = input
    round = 0
    AddRoundKey (state, roundkey[round])
    for round = 1 to 9 in steps of 1
        SubBytes(state)
        ShiftRows(state)
        MixColumns(state)
        AddRoundKey(state,roundkey[round])
    end for
    SubBytes(state)
    ShiftRows(state)
    AddRoundKey(state, roundkey[round+1])
    output = state
end
```

The pseudocode shows the four distinct operations in the encryption process (see also Figure 7.8):

1. **In the *SubBytes* operation**, every byte in the state S is substituted with another byte, using a look-up table called the S-box. The S-box used is derived from the inverse function over $GF(2^8)$,[11] known to have good nonlinearity properties. This operation is the only one that provides nonlinearity for this encryption. Although the S-table can be mathematically derived, most implementations simply have the substitution table stored in memory.
2. **In the *ShiftRows* operation**, each row is shifted cyclically a fixed number of steps. Specifically, the elements of the first row are left as is, the elements of the second row are shifted left by one column, the elements of the third row are shifted left by two columns, and the elements of the last row are shifted left by three columns. This operation ensures that each column of the output state of this step is composed of bytes from each column of the input state.
3. **In the *MixColumns* operation**, each column is linearly transformed by multiplying it with a matrix in finite field. More precisely, each column is treated as a polynomial over $GF(2^8)$ and is then multiplied modulo $x^4 + 1$ with a fixed polynomial $c(x) = 3x^3 + x^2 + x + 2$. This invertible linear transformation, along with the ShiftRows operation, provides diffusion in the cipher.
4. **In the *AddRoundKey* operation**, each byte in the state is XORed with a round key. The AES process includes deriving 11 round keys from the cipher key delivered to the encryption engine. The delivered cipher key itself would be the result of a number of transformations, such as hashing, done on the original master secret key. The 11 round keys are derived from

---

11. Galois, or finite, field.

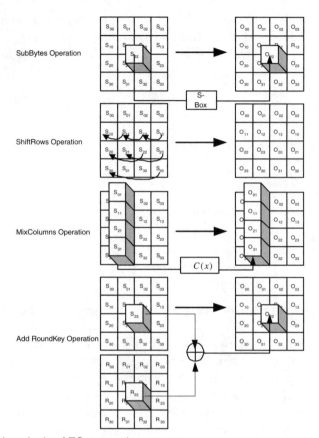

Figure 7.8 Operations in the AES encryption process

the cipher key, using a computationally simple algorithm.

In order to use a block cipher, such as AES, a reversible mechanism is needed to convert an arbitrary-length message into a sequence of fixed-size blocks prior to encryption. The method to convert between messages and blocks is referred to as the cipher's *mode of operation*, several of which are proposed for AES. The mode of operation needs to be carefully chosen so that is does not create any security holes and with implementation considerations in mind. The mode used in WiMAX is called the *counter mode*, an example of which is illustrated in Figure 7.9.

In counter mode, instead of directly encrypting the plain text, an arbitrary block, called the counter, is encrypted using the AES algorithm, and the results are XORed with the plain text to produce the ciphertext. The arbitrary block is called the counter because it is generally incremented by 1 for each successive block processed. In Figure 7.9, the counter starts at 1, but in practice, it can be any arbitrary value. By changing the value of the counter for every block, the

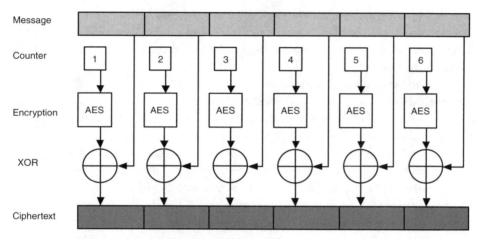

Figure 7.9 An AES counter operating mode

In addition to providing this additional protection, the counter mode has the remarkable property of making the decryption process exactly the same as encryption, since XORing the same value twice produces the original value, making the implementation easier. Counter mode is also suitable for parallel encryption of several blocks. Further, if the message doesn't break into an exact number of blocks, this mode allows you to take the last short block and XOR it with the encrypted block and simply send the required number of bits from the output. These interesting properties make counter mode a popular choice for AES implementation. Both Wi-Fi and WiMAX systems specify the use of AES in counter mode with cipher-block chaining message-authentication code (CBC-MAC). CBC-MAC, a protocol defined in RFC 3610, uses the same encryption key for deriving a message-integrity-check value.

7.3.2 Public Key Infrastructure

With symmetric key encryption, both the transmitter and the receiver need to use the same key, which raises the question of how the key itself can be securely transmitted. One way to do this is to establish the shared secret key a priori via an out-of-band mechanism. For example, a shared secret password could be hardcoded into both the transmitter and the receiver; alternatively, a service provider could give the key to a subscriber at the time of signing up for service. This approach, however, does not scale very well for widespread use. For example, it becomes impossible to generate millions of individual unique keys and deliver them to each person. Also, relying on out-of-band mechanisms is cumbersome, prone to errors, and often not very practical.

Asymmetric key encryption is an elegant solution to the key-distribution problem. Asymmetric key encryption uses two keys: a *public key* and a *private key*. When a ciphertext is encrypted using one of the two keys, it can be decrypted only by the other key. Both the keys are generated simultaneously using the same algorithm—RSA [52]—and the public key is disclosed widely and the private key is kept secret (see Sidebar 7.2). The *public key infrastructure*

are generated simultaneously using the same algorithm—RSA [52]—and the public key is disclosed widely and the private key is kept secret (see Sidebar 7.2). The *public key infrastructure* (PKI), which is widely used to secure a variety of Internet transactions, is built on this idea of using asymmetric keys.

Asymmetric keys are useful for a variety of security applications.

Authentication: Here, we need a mechanism to ensure that a given user or device is as stated. For example, to ensure that the data received is really from user B, user A can use the process illustrated in Figure 7.10, using public and private keys, along with a random number. If B returns A's random number, A can be assured that the message was sent by B and no one else. Similarly, B can be assured that A received the message correctly. The message could not have been read by anyone else and could not have been generated by anyone else, since no other user has the private key or the correct random number.

Shared secret key distribution: To securely send data to user B, user A can do so by using the public key of user B to encrypt the data. Since it now can be decrypted only by the private key of user B, the transaction is secured from everyone else. This secure transaction can now be used to distribute a shared secret key, which can then be used to encrypt the rest of the communication, using a symmetric key algorithm, such as AES. Figure 7.10 also shows how, after mutual authentication, a shared key is established for encrypting the rest of the session.

Nonrepudiation and message integrity: Asymmetric keys and PKI can also be used to prove that someone said something. This *nonrepudiation* is the role often played by signatures on a standard letter. In order to establish nonrepudiation, it is not necessary to encrypt the entire text, which is sometimes computationally expensive and unnecessary. An easier way to guarantee that the text came from the sender and has not been tampered with is to create a message digest from the message and then encrypt the digest, using the private key of the sender. A *message digest* is a short fixed-length string that can be generated from an arbitrarily long message. It is very unlikely that two different messages generate the same digest, especially when at least 128-bit message digests are used. MD-5 [51] and SHA [22] are two algorithms used for computing message digests, both of which are much faster and easier to implement than encryption. By sending the unencrypted text along with an encrypted digest, it is possible to establish nonrepudiation and message integrity.

Digital certificates: Digital certificates are a means of certifying the authenticity and validity of public keys. As part of the public key infrastructure, a certification authority, which essentially is a trusted independent organization, such as VeriSign, certifies a set of public and private keys for use with PKI transactions. The certification authority issues digital certificates that contain the user's name, the expiry date, and the public key. This certificate itself is digitally signed by the certification authority using its private key. The public key of Certification Authorities are widely distributed and known; for example, every browser knows them. In the context of broadband wireless services, subscriber terminals may be issued individual digital certificates that are hardcoded into the device, and can be used for device authentication.

Sidebar 7.2 The Math Behind Asymmetric Key Encryption: RSA Algorithm

Asymmetric key encryption is based on the simple fact that it is quite easy to multiply two large prime numbers but computationally very intensive to find the two prime factors of a large number. In fact, even using a supercomputer, it may take millions of years to do prime factorization of large numbers, such as a 1,024-bit number. It should be noted that although no computationally efficient algorithms are known for prime factorization, it has not been proved that such algorithms do not exist. If someone were to figure out an easy way to do prime factorization, the entire PKI encryption system would collapse.

Here are the steps the RSA (Rivest-Shamin-Adleman) algorithm uses for public/private key encryption [52].

1. Find two large prime numbers p and q such that $N = pq$. N is often referred to as the *modulus*.
2. Choose E, the *public exponent*, such that $1 < E < N$, and E and $(p-1)(q-1)$ are relatively prime. Two numbers are said to be relatively prime if they do not share a common factor other than 1. N and E together constitute the *public key*.
3. Compute D, the *private key*, or *secret exponent*, such that $(DE - 1)$ is evenly divisible by $(p-1)(q-1)$. That is, $DE = 1\{\mod[(p-1)(q-1)]\}$. This can be easily done by finding an integer X that causes $D = (X(p-1)(q-1) + 1)/E$ to be an integer and then using that value of D.
4. Encrypt given message M to form the ciphertext C, using the function $C = M^E[\mod(N)]$, where the message M being encrypted must be less than the modulus N.
5. Decrypt the ciphertext by using the function $M = C^D[\mod(N)]$. To crack the private key D, one needs to factorize N.

7.3.3 Authentication and Access Control

Access control is the security mechanism to ensure that only valid users are allowed access to the network. In the most general terms, an access control system has three elements: (1) an entity that desires to get access: the *supplicant*, (2) an entity that controls the access gate: the *authenticator*, and (3) an entity that decides whether the supplicant should be admitted: the *authentication server*.

Figure 7.11 shows a typical access control architecture used by service providers. Access control systems were first developed for use with dial-up modems and were then adapted for broadband services. The basic protocols developed for dial-up services were PPP (point-to-point protocol) [60] and remote dial-in user service (RADIUS) [50]. PPP is used between the

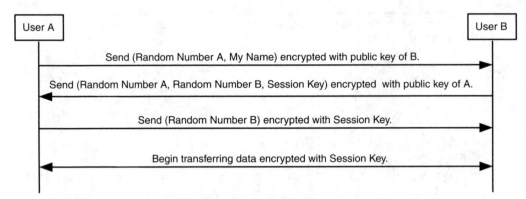

Figure 7.10 Mutual authentication and shared key distribution using PKI

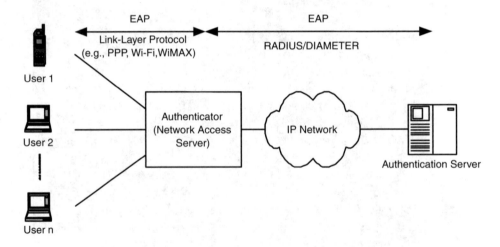

Figure 7.11 Access control architecture

supplicant and the authenticator, which in most cases is the edge router or network access server (NAS), and RADIUS is used between the authenticator and the authentication server.

PPP originally supported only two types of authentication schemes: PAP (password authentication protocol) [37] and CHAP (challenge handshake authentication protocol) [65], both of which are not robust enough to be used in wireless systems. More secure authentication schemes can be supported by PPP using EAP (extensible authentication protocol) [38].

7.3.3.1   Extensible Authentication Protocol

EAP, a flexible framework created by the IETF (RFC 3748), allows arbitrary and complicated authentication protocols to be exchanged between the supplicant and the authentication server. EAP is a simple encapsulation that can run over not only PPP but also any link, including the WiMAX link. Figure 7.12 illustrates the EAP framework.

EAP includes a set of negotiating messages that are exchanged between the client and the authentication server. The protocol defines a set of request and response messages, where the authenticator sends requests to the authentication server; based on the responses, access to the client may be granted or denied. The protocol assigns type codes to various authentication methods and delegates the task of proving user or device identity to an auxiliary protocol, an *EAP method*, which defines the rules for authenticating a user or a device. A number of EAP methods have already been defined to support authentication, using a variety of credentials, such as passwords, certificates, tokens, and smart cards. For example, protected EAP (PEAP) defines a password-based EAP method, EAP-transport-layer security (EAP-TLS) defines a certificate-based EAP method, and EAP-SIM (subscriber identity module) defines a SIM card–based EAP method. EAP-TLS provides strong mutual authentication, since it relies on certificates on both the network and the subscriber terminal.

In WiMAX systems, EAP runs from the MS to the BS over the PKMv2 (Privacy Key Management) security protocol defined in the IEEE 802.16e-2005 air-interface. If the authenticator is not in the BS, the BS relays the authentication protocol to the authenticator in the access service network (ASN). From the authenticator to the authentication server, EAP is carried over RADIUS.

7.3.3.2 RADIUS

The most widely used standard for communication between the authenticator and the authentication server, RADIUS, is an IETF standard [50] that defines the functions of the authentication server and the protocols to access those functions. RADIUS is a client/server UDP application that runs over IP. The authentication server is the RADIUS server, and the authenticator is the RADIUS client. In addition to authentication, RADIUS supports authorization and accounting functions, such as measuring session volume and duration, that can be used for charging and billing purposes. The authentication, authorization, and accounting functions are collectively referred to as AAA functions. Numerous extensions to RADIUS have been defined to accommodate a variety of needs, including supporting EAP.

RADIUS, however, does have a number of deficiencies that cannot be easily overcome by modifications. Recognizing this, the IETF has developed a new standard for AAA functions: DIAMETER [13]. Although not backward compatible with RADIUS, DIAMETER does provide an upgrade path to it. DIAMETER has greater reliability, security, and roaming support than RADIUS does.

7.4 Mobility Management

Two basic mechanisms are required to allow a subscriber to communicate from various locations and while moving. First, to deliver incoming packets to a mobile subscriber, there should be a mechanism to locate all mobile stations (MS)—including idle stations—at any time, regardless of where they are in the network. This process of identifying and tracking a MS's current point of attachment to the network is called *location management*. Second, to maintain an ongoing session as the MS moves out of the coverage area of one base station to that of another, a mechanism to seamlessly transition, or hand off, the session is required. The set of procedures to

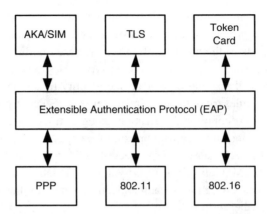

Figure 7.12 IEAP architecture

manage this is called *handoff management*. Location management and handoff management together constitute *mobility management*.

7.4.1 Location Management

Location management involves two processes. The first process is called *location registration*, or *location update*, in which the MS periodically informs the network of its current location, which leads the network to authenticate the user and update its location profile in a database. The databases are usually placed in one or more centralized locations within the network. The location is typically defined by an area that encompasses the coverage area of one or more base stations. Making the location area large reduces the number of location updates. Having every MS, including idle MS, report to the network every time it moves from the coverage range of one BS to another could cause an unacceptable signaling load on the network, particularly when the base stations are microcells and when the number of subscribers is very large. To lighten this burden, service providers typically define larger location areas that cover several base stations. The frequency of location update is also an important consideration. If location update is done infrequently, the MS risks moving out of its current location area without the network being notified, which leads to the network having inaccurate location information about the mobile. To support global roaming, location management must be done not only within a single operator's network but also across several operators tied through roaming agreements.

The second process related to location management is called *paging*. When a request for session initiation, e.g., incoming call, arrives at the network, it looks up the location database to determine the recipient's current location area and then pages all the base stations within and around that area for the subscriber. Obviously, the larger the number of base stations within a defined location area, the greater the paging resources required in the network. Network operators need to make the trade-off between using resources for location update signaling from all the mobile stations versus paging over a large area.

7.4.2 Handoff Management

Compared to location management, handoff management has a much tighter real-time performance requirement. For many applications, such as VoIP, handoff should be performed seamlessly without perceptible delay or packet loss. To support these applications, WiMAX requires that for the full mobility—up to 120kmph scenario—handoff latency be less than 50ms with an associated packet loss that is less than 1 percent.

The handoff process can be thought of as having two phases. In the first phase, the system detects the need for handoff and makes a decision to transition to another BS. In the second phase, the handoff is executed, ensuring that the MS and the base stations involved are synchronized and all packets are delivered correctly, using appropriate protocols.

Handoff decisions may be made by either the MS or the network, based on link-quality metrics. In WiMAX, the MS typically makes the final decision, whereas the BS makes recommendations on candidate target base stations for handoff. The decision is based on signal-quality measurements collected and periodically reported by the MS. The MS typically listens to a beacon or a control signal from all surrounding base stations within range and measures the signal quality. In WiMAX, the base station may also assist in this process by providing the MS with a neighbor list and associated parameters required for scanning the neighboring base stations. The received signal strength (RSS) or signal-to-interference plus noise ratio (SINR) may be used as a measure of signal quality. SINR is a better measure for high-density cellular deployments but is more difficult to measure than is RSS.

Figure 7.13 shows a simple case involving two base stations and an MS moving away from base station A (serving base station) toward base station B (target base station). The minimum signal level (MSL) is the point below which the quality of the link becomes unacceptable and, absent a handoff, will lead to excessive packet loss and the session being dropped. It should be noted that the MSL may vary, depending on the particular QoS needs of the application within the session. For example, a higher-throughput application may have a higher MSL when compared to a low-data-rate application.

Typically, handoff procedures are initiated when the signal drops below a handoff threshold, which is set to be Δ higher than the MSL. Also, handoff is typically executed only if there is another BS for which the received signal quality is at least Δ higher than the MSL. Using a larger Δ will minimize the likelihood of signal dropping below MSL while handoff is in execution.

A good handoff algorithm should minimize handoff failures and avoid unnecessary handoffs. Two metrics often used to assess the performance of handoff algorithms are *dropping probability* and *handoff rate*. Dropping probability quantifies handoff failures, which occur when the signal level drops below the MSL for a duration of time. The handoff rate quantifies how often handoff decisions are made, which depends in part on how frequently measurements are taken and reported back to the network. Measurements, however, consume radio resources and hence reduce the available capacity.

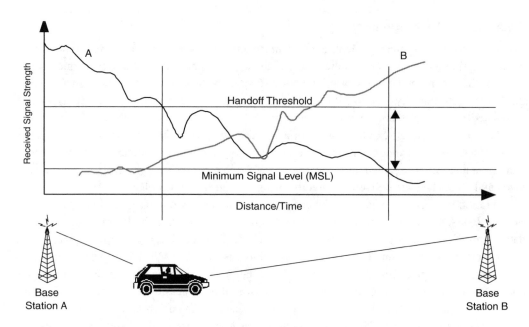

Figure 7.13 Handoff detection based on signal strength

To minimize the dropping probability, the handoff procedures need to be executed quickly, and the Δ set higher so that the likelihood of the signal's dropping below the MSL before handoff execution is minimized. Obviously, setting a large Δ implies a costlier cellular design with larger overlap among cells. Being too quick to hand over may also lead to excessive and unnecessary handoffs, particularly when there is significant signal fluctuations.

Clearly, there is a trade-off between dropping probability and handoff rate. Too few handoffs may lead to dropped calls, and too many handoffs may cause signaling overload and degrade service quality. The nature of the trade-off between dropping probability and handoff rate depends on the signal-fluctuation model and the handoff decision algorithm used. For example, Table 7.4 shows the results of a simulation study reported in [67]. The table illustrates the trade-off between dropping probability and handoff rate for three algorithms, which, respectively, base handoff decisions on (1) instantaneous value of signal level, (2) signal level averaged over past ten samples, and (3) true expected value of signal level. As reported in [67], these results are based on Matlab simulations of an MS moving at 20m/s from one BS to another separated by 1km. A fourth-power exponential decay and lognormal shadow fading with a correlation distance of 50m[12] is used to model the signal. Signal samples are assumed to be taken every 0.5 seconds.

---

12. Defined as the distance at which the signal correlation drops to 0.5.

Table 7.4 Trade-off Between Dropping Probability and Handoff Rate [67]

Handoff Decision Based On:	Dropping Probability with $\Delta = 10$ dB	Dropping Probability with $\Delta = 5$ dB	Dropping Probability with $\Delta = 0$ dB	Number of Handoffs
Instantaneous value of signal level	0.003	0.024	0.09	7.6
Average signal level measured over ten samples	0.014	0.05	0.13	1.8
True expected value of signal level	0.02	0.06	0.14	1

Table 7.4 shows that selecting a BS based on strongest instantaneous value offers the best dropping probability at the cost of large number of handoffs. On the other hand, making handoff decisions based on true expected value leads to increased dropping probability but keeps the number of handoffs at a minimum. While the results shown here are for simple signal-level averaging, more complex schemes that use knowledge of the fading environment to predict impending signal loss may provide better handoff performance. It is, however, quite challenging to make a generalized fading model that fits a variety of environments.

Another common technique used to minimize handing off back and forth between base stations under rapid fading conditions is to build a signal-quality hysteresis into the algorithm. Once a handoff occurs from base station A to B, the handoff threshold to initiate a handoff from B back to A is typically set at a higher value.

Handoff decision making also needs to take into account whether radio resources are available in the target BS to handle the session. To minimize the probability of dropping sessions owing to lack of resources at the target BS, some system designs may reserve a fraction of network resources solely for accepting handoff sessions. Handoff sessions are often given higher priority over new sessions from an admission control standpoint. Providing reservation or prioritization for handoff sessions, however, consumes additional radio resources and leads to a decreased spectral efficiency. A better approach—especially in dense deployments, in which there is often more than one candidate BS to receive a given handoff—is to incorporate radio resource information in the handoff decision. By devising a scheme that favors base stations that have acceptable signal quality and more available resources over the one with the best signal quality and limited resources, it may be possible to mitigate the loss in spectral efficiency. In WiMAX, base stations may communicate their resource availability to one another over the backbone, and this may be used to help the MS select the appropriate target for handover.

Once a decision to hand off an ongoing session to a target BS is made, a number of steps need to be completed to fully execute the handoff. These steps include establishing physical connectivity to the new BS—ranging, synchronization, channel acquisition, and so on—performing the necessary security functions for reassociation with a new BS, and transferring the MAC state from the old BS to the target BS. To make the handover seamless—that is, fast and error free—a number of mechanisms could be used.

- Performing initial ranging and synchronization with neighboring base stations prior to handoff.
- Establishing physical-layer connections with more than one BS at a time so that data transfer can be switched from one to the other without the need for executing a full set of handoff signaling procedures. IEEE 802.16e-2005 supports this functionality, which is called fast base station switching (FBSS). In this case, if all the base stations with which the MS has a connection receive downstream packets from the network destined for the MS, packet loss when switching can be significantly reduced.
- Transferring all undelivered MAC layer packets in the queue of the old BS to the target BS via the backbone to reduce packet loss and/or the need for higher-layer retransmissions (delay). Transferring MAC-layer ARQ states to the target BS can also reduce unnecessary MAC-layer retransmissions.

7.4.3 Mobile IP

The discussion of mobility management thus far has assumed that when the MS moves from the coverage area of one BS to another, all that is needed is to maintain the physical connection such that packets can continue to flow. This, however, is not sufficient. For an application session to stay intact, the IP address of the MS must remain unchanged throughout an application session. If the entire wireless network is architected such that it belongs to a single IP subnet, the IP address of the MS could indeed remain the same across the entire network. In a strictly flat IP architecture, however, we would have the BS itself act as an IP router, and moving across them would mean a change in IP subnets. Even when the base station is not architected to be an IP router, one may move across two BSs that belong to different IP subnets. This is indeed the case when moving across different access service networks[13] in a WiMAX network. When that happens, the IP address of the MS is forced to change. Then the IP connection breaks down, and ongoing application sessions are terminated, though physical connectivity over the air is maintained seamlessly. Movement across subnets will also happen when dealing with heterogeneous wireless networks—for example, when moving from a WiMAX network to a Wi-Fi network or a 3G cellular network. Therefore, a solution is needed to keep an ongoing session intact even when the MS moves across subnet boundaries. Mobile IP (MIP) is the current IETF solution for this problem of IP mobility [46].

Mobile IP is specifically designed as an overlay solution to Internet Protocol version 4 (IPv4) to support user mobility from one IP subnet to another. Mobile IP is designed to be transparent to the application in the sense that applications do not have to know that the user has

---

13. See Chapter 10 for a description of WiMAX network architecture.

moved to a new IP subnet. Mobile IP is also transparent to the network in the sense that the routing protocols or routers need not be changed.

7.4.3.1 Components

Figure 7.14 shows the basic components of mobile IP. The MIP client is implemented in the terminal that is moving (MS in WiMAX) and is referred to as the *mobile node* (MN). The IP host with which the MN is communicating is called the *correspondent node* (CN). Mobile IP defines two addresses for each MN. The first address, the address issued to the MN by its home network, is called the *home address* (HoA). This IP address can be thought of as identifying the mobile to the IP network. The second address, the *care-of address* (CoA), is a temporary IP address that is assigned to the MN by the visited network. This IP address can be thought of as providing information about the current logical location of the MN.

In order to manage mobility, dynamic mapping is needed between the fixed identifier IP address and the CoA. This need is met through the use a mobility agent, the *home agent* (HA), located in the home network, working with another mobility agent, the *foreign agent* (FA), located in the visited network. Both of these mobility agents can be thought of as specialized routers.

There is also the option of colocating the FA with the MN itself; this scenario is referred to as a *colocated foreign agent*. The CoA of the MN is the address of the FA. Whenever the MN moves away from the home network to a visited network, this movement is detected through the use of location-discovery protocols that are based on extensions to ICMP (Internet control message protocol) router discovery protocol [18]. Mobility agents advertise their presence to enable discovery by the MN. Once in a visited network, the MN obtains a new address and sends an update message to the HA, informing it of the address of the new FA. This update registration can be done directly by the MN for a colocated CoA or is relayed by the FA if the visited subnet FA address is used as the CoA.

Once the HA is updated with the new CoA, all packets destined to the MN that arrive at the home network are forwarded to the appropriate FA CoA by encapsulating them in a tunneling protocol. IP-in-IP encapsulation as defined in RFC 2003 [45] is used for this tunneling. Minimal encapsulation (RFC 2004) [47] or GRE (generic routing encapsulation) tunneling (RFC 1701) [56] may optionally be supported as well. The FA decapsulates the packets and delivers them to the MN. By having the HA act as the anchor point for all packets destined to the MN, mobile IP is able to deliver all packets to the MN regardless of its location.

Mobile IP is required only for delivering packets destined to the MN. Packets from the MN can be carried directly without the need for mobile IP, except, of course, if the CN is also mobile, in which case it will have to go through the HA of the CN.

Clearly, packets destined for the MN take a different path from those originating from the MN. This *triangular routing* is illustrated in Figure 7.15. Triangular routing causes some problems and is one of the key limitations of mobile IP.

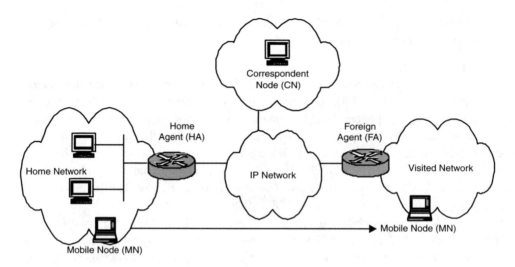

Figure 7.14 Mobile IP components

7.4.3.2   Limitations and Work-arounds

Mobile IP has a number of limitations, most stemming from the use of triangular routing, that make it a suboptimal mobility-management solution. These limitations are as follows:

- **Inefficient routing.** Triangular routing can be extremely wasteful if the mobile roams to a location that is topologically far away from the home network. For example, if a person with a U.S. home network were to access a Web site in Korea while in Korea, packets from the Web site in Korea would have to go the U.S. home agent before being tunneled back and delivered to the user in Korea. One proposed solution to this problem is the optional extension to mobile IP, called *route optimization*, which allows the CN to send packets directly to the MN, in response to a *binding update* from the HA, informing the CN of the MN's new CoA. Implementing route optimization, however, requires changes to the CN's protocol stack, which is not practical in many scenarios.

- **Ingress filtering issues.** Many firewalls do not allow packets coming from a topologically incorrect source address. Since the MN uses its home address as its source address even when in a visited network, firewalls in the visited network may discard packets from the MN. A solution to this problem is a technique called *reverse tunneling*, which is another optional extension to mobile IP [40]. This solution requires that the MN establish a tunnel from the CoA to the HA, where it can be decapsulated and forwarded to the correct CN. (Figure 7.16).

- **Private address issues.** Mobile IP does not allow for the use of private addressing using network address translation (NAT), wherein one public IP address is shared by many nodes using different port numbers. Since packets are tunneled from the HA (and CN, in the case of route optimization) using IP-in-IP encapsulation to the MN's publicly routable

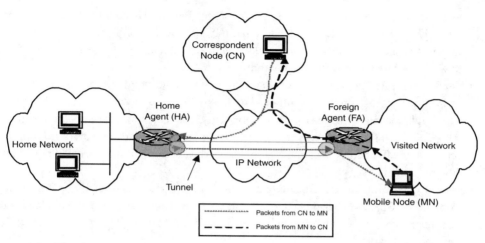

Figure 7.15 Triangular routing

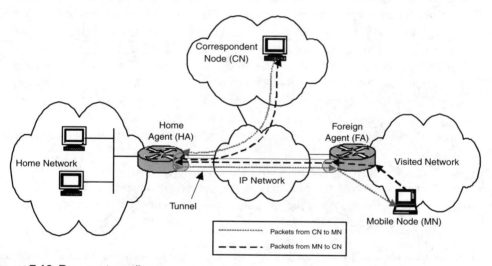

Figure 7.16 Reverse tunneling

care-of address, a NAT server will not be able to translate this to the private care-of-address, since the port number information is lost. One proposed solution is to use IP-in-UDP encapsulation instead, whereby the UDP header can carry the port number information [36].

- **Address shortage.** Another issue with mobile IP relates to the IPv4 address shortage. Mobile IP requires that every MN be given a permanent home IP address, which is wasteful of scarce IP addresses. In the visited network, the MN may be assigned a DHCP address or private address if there is an FA in the network that can connect to various mobiles. In the case of colocated FA, each mobile node will need a unique public IP address, since it will have to decapsulate the IP-in-IP tunnel. In addition to the need for public IP addresses, a colocated FA also has the disadvantage of having to tunnel through the wireless air interface, which introduces an overhead that is better to avoid in wireless networks, where bandwidth is at a premium.

- **Need for an FA.** The fact that FAs are required to support mobile IP implies that every network that the MN moves to will need to deploy FAs, and those FAs need to have a trusted relationship with the HA. This is not easily realizable in practice. The alternative of using colocated FAs on the MN themselves suffer from the disadvantages of needing publicly routable IP address, creating additional overhead on the air interface and potentially slowing the handover process.

- **Loss of QoS information.** Tunneling used in mobile IP also makes QoS implementation problematic. Since IP packet headers that provide QoS information may be hidden inside the tunnel, intermediate routers may not be able to implement the QoS requirements of the tunneled packet.

- **Issues with certain IP applications.** Since traffic to each MN has to be tunneled individually, multicasting is problematic in mobile IP. The same is true for Web caching. Web caching can be done only outside the tunnel, and therefore using mobile IP reduces an operator's flexibility in terms of how the caches are positioned.

- **Signaling overhead.** Mobile IP requires notification of the HA every time the terminal moves from one IP subnet to another. This can create a large signaling overhead, especially if the movements happen frequently, as would be the case when moving between microcell BS routers and when the HA is far away from the visited network. This issue can be mitigated by using local proxy mobility agents such that signaling messages remain regionalized.

- **Slow handover.** Since the MN must notify its change of CoA to the HA, the handover process from one network to the other may be slow, especially if the HA is far away from the visited network. The handover process may also lead to loss of any in-transit packets that are delivered while the binding update is being sent to the HA.

Although it provides a good mechanism for IP packets to be delivered to a device that moves from one network to another, mobile IP by itself is not sufficient to guarantee seamless session continuity. Mobile IP was conceived as a solution for slow macromobility, where handover is expected to be very infrequent, and the speed and smoothness of handover is not critical. It does not perform well for applications that require frequent, fast, and smooth handovers. However, a variety of tricks can be used with mobile IP to enable seamless handover.

One method to reduce handover delays is to figure out when a handover is imminent and to take proactive action to initiate a second connection with the target network before executing the handoff. These connections could be based on link-layer primitives, such as power measurements. The idea would be to acquire a new CoA as soon as possible and, if the mobile node can listen to two links at once, it can hold on to its current CoA for a short while after the handover. This can stop any packets from being lost while the binding update messages are being sent.

The other method would be to have two simultaneous bindings in the HA, which can then bicast all packets to the mobile on both the CoAs. Another approach is to set up a temporary tunnel between the previous CoA and the new CoA. The latter approach is supported in the WiMAX network architecture.

In summary, mobile IP, if implemented properly with all the optional fixes and coupled with effective network detection and selection mechanisms, can be an effective macromobility solution. The handover latency may be an issue for delay-sensitive applications, such as VoIP, but for several data applications, this overlay solution may perform satisfactorily.

7.4.3.3 Proxy Mobile IP

Mobile IP as defined in RFC 3344 requires a mobile IP client or MN functionality in every mobile station. This is a challenging requirement since most IP hosts and operating systems currently do not have support for a mobile IP client. One way to get around this problem is to have a node in the network that acts as a proxy to the mobile IP client. This *mobility proxy agent* (MPA) could perform registration and other MIP signalling on behalf of the MN. Like in the case of client-based mobile IP (CMIP), the MPA may include a colocated FA functionality or work with an external FA entitiy. This network-based mobility scheme, called *proxy mobile IP* (PMIP), offers a way to support IP mobility without requiring changes to the IP stack of the end-user device and has the added advantage of eliminating the need for MIP related signaling over the bandwidth-challenged air-interface [68]. PMIP requires only incremental enhancements to the traditional client-based mobile (CMIP) and is designed to coexist well with CMIP. The network architecture defined by the WiMAX Forum supports both PMIP and CMIP.

7.4.3.4 Mobile IP for IPv6

Unlike in IPv4, IPv6 designers considered mobility from the beginning, not as an afterthought. As a result, Mobile IPv6 [33] does have several advantages over mobile IP for IPv4. The primary advantage is that route optimization is built into IPv6; therefore, packets do not have to travel through the HA to get to the MN. Route-optimization binding updates are sent to CNs by the MN rather than by the HA. Mobile IPv6 also supports secure route optimization [4, 42, 44]. Other advantages include

- **No foreign agents.** Owing to the increased address space, IPv6 requires only the colocated CoA to be used and does not require the use of an FA CoA. Enhanced features in IPv6, such as neighbor discovery, address autoconfiguration, and the ability of any router to send router advertisements, eliminate the need for foreign agents.
- **No ingress filtering issues.** The MN's home address is carried in a packet in the home address destination option. This allows an MN to use its CoA as the source address in the IP header of packets it sends; therefore, packets pass normally through firewalls, without resorting to reverse tunneling.

- **No tunnelling.** In IPv6, the MN's CoA is carried by the routing-header option that is added to the original packet. This eliminates the need for encapsulation, thereby reducing overhead and keeping any QoS information in the packet visible.
- **Reduced signaling overhead.** There is no need for separate control packets, because the destination option in the IPv6 allows control messages to be piggybacked onto any IPv6 packet.

Although mobility support in IPv6 has a number of advantages, the question of its deployment remains uncertain. It should be noted that although some of the benefits listed here require IPv6 support in the CN, it is possible to use mobile IPv6 with IPv4 CNs as well. In this case, however, the mechanism reverts to a bidirectional tunneling mode. A number of new protocols are being developed to improve the performance of mobile IPv6. Among them are protocols for supporting fast handovers [35] and hierarchical mobility management [62].

7.5 IP for Wireless: Issues and Potential Solutions

The Internet Protocol is a network-layer protocol following a modular design that allows it to run over any link layer and supports carrying a variety of applications over it. The modularity and simplicity of IP design have led to a remarkable growth in the number of applications developed for it. The remarkable success of the Internet has made IP the network-layer protocol of choice for all modern communication systems; not only for data communications but also voice, video, and multimedia communications. WiMAX has chosen IP as the protocol for delivering all services.

IP's modularity and simplicity are achieved by making a number of assumptions about the underlying network. IP assumes that the link layers in the network are generally reliable and introduce very few errors. IP does not strive for efficient use of network resources; rather, it assumes that the network has sufficient resources. Some of these assumptions do not hold well in a wireless network; as a result running IP over wireless networks introduces problems that need to be addressed. In this section, we cover two such problems. The first problem results from the error-prone nature of wireless links, the second, from the bandwidth scarcity of wireless links.

7.5.1 TCP in Wireless

The transport control protocol (TCP) is used by a large number of IP applications, such as e-mail, Web services, and TELNET. As a connection-oriented protocol, TCP ensures that data is transferred reliably from a source to a destination. TCP divides data from the application layer into segments and ensures that every segment is delivered reliably, by including a sequence number and checksum in its header. Every TCP segment received correctly is acknowledged by sending back an ACK packet with the sequence number of the next expected packet. The receiver also provides flow control by letting the transmitter know how many data bytes it can handle without buffer overflow—this is called the *advertised window*—and the transmitter adjusts its transmission rate to ensure that the number of segments in transit is always less than the advertised window.

TCP also manages network buffer overflows. Since TCP is transparent to the intermediate routers, the sender has to indirectly figure out network buffer overflows by keeping a timer that estimates the round-trip time (RTT) for TCP segments. If it does not receive an ACK packet before its timer expires, a sender will assume that the packet was lost owing to network congestion and will retransmit the packet.

Figure 7.17 illustrates how TCP manages network congestion. TCP maintains two variables: a *congestion window* and a *slow-start threshold*. The congestion window determines the number of segments that is transmitted within an RTT. At the start of a TCP session, the congestion window is set to 1, and the transmitter sends only one segment and waits for an acknowledgment. When an ACK is received, the congestion window is doubled, and two segments are transmitted at a time. This process of doubling the congestion window continues until it reaches the maximum indicated by the advertised window size or until the sender fails to get an acknowledgment before the timer expires. At this point, TCP infers that the network is congested and begins the recovery process by dropping the congestion window back to one segment. Resetting the congestion window to one segment allows the system to clear all packets in transit. Now, if a retransmission also fails, the TCP sender will also exponentially back off its retransmission time, providing more time for the system to clear the congestion. If transmission is successful after restart, the process of doubling the congestion window size after every transmission continues until the contention window size reaches half the size at which it detected the previous congestion. This is called the *slow-start threshold*. Once at this threshold, the congestion window is increased only linearly—that is, by one segment size at a time—in what is called the *congestion-avoidance* algorithm. This process continues as shown in Figure 7.17.

Network congestion may also be detected by receiving one or more—typically, three— duplicate ACK packets, which are sent when packets are received out of order. When that happens, TCP performs a *fast retransmit*—retransmit the missing packet without waiting for the timeout to expire—and *fast recovery*—that is follow the congestion-avoidance mechanism without resetting the congestion window back to 1—operation [63].

Clearly, TCP provides a mechanism for reliable end-to-end transmission without requiring any support from intermediate nodes. This is done, however, by making certain assumptions about the network. Specifically, TCP assumes that all packet losses, or unacknowledged packets and delays are caused by congestion and that the loss rate is small. This assumption is not valid in a wireless network, where packet errors are very frequent and caused mostly by poor channel conditions. Responding to packet errors by slowing down does not solve the problem if the errors are not caused by congestion. Instead, it serves only to unnecessarily reduce the throughput. Frequent errors will lead to frequent initiation of slow-start mechanisms, keeping TCP away from achieving steady state throughput.

Further, in the presence of frequent losses, TCP throughput can be shown to be inversely proportional to round-trip time. This makes intuitive sense, since transmission rates are increased

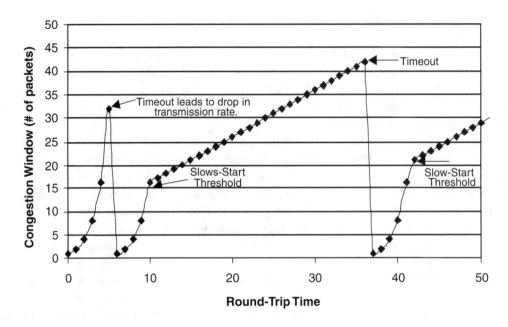

Figure 7.17 TCP congestion control

only every round-trip time. When wireless networks have large latencies, this also leads to throughput reduction. Large latencies coupled with high data rates can also mean that at a given time, large amount of data is in transit. This can lead to TCP's assuming that the receiver buffer is full and slowing its transmission rate.

TCP performance is particularly bad under conditions of burst errors. Fast retransmit and fast recovery improve the throughput of TCP under sporadic random losses only if such losses occur only once within an RTT. Consecutive failed attempts will cause the TCP sender to exponentially back off its retransmission timer. A loss of a series of packets can therefore cause the timer to be set very long, leading to long periods of inactivity and underutilization of the available link bandwidth [64].

Clearly, running TCP over wireless channel leads to unnecessary degradation in throughput, inefficient utilization of scarce resources, and excessive interruptions in data transmissions. In mobile systems, these problems are exacerbated during handover. Given these problems, a lot of research into methods to improve TCP performance in wireless networks has occurred over the past decade or so.

A number of simple tricks can be used to improve TCP performance in wireless networks. For example, increasing the maximum allowed window size, using selective repeat-ARQ instead of the go-back-N ARQ for retransmission [23, 39], and using an initial window size larger than one segment [3] have all been shown to improve TCP performance over wireless links. Although these optimization methods do provide marginal improvements, they do not mitigate all the problems of TCP in wireless.

Broadly speaking, there are two approaches to mitigating the TCP issues in wireless. One approach is to make TCP aware of the wireless links and make it change its behavior. These schemes typically attempt to differentiate between congestion-based losses and channel-induced losses so that TCP congestion-control behavior is not activated when packets are lost owing to channel-induced errors. Examples of such schemes are TCP–Santa Cruz, Freeze-TCP, and others [12, 25, 43, 66] introduced in the late 1990s. Since many of these schemes requires making changes to the TCP stacks at every host, they are not considered practical.

To mitigate this concern, proposals have been made to break the TCP link into two pieces—one that is between the wireless node and the base station or edge router in the wireless network, and the other between the wireless network and the fixed host—and to implement a "wireless-aware" TCP stack only for the wireless link. Examples of these include indirect TCP (I-TCP) [6] and the Berkeley snoop module [8]. These schemes, however, break the end-to-end semantics of TCP and can cause problems for some applications, for instance, when there is end-to-end encryption.

The alternative approach is to make the wireless link adapt to the needs of TCP. An obvious way to make TCP work well in wireless networks is to make the link more reliable. To an extent, this could be done by using strong error correction and link-layer retransmission schemes (ARQ). Most modern wireless broadband networks, including WiMAX, do have link-layer ARQ. WiMAX support hybrid-ARQ at the physical layer in addition to the standard ARQ at the MAC layer.

ARQ at the link level can make the wireless link look relatively error free to TCP, but does introduce the problem of variable delays in packet delivery, which can cause incorrect estimation of round-trip delays and hence inaccurate setting of the TCP timeouts. As a result, it is likely, for example, that TCP may assume that a packet is lost while it is correctly being retransmitted using a link-layer ARQ process. This again is wasteful of wireless bandwidth. By having a closer coordination between the link layer and the TCP layer, however, this solution can potentially be made effective. Cross-layer design to improve interaction between the link layer and higher layers is an active area of research in wireless networks and has the potential to offer significant performance improvements [58].

7.5.2 Header Compression

In such IP applications as VoIP, messaging, and interactive gaming, the payload sizes of packets tend to be fairly small. For these packets, the size of the header becomes a large fraction of the total packet size. For example, voice packets are typically 20–60 bytes long, whereas the associated header is 40 bytes long. Since the headers, which contain the source and destination IP and port addresses, sequence numbers, protocol identifiers, and so on, have very little variation from one packet to another for a given flow, it is possible to compress them heavily and to save more than 80 percent bandwidth (see Table 7.5). In addition to bandwidth savings, header compression can reduce packet loss, since smaller packets are less likely to suffer from bit errors for a given BER, and improve the interactive response times.

Table 7.5 Gains Achievable Through Header Compression

Protocol Type	Header Size (bytes)	Minimum Compressed Header Size (bytes)	Bandwidth Savings (%)
IPv4/TCP	40	4	90
IPv4/UDP	28	1	96.4
IPv4/UDP/RTP	40	1	97.5
IPv6/TCP	60	4	93.3
IPv6/UDP	48	3	93.75
IPv6/UDP/RTP	60	3	95

Header compression uses the concept of *flow context*, a collection of information about static and dynamic fields and change patterns in the packet header. This context is used by the compressor and the decompressor to achieve maximum compression. The first few packets of a flow are sent without compression and are used to build the context on both sides. The number of initial uncompressed packets is determined based on link BER and round-trip time. Using periodic feedback about link conditions, the amount of compression can also be varied. Once a context is established, compressed packets are sent with a context identifier prefixed to it.

Several header-compression techniques have been developed over the years [15, 19, 31]. We discuss only one of them, *robust header compression* (ROHC) [10], which is supported in WiMAX. ROHC is a more complex technique, but works well under conditions of high BER and long round-trip times and can reduce the header size to a minimum of 1 byte. An extensible framework for compression, ROHC can be used on a variety of headers, including IP/UDP/RTP for VoIP and IP/ESP for VPN.

At the beginning of a flow, a static update message that contains all the fields not expected to change such as IP source and destination address, is sent. Dynamic fields are sent uncompressed in the beginning and when there is a failure. Otherwise, dynamic fields are sent compressed, using a window-based least-significant bits encoding. ROHC includes an error-recovery process at the decompressor, as shown in Figure 7.18. A CRC that is valid for the uncompressed header is sent with each compressed header. If the CRC fails after decompression, the decompressor tries to interpolate from the previous headers the missing data and checks again. This is tried a few times; and if unsuccessful, a context update is requested. The compressor then sends enough information to fix the context. This error-recovery mechanism is what makes ROHC compression scheme robust. ROHC is widely recognized as a critical piece of any wireless IP network, and the IETF has a charter dedicated to continually making additions and enhancements to ROHC [53].

One negative consequence of using header compression over the air link is that the bandwidth requirements of a particular application become different over the air and in the rest of the

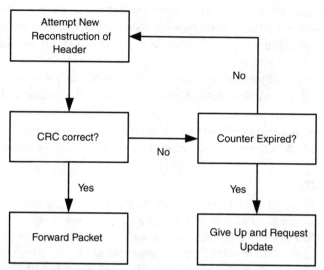

Figure 7.18 ROHC decompressor recovery process

network. This makes it difficult for an application to make the correct end-to-end bandwidth requests for QoS.

7.6 Summary and Conclusions

In this chapter, we provided a brief overview of the various end-to-end aspects of broadband wireless networks.

- QoS is of two types: one based on per flow handling and one based on aggregate handling. Per flow handling offers better QoS but has scalability issues. Most IP networks today rely on aggregate handling.
- The IETF has developed a number of architectures and protocols for providing QoS in an IP network. Three major emerging IP QoS technologies are integrated services, differentiated services, and multiprotocol label switching.
- The session initiation protocol (SIP), a simple, flexible, and powerful text-based protocol, has rapidly established itself as the protocol of choice for multimedia session control in IP networks.
- Wireless network designs should include support for basic security mechanisms, such as encryption, authentication, and access control. The IEEE 802.16e-2005, along with the WiMAX architecture, has support for robust and flexible security mechanisms.
- To support mobile users, broadband wireless networks should incorporate mechanisms for location management and handoff management. Developing good handoff mechanisms is critical to the performance of mobile networks.

- In addition to physical-layer and MAC-layer mechanisms to support handoff, there is a need to deploy mobile IP to support transfer of ongoing IP connections across subnets in a WiMAX networks. Enhancements to mobile IP are needed to achieve seamless session continuity.
- TCP was not designed for running over noisy and bandwidth constrained links and hence performs poorly over wireless links. A number of potential solutions to this problem are available.
- Header compression can improve throughput efficiency of bandwidth-constrained wireless links. The WiMAX standard has support for robust header compression.

7.7 Bibliography

[1] I. F. Akyildiz, J. McNair, J. S. M. Ho, H. Uzunalioglu, and W. Wang. Mobility management in next generation wireless systems. *Proceedings of the IEEE*, 87(8): August 1999.

[2] I. F. Akyildiz, J. Xie, and S. Mohanty. A survey of mobility management in next-generation all-IP-based wireless systems. *IEEE Wireless Communications Magazine*, 11(4):16–28, August 2004.

[3] M. Allman. Increasing TCP's initial window. *IETF RFC 2414*. September 1998.

[4] J. Arkko et al. Using IPSec to protect mobile IPv6 signalling between mobile nodes and home agents. *IETF RFC 3776*. June 2004.

[5] G. Armitage. *Quality of Service in IP Networks*. Sams, 2001.

[6] A. Bakre and B.R. Badrinath. I-TCP: Indirect TCP for mobile hosts. *Proceedings of the 15th International Conference on Distributed Computing Systems (ICDCS)*. May 1995.

[7] H. Balakrishna, et al. A comparison of mechanisms for improving TCP performance over wireless links. *Proceedings of ACM/IEEE Mobicom*, pp. 77–89, September 1997.

[8] H. Balakrishna et al. Improving TCP/IP performance over wireless networks. *Proceedings of ACM/IEEE Mobicom*. November 1995.

[9] L. Blunk and J. Vollbrecht. PPP extensible authentication protocol (EAP). *IETF RFC 2284*. March 1998.

[10] C. Borman et al. Robust header compression (ROHC): Framework and four profiles: RTP, UDP, ESP, uncompressed. *IETF RFC 3095*. July 2001.

[11] R. Braden et al. Resource reservation protocol. *IETF RFC 2205*. September 1997.

[12] L. Brakmo and L. Peterson. TCP Vegas: End to end congestion avoidance on a global internet. *IEEE Journal on Selected Areas in Communication*, 13(8):1465–1480, October 1995.

[13] P. Calhoun et al. Diameter in use. *IETF RFC 3588*. September 2003.

[14] A. T. Campbell et al. Comparison of IP micromobility protocols. *IEEE Wireless Communications Magazine*, 9(1):72-82, February 2002.

[15] S. Casner. Compressing IP/UDP/RTP headers for low speed serial links. *IETF RFC 2508*. February 1999.

[16] F. M. Chiussi, D.A. Khotimsky, and S. Krishnan. Mobility management in third generation all-IP Networks. *IEEE Communications Magazine*, 40(9):124–135, September 2002.

[17] J. Daemen and V. Rijmen, *The Design of Rijndael: AES—The Advanced Encryption Standard*. Springer-Verlag, 2002.

[18] S. Deering. ICMP router discovery messages. *IETF RFC 1256*. September 1991.

[19] M. Degermark. IP header compression. *IETF RFC 2507*. February 1999.

[20] T. Dierks. TLS protocol version 1.0. *IETF RFC 2246.* January 1999.

[21] D. Durham. The common open policy service (COPS) protocol. *IETF RFC 2748*, January 2000.

[22] D. Eastlake and P. Jones. The US secure hash algorithm 1 (SHA1). *IETF RFC 3174.* September 2001.

[23] K. Fall and S. Floyd. Simulation based comparison of Tahoe, Reno, and SACK TCP. *Computer Communications Review*, 1996.

[24] G. J. Foschini, B. Gopinath, and Z. Miljanic. Channel cost of mobility. *IEEE Transactions on Vehicular Technology*, 42(4):414-424, November 1993.

[25] T. Goff et al. Freeze-TCP: A true end-to-end enhancement mechanism for mobile environments. *Proceedings of the IEEE Infocom 2000*, pp. 1537–1545, 2000.

[26] H. Gossain et al. Multicast: Wired to wireless. *IEEE Communications Magazine*, 40(6):116–123, June 2002.

[27] M. Gudmundson. Analysis of handover algorithm. *Proceedings of the IEEE Vehicular Technology Conference*, May 1991.

[28] M. Gudmundson. A correlation model for shadow fading in mobile radio. *Electronics Letters*, 27(23): 146–2147, November 1991.

[29] J. Heinanen et al. Assured Forwarding PHB Group. *IETF RFC 2597.* June 1999.

[30] J. Hodges and R. Morgan. Lightweight directory access protocol (v3): Technical specifications. *IETF RFC 3377*, September 2002.

[31] V. Jacobson. Compressing TCP/IP headers for low-speed serial links. *IETF RFC 1144.* February 1990.

[32] V. Jacobson et al. An expedited forwarding PHB. *IETF RFC 2598.* June 1999.

[33] D. Johnson, C. Perkins, and J. Arkko. Mobility Support for IPv6. *IETF RFC 3775.* June 2004.

[34] A. B. Johnston. *Understanding the Session Initiation Protocol*, 2nd edition. Artech House, 2004.

[35] R. Koodli, ed. Fast handovers for Mobile IPv6. *IETF RFC 4068.* July 2005.

[36] Levkowetz. Mobile IP traversal of network address translation devices. *IETF RFC 3519.* April 2003.

[37] B. Lloyd and W. Simpson. PPP authentication protocols. *IETF RFC 1334.* October 1992.

[38] L. Mamakos et al. A method for transmitting PPP over Ethernet (PPPoE). *IETF RFC 2516.* February 1999.

[39] M. Mathis et al. TCP selective acknowledgment options. *IETF RFC 2018.* 1996.

[40] G. Montenegro. Reverse tunneling for mobile IP revised. *IETF RFC 3024.* January 2001.

[41] National Institute of Standards and Technology (NIST). Federal Information Processing Standard (FIPs) 197. csrc.nist.gov/encryption/aes/index.htm.

[42] P. Nikander et al. Mobile IP version 6 route optimization security design background. *IETF RFC 4225.* December 2005.

[43] C. Parsa and J.J. Garcia-Lana-Aceves. Differentiating congestion versus random loss: A method for improving TCP performance over wireless links. *Proceedings of 2nd IEEE Wireless Communications and Networking Conference* (WCNC); September 2000.

[44] A. Patel et al. Authentication protocol for mobile IPv6. *IETF RFC 4285.* January 2006.

[45] C. Perkins. IP Encapsulation within IP. *IETF RFC 2003.* October 1996.

[46] C. Perkins. IP mobility support for IPv4. *IETF RFC 3344.* August 2002.

[47] C. Perkins. Minimum Encapsulation within IP. *IETF RFC 2004.* October 1996.

[48] M. Poikselka. *The IMS: IP Multimedia Concepts and Services*, 2nd edition. Wiley, 2006.

[49] M. Rhodes-Ousley. *Network Security: The Complete Reference*. McGraw-Hill. 2003.

[50] C. Rigney et al. Remote dial-in user service (RADIUS). *IETF RFC 2865.* June 2000.

[51] R. Rivest. The MD5 message-digest algorithm. *IETF RFC 1321*. April 1992.

[52] R. Rivest, A. Shamir, and L. Adleman. A method for obtaining digital signatures and public-key cryptosystems. *Communications of the ACM*, 21(2): 120–126, February 1978.

[53] Robust Header Compression. http://www.ietf.org/html.charters/rohc-charter.html.

[54] Rosen et al. Multiprotocol label switching architecture. *IETF RFC 3031*. January 2001.

[55] J. Rosenberg et al. SIP: Session initiation protocol. *IETF RFC 3261*. June 2002.

[56] S. Sanks et al. Generic routing encapsulation (GRE). *IETF RFC 1701*. October 1994.

[57] H. Schulzrinne. et al. A transport protocol for real-time applications. *IETF RFC 1889*. January 1996.

[58] S. Shakkottai, T. S. Rappaport, and T. S. Karlsson. Cross-layer design for wireless networks. *IEEE Communications Magazine*, 41(10):74-80, October 2003.

[59] S. Shenker and J. Wroclawski. General characterization parameters for integrated service network elements. *IETF RFC 2215*. September 1997.

[60] W. Simpson. The point-to-point protocol (PPP). *IETF RFC 1661*. July 1994.

[61] W. Simpson. PPP challenge handshake authentication protocol (CHAP). *IETF RFC 1994*. August 1996.

[62] H. Soliman et al. Hierarchical mobile IPv6 mobility management (HMPIv6). *IETF RFC 4140*. August 2005.

[63] W. Stevens. TCP slow start, congestion avoidance, fast retransmit and fast recovery algorithms. *IETF RFC 2001*. January 1997.

[64] Y. Tian, K. Xu, and N. Ansari. TCP in wireless environments: Problems and solutions. *IEEE Communications Magazine*, 43(3):S27–S32, March 2005.

[65] Z. Wang. *Internet QoS: Architectures and Mechanisms for Quality of Service*. Academic Press, 2001.

[66] K. Xu et al. TCP—Jersey for wireless IP communications. *IEEE Journal on Selected Areas in Communications*, 22(4):747–756, May 2004.

[67] J. Zander, and SL. Kim. *Radio Resource Management for Wireless Networks*. Artech House, 2001.

[68] K. Leung et al. Mobility management using proxy mobile IPv4. draft-leung-mip4-proxy-mode-01.txt, Internet Draft, June 2006.

PART III

Understanding WiMAX and Its Performance

PHY Layer of WiMAX

The physical (PHY) layer of WiMAX is based on the IEEE 802.16-2004 and IEEE 802.16e-2005 standards and was designed with much influence from Wi-Fi, especially IEEE 802.11a. Although many aspects of the two technologies are different due to the inherent difference in their purpose and applications, some of their basic constructs are very similar. Like Wi-Fi, WiMAX is based on the principles of orthogonal frequency division multiplexing (OFDM) as previously introduced in Chapter 4, which is a suitable modulation/access technique for non–line-of-sight (LOS) conditions with high data rates. In WiMAX, however, the various parameters pertaining to the physical layer, such as number of subcarriers, pilots, guard band and so on, are quite different from Wi-Fi, since the two technologies are expected to function in very different environments.

The IEEE 802.16 suite of standards (IEEE 802.16-2004/IEEE 802-16e-2005) [3, 4] defines within its scope four PHY layers, any of which can be used with the media access control (MAC) layer to develop a broadband wireless system. The PHY layers defined in IEEE 802.16 are

- *WirelessMAN SC*, a single-carrier PHY layer intended for frequencies beyond 11GHz requiring a LOS condition. This PHY layer is part of the original 802.16 specifications.
- *WirelessMAN SCa*, a single-carrier PHY for frequencies between 2GHz and 11GHz for point-to-multipoint operations.
- *WirelessMAN OFDM*, a 256-point FFT-based OFDM PHY layer for point-to-multipoint operations in non-LOS conditions at frequencies between 2GHz and 11GHz. This PHY layer, finalized in the IEEE 802.16-2004 specifications, has been accepted by WiMAX for fixed operations and is often referred to as *fixed WiMAX*.
- *WirelessMAN OFDMA*, a 2,048-point FFT-based OFDMA PHY for point-to-multipoint operations in NLOS conditions at frequencies between 2GHz and 11GHz. In the IEEE

802.16e-2005 specifications, this PHY layer has been modified to SOFDMA (scalable OFDMA), where the FFT size is variable and can take any one of the following values: 128, 512, 1,024, and 2,048. The variable FFT size allows for optimum operation/implementation of the system over a wide range of channel bandwidths and radio conditions. This PHY layer has been accepted by WiMAX for mobile and portable operations and is also referred to as *mobile WiMAX*.

Figure 8.1 shows the various functional stages of a WiMAX PHY layer. The first set of functional stages is related to forward error correction (FEC), and includes channel encoding, rate matching (puncturing or repeating), interleaving, and symbol mapping. The next set of functional stages is related to the construction of the OFDM symbol in the frequency domain. During this stage, data is mapped onto the appropriate subchannels and subcarriers. Pilot symbols are inserted into the pilot subcarriers, which allows the receiver to estimate and track the channel state information (CSI). This stage is also responsible for any space/time encoding for transmit diversity or MIMO, if implemented. The final set of functions is related to the conversion of the OFDM symbol from the frequency domain to the time domain and eventually to an analog signal that can be transmitted over the air. Although Figure 8.1 shows only the logical components of a transmitter, similar components also exist at the receiver, in reverse order, to reconstruct the transmitted information sequence. Like all other standards, only the components of the transmitter are specified; the components of the receiver are left up to the equipment manufacturer to implement.

In the first section of this chapter, we describe the various components of the channel encoding and symbol-mapping stages as defined in the IEEE 802.16e-2005 standard. The various mandatory and optional channel coding and modulation schemes are discussed. Next, we describe the construction of the OFDM symbol in the frequency domain. This stage is very critical and unique to IEEE 802.16e-2005, since various subcarrier permutations and mappings are allowed within the standard, allowing adaptation based on environmental, network, and spectrum related parameters. We then discuss the optional multiantenna features of IEEE 802.16e-2005 for various modes, such as transmit diversity and spatial multiplexing. Finally, we describe the various physical-layer control mechanisms, such as power control and measurement reporting.

8.1 Channel Coding

In IEEE 802.16e-2005, the channel coding stage consists of the following steps: (1) data randomization, (2) channel coding, (3) rate matching, (4) HARQ, if used, (5) and interleaving. Data randomization is performed in the uplink and the downlink, using the output of a maximum-length shift-register sequence that is initialized at the beginning of every FEC block. This shift-register sequence is modulo 2, added with the data sequence to create the randomized data. The purpose of the randomization stage is to provide layer 1 encryption and to prevent a rogue receiver from decoding the data. When HARQ is used, the initial seed of the shift-register

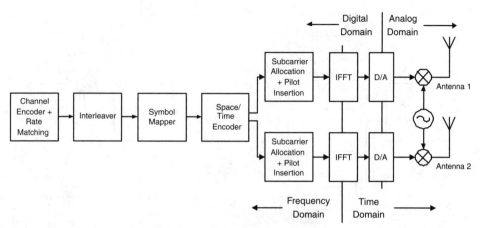

Figure 8.1 Functional stages of WiMAX PHY

sequence for each HARQ transmission is kept constant in order to enable joint decoding of the same FEC block over multiple transmissions.

Channel coding is performed on each FEC block, which consists of an integer number of *subchannels*. A subchannel is the basic unit of resource allocation in the PHY layer and comprises several data and pilot subcarriers. The exact number of data and pilot subcarriers in a subchannel depends on the *subcarrier permutation* scheme, which is explained in more detail later. The maximum number of subchannels in an FEC block is dependent on the channel coding scheme and the modulation constellation. If the number of subchannels required for the FEC block is larger than this maximum limit, the block is first segmented into multiple FEC subblocks. These subblocks are encoded and rate matched separately and then concatenated sequentially, as shown in Figure 8.2, to form a single coded data block. Code block segmentation is performed for larger FEC blocks in order to prevent excessive complexity and memory requirement of the decoding algorithm at the receiver.

8.1.1 Convolutional Coding

The mandatory channel coding scheme in IEEE 802.16e-2005 is based on binary nonrecursive convolutional coding (CC). The convolutional encoder uses a constituent encoder with a constraint length 7 and a native code rate 1/2, as shown in Figure 8.3. The output of the data randomizer is encoded using this constituent encoder. In order to initialize the encoder to the 0 state, each FEC block is padded with a byte of 0x00 at the end in the OFDM mode. In the OFDMA mode, tailbiting is used to initialize the encoder, as shown in Figure 8.3. The 6 bits from the end of the data block are appended to the beginning, to be used as flush bits. These appended bits flush out the bits left in the encoder by the previous FEC block. The first 12 parity bits that are generated by the convolutional encoder which depend on the 6 bits left in the encoder by the previous FEC block are discarded. Tailbiting is slightly more bandwidth efficient than using flush bits since the FEC blocks are not padded unneccessarily. However, tailbiting requires a more complex decoding

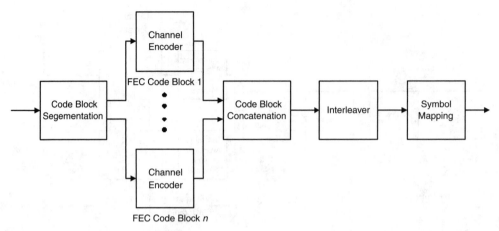

Figure 8.2 Code block segmentation

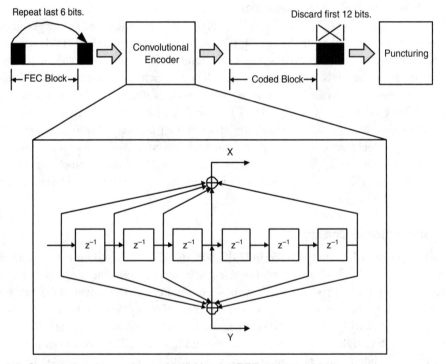

Figure 8.3 Convolutional encoder and tailbiting in IEEE 802.16e-2005

algorithm, since the starting and finishing states of the decoder are no longer known.[1] In order to achieve code rates higher than 1/2, the output of the encoder is punctured, using the puncturing pattern shown in Table 8.1.

In the downlink of the OFDM mode, where subchannelization is not used, the output of the data randomizer is first encoded using an outer systematic Reed Solomon (RS) code and then encoded using an inner rate 1/2 binary convolutional encoder. The RS code is derived from a systematic RS ($N = 255$, $K = 239$, $T = 8$) code using GF(2^8). The total DL and UL PHY data rates for the allowed modulation and code rates are shown in Table 8.2 for a 10MHz channel bandwidth with an FFT size of 1,024, an oversampling rate of 8/7, and a frame length of 5msec.

8.1.2 Turbo Codes

Apart from the mandatory channel coding schemes mentioned in the previous section, several optional channel coding schemes such as block turbo codes, convolutional turbo codes, and low density parity check (LDPC) codes are defined in IEEE 802.16e-2005. Of these optional channel coding modes, the convolutional turbo codes (CTC) are worth describing because of their superior performance and high popularity in other broadband wireless systems, such as HSDPA, WCDMA, and 1xEV-DO. As shown in Figure 8.4, WiMAX uses duobinary turbo codes with a constituent recursive encoder of constraint length 4. In duo binary turbo codes two consecutive bits from the uncoded bit sequence are sent to the encoder simultaneously. Unlike the binary turbo encoder used in HSDPA and 1xEV-DO, which has a single generating polynomial for one party bit, the duobinary convolution encoder has two generating polynomials, $1+D^2+D^3$ and $1+D^3$ for two parity bits. Since two consecutive bits are used as simultaneous inputs, this encoder has four possible state transitions compared to two possible state transitions for a binary turbo encoder.

Duobinary turbo codes are a special case of nonbinary turbo codes, which have many advantages over conventional binary turbo codes [2]:

- **Better convergence:** The better convergence of the bidimensional iterative process is explained by a lower density of the erroneous paths in each dimension, reducing the correlation effects between the component decoders.

- **Larger minimum distances:** The nonbinary nature of the code adds one more degree of freedom in the design of permutations (interleaver)—intrasymbol permutation—which results in a larger minimum distance between codewords.

- **Less sensitivity to puncturing patterns:** In order to achieve code rates higher than 1/3 less redundancy, bits need to be punctured for nonbinary turbo codes, thus resulting in better performance of punctured codes.

---

1. In the case of a conventional Viterbi decoder, the start and end states of the trellis are the 0 state.

Table 8.1 Puncturing for Convolutional Codes[a]

Code Rate	R 1/2	R 2/3	R 3/4	R 5/6
d_{free}	10	6	5	4
Parity 1 (X)	11	10	101	10101
Parity 2 (Y)	11	11	110	11010
Output	X_1Y_1	$X_1Y_1Y_2$	$X_1Y_1Y_2X_3$	$X_1Y_1Y_2X_3Y_4X_5$

a. The R5/6 puncturing is used when the convolutional encoder is used with Reed Solomon codes. The R5/6 convolutional encoder with the RS encoder provides an overall coding rate of 3/4.

Table 8.2 Data Rate in Mbps for the Mandatory Coding Modes

DL:UL Ratio	1:1				3:1			
Cyclic Prefix	1/4	1/8	1/16	1/32	1/4	1/8	1/16	1/32
QPSK R1/2 DL	2.880	3.312	3.456	3.600	4.464	4.896	5.328	5.472
QPSK R1/2 UL	2.352	2.576	2.800	2.912	1.120	1.344	1.344	1.456
QPSK R3/4 DL	4.320	4.968	5.184	5.400	6.696	7.344	7.992	8.208
QPSK R3/4 UL	3.528	3.864	4.200	4.368	1.680	2.016	2.016	2.184
16 QAM R1/2 DL	5.760	6.624	6.912	7.200	8.928	9.792	10.656	10.944
16 QAM R1/2 UL	4.704	5.152	5.600	5.824	2.240	2.688	2.688	2.912
16 QAM R3/4[a] DL	8.640	9.936	10.368	10.800	13.392	14.688	15.984	16.416
16 QAM R3/4 UL	7.056	7.728	8.400	8.736	3.360	4.032	4.032	4.368
64 QAM R2/3 DL	11.520	13.248	13.824	14.400	17.856	19.584	21.312	21.888
64 QAM R2/3 UL	9.408	10.304	11.200	11.648	4.480	5.376	5.376	5.824
64 QAM R3/4 DL	12.960	14.904	15.552	16.200	20.088	22.032	23.976	24.624
64 QAM R3/4 UL	10.584	11.592	12.600	13.104	5.040	6.048	6.048	6.552

a. 16 QAM R3/4 and 64 QAM R1/2 have the same data rate.

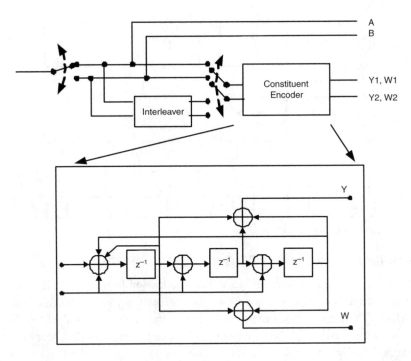

Constituent Encoder

Figure 8.4 Turbo Encoder in IEEE 802.16e-2005

- **Robustness of the decoder:** The performance gap between the optimal MAP decoder and simplified suboptimal decoders, such as log-MAP and the soft input soft output (SOVA) algorithm, is much less in the case of duobinary turbo codes than in binary turbo codes.

The output of the native R1/3 turbo encoder is first separated into the six sub blocks (A, B, Y1, Y2, W1, and W2), where A and B contain the systematic bits, Y1 and W1 contain the parity bits of the encoded sequence in natural order, and Y2, and W2 contain the parity bits of the interleaved sequence. Each of the six subblocks is independently interleaved, and the subblocks containing the parity bits are punctured to achieve the target code rate as shown in Figure 8.5. The subblock interleaver consists of two stages: (1) The first stage of the interleaver flips bits contained in the alternating symbol.[2] (2) The second stage of the subblock interleaver permutates the positions of the symbols. In order to achieve the target code rate, the interleaved subblocks Y1, Y2, W1, and W2 are punctured using a specific puncturing pattern. When HARQ (hybrid-ARQ) is used, the puncturing pattern of the parity bits can change from one transmission to the next, which allows the receiver to generate log likelihood ratio (LLR) estimates of more parity bits with each new retransmission.

---

2. Here, each symbol refers to a pair of consecutive bits. This is a common nomenclature for duobinary turbo codes, which process 2 bits at a time.

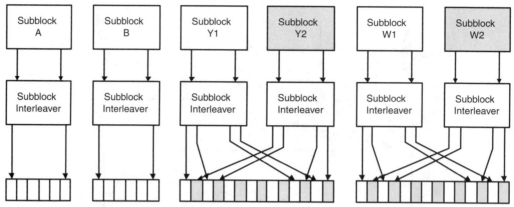

Figure 8.5 Subblock interleaving

8.1.3 Block Turbo Codes and LDPC Codes

Other channel coding schemes, such as block turbo codes and LDPC codes, have been defined in WIMAX as optional channel coding schemes but are unlikely to be implemented in fixed or mobile WiMAX. The reason is that most equipment manufacturers have decided to implement the convolutional turbo codes for their superior performance over other FEC schemes. The block turbo codes consist of two binary extended Hamming codes that are applied on natural and inter-leaved information bit sequences, respectively. The LDPC code, as defined in IEEE 802.16e-2005, is based on a set of one or more fundamental LDPC codes, each of the fundamental codes is a systematic linear block code that can accommodate various code rates and packet sizes.The LDPC code can flexibly support various block sizes for each code rate through the use of an expansion factor.

8.2 Hybrid-ARQ

IEEE 802.16e-2005 supports both type I HARQ and type II HARQ. In type I HARQ, also referred to as *chase combining*, the redundancy version of the encoded bits is not changed from one transmission to the next: The puncturing pattern remains same. The receiver uses the current and all previous HARQ transmissions of the data block in order to decode it. With each new transmission, the reliability of the encoded bits improves thus reducing the probability of error during the decoding stage. This process continues until either the block is decoded without error—passes the CRC check—or the maximum number of allowable HARQ transmissions is reached. When the data block cannot be decoded without error and the maximum number of HARQ transmissions is reached, a higher layer, such as MAC or TCP/IP, retransmits the data block. In that case, all previous transmissions are cleared, and the HARQ process start over.

In the case of type II HARQ, also referred to as *incremental redundancy*, the redundancy version of the encoded bits is changed from one transmission to the next, as shown in Figure 8.6. Thus, the puncturing pattern changes from one transmission to the next, not only improving the

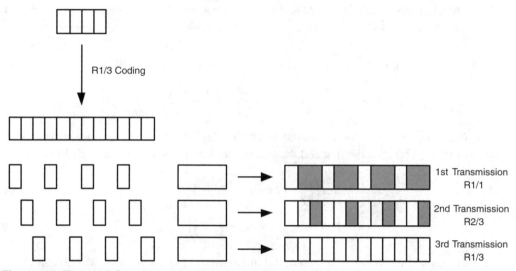

Figure 8.6 The HARQ process with incremental redundancy

LLR of parity bits but also reducing the code rate with each additional transmission. Incremental redundancy leads to lower bit error rate (BER) and block error rate (BLER) than in chase combining. The puncturing pattern to be used for a given HARQ transmission is indicated by the subpacket identity (SPID). By default, the SPID of the first transmission is always 0, which ensures that all the systematic bits are sent, as only the parity bits are punctured, and the transmission is self-decodable. The SPIDs of the subsequent transmission can be chosen by the system at will. Note that although the SPIDs of the various transmissions can be in natural increasing order—0, 1, 2—this is not necessary. Any order of SPIDs is allowed, as long as long as it starts with 0.

8.3 Interleaving

After channel coding, the next step is interleaving. The encoded bits are interleaved using a two-step process. The first step ensures that the adjacent coded bits are mapped onto nonadjacent subcarriers, which provides frequency diversity and improves the performance of the decoder. The second step ensures that adjacent bits are alternately mapped to less and more significant bits of the modulation constellation. It should be noted that interleaving is performed independently on each FEC block. As explained in Section 8.6, the separation between the subcarriers, to which two adjacent bits are mapped onto, depends on the subcarrier permutation schemes used. This is very critical, since for 16 QAM and 64 QAM constellations, the probability of error for all the bits is not the same. The probability of error of the most significant bit (MSB) is less than that of the least significant bit (LSB) for the modulation constellations.

Equation (8.1) provides the relation between k, m_k, and j_k, the indices of the bit before and after the first and second steps of the interleaver, respectively, where N_c is the total number of bits in the block, and s is $M/2$, where M is the order of the modulation alphabet (2 for QPSK, 4 for 16 QAM, and 6 for 64 QAM), and d is an arbitrary parameter whose value is set to 16:

$$n_k = \left(\frac{N_c}{d}\right) k_{mod(d)} + floor\left(\frac{k}{d}\right) \tag{8.1}$$

$$k = s \cdot floor\left(\frac{m_k}{s}\right) + \left(m_k + N_c - floor\left(\frac{d \cdot m_k}{N_c}\right)\right)_{mod(d)}.$$

The deinterleaver, which performs the inverse of this operation, also works in two steps. The index of the *jth* bit after the first and the second steps of the deinterleaver is given by

$$n_j = s \cdot floor\left(\frac{j}{s}\right) + \left(j + floor\left(\frac{d \cdot j}{N_c}\right)\right)_{mod(d)} \tag{8.2}$$

$$_j = dm_j - \left((N_c - 1) \cdot floor\left(\frac{d \cdot m_j}{N_c}\right)\right).$$

When convolutional turbo codes are used, the interleaver is bypassed, since a subblock interleaver is used within the encoder, as explained in the previous section.

8.4 Symbol Mapping

During the symbol mapping stage, the sequence of binary bits is converted to a sequence of complex valued symbols. The mandatory constellations are QPSK and 16 QAM, with an optional 64 QAM constellation also defined in the standard, as shown in Figure 8.7. Although the 64 QAM is optional, most WiMAX systems will likely implement it, at least for the downlink.

Each modulation constellation is scaled by a number c, such that the average transmitted power is unity, assuming that all symbols are equally likely. The value of c is $\sqrt{1/2}$, $\sqrt{1/10}$, and $\sqrt{1/42}$ for the QPSK, 16 QAM, and 64 QAM modulations, respectively. The symbols are further multiplied by a pseudorandom unitary number to provide additional layer 1 encryption:

$$s_k = 2\left(\frac{1}{2} - w_k\right)s_k, \tag{8.3}$$

where k is the subcarrier index, and w_k is a pseudorandom number generated by a shift register of memory order 11. Preamble and midamble symbols are further scaled by $2\sqrt{2}$, which signifies an eight fold boost in the power and allows for more accurate synchronization and various parameter estimations, such as channel response and noise variance.

8.5 OFDM Symbol Structure

As discussed in Chapter 4, in an OFDM system, a high-data-rate sequence of symbols is split into multiple parallel low-data rate-sequences, each of which is used to modulate an orthogonal tone, or subcarrier. The transmitted baseband signal, which is an ensemble of the signals in all the subcarriers, can be represented as

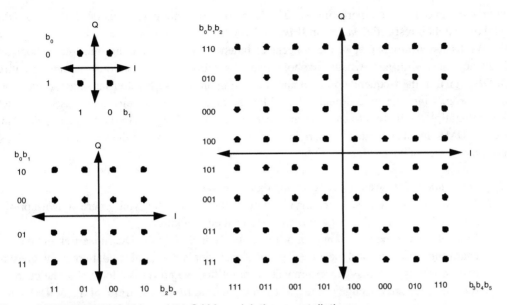

Figure 8.7 QPSK, 16 QAM, and 64 QAM modulation constellations

$$x(t) = \sum_{i=0}^{L-1} s[i] e^{-2\pi j (\Delta f + iB_c)t} \quad 0 \le t \le T, \tag{8.4}$$

where $s[i]$ is the symbol carried on the ith subcarrier; B_c is the frequency separation between two adjacent subcarriers, also referred to as the subcarrier bandwidth; Δf is the frequency of the first subcarrier; and T' is the total useful symbol duration (without the cyclic prefix). At the receiver, the symbol sent on a specific subcarrier is retrieved by integrating the received signal with a complex conjugate of the tone signal over the entire symbol duration T'. If the time and the frequency synchronization between the receiver and the transmitter is perfect, the orthogonality between the subcarriers is preserved at the receiver. When the time and/or frequency synchronization between the transmitter and the receiver is not perfect,[3] the orthogonality between the subcarriers is lost, resulting in *intercarrier interference* (ICI). Timing mismatch can occur due to misalignment of the clocks at the transmitter and the receiver and propagation delay of the channel. Frequency mismatch can occur owing to relative drift between the oscillators at the transmitter and the receiver and nonlinear channel effects, such as Doppler shift. The flexibility of the WiMAX PHY layer allows one to make an optimum choice of various PHY layer parameters, such as cyclic prefix length, number of subcarriers, subcarrier separation, and preamble interval, such that the performance degradation owing to ICI and ISI (intersymbol interference) is minimal

---

3. Time synchronization is not as critical as frequency synchronization, as long as it is within the cyclic prefix window.

without compromising the performance. The four primitive parameters that describe an OFDM symbol, and their respective values in IEEE 802.16e-2005, are shown in Table 8.3.

As discussed in Chapter 4, the concept of independently modulating multiple orthogonal frequency tones with narrowband symbol streams is equivalent to first constructing the entire OFDM signal in the frequency domain and then using an inverse fast fourier transform to convert the signal into the time domain. The IFFT method is easier to implement, as it does not require multiple oscillators to transmit and receive the OFDM signal. In the frequency domain, each OFDM symbol is created by mapping the sequence of symbols on the subcarriers. WiMAX has three classes of subcarriers.

1. *Data subcarriers* are used for carrying data symbols.
2. *Pilot subcarriers* are used for carrying pilot symbols. The pilot symbols are known a priori and can be used for channel estimation and channel tracking.
3. *Null subcarriers* have no power allocated to them, including the DC subcarrier and the guard subcarriers toward the edge. The DC subcarrier is not modulated, to prevent any saturation effects or excess power draw at the amplifier. No power is allocated to the guard subcarrier toward the edge of the spectrum in order to fit the spectrum, of the OFDM symbol within the allocated bandwidth and thus reduce the interference between adjacent channels.

Figure 8.8 shows a typical frequency domain representation of an IEEE 802.16e-2005 OFDM symbol containing the data subcarriers, pilot subcarriers, and null subcarriers. The power in the pilot subcarriers, as shown here, is boosted by 2.5 dB, allowing reliable channel tracking even at low-SNR conditions.

Table 8.3 Primitive Parameters for OFDM Symbol[a]

Parameter	Value (MHz)	Definition
B	Variable (1.25, 1.75, 3.5, 5, 7, 8.75, 10, 14, 15[b])	Nominal channel bandwidth
L	256 for OFDM; 128, 512, 1,024, 2,048 for SOFDMA	Number of subcarriers, including the DC subcarrier pilot subcarriers and the guard subcarriers
n	8/7, 28/25	Oversampling factor
G	1/4, 1/8, 1/16, and 1/32	Ratio of cyclic prefix time to useful symbol time

a. Not all values are part of the initial WiMAX profile.

b. The 8.75MHz channel bandwidth is for WiBro.

8.6 Subchannel and Subcarrier Permutations

In order to create the OFDM symbol in the frequency domain, the modulated symbols are mapped on to the subchannels that have been allocated for the transmission of the data block.

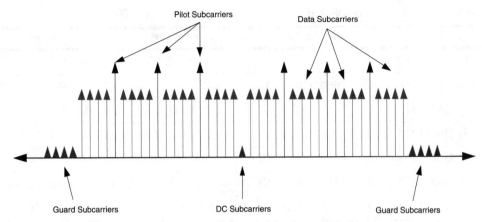

Figure 8.8 Frequency-domain representation of OFDM symbol

A subchannel, as defined in the IEEE 802.16e-2005 standard, is a logical collection of subcarriers. The number and exact distribution of the subcarriers that constitute a subchannel depend on the *subcarrier permutation* mode. The number of subchannels allocated for transmitting a data block depends on various parameters, such as the size of the data block, the modulation format, and the coding rate. In the time and frequency domains, the contiguous set of subchannels allocated to a single user—or a group of users, in case of multicast—is referred to as the *data region* of the user(s) and is always transmitted using the same *burst profile*. In this context, a burst profile refers to the combination of the chosen modulation format, code rate, and type of FEC: convolutional codes, turbo codes, and block codes. The allowed uplink and downlink burst profiles in IEEE 802.16e-2005 are shown in Table 8.4.

The BPSK R1/2 burst profile, used only for broadcast control messages, is not an allowed burst profile for transmission of data or dedicated control messages in the OFDMA mode. However, in the OFDM mode, the BPSK R1/2 is an allowed burst profile for data and dedicated control messages.

It is important to realize that in WiMAX, the subcarriers that constitute a subchannel can either be adjacent to each other or distributed throughout the frequency band, depending on the subcarrier permutation mode. A distributed subcarrier permutation provides better frequency diversity, whereas an adjacent subcarrier distribution is more desirable for beamforming and allows the system to exploit multiuser diversity. The various subcarrier permutation schemes allowed in IEEE 802.16e-2005 are discussed next.

8.6.1 Downlink Full Usage of Subcarriers

In the case of DL FUSC, all the data subcarriers are used to create the various subchannels. Each subchannel is made up of 48 data subcarriers, which are distributed evenly throughout the entire frequency band, as depicted in Figure 8.9. In FUSC, the pilot subcarriers are allocated first, and

Table 8.4 Uplink and Downlink Burst Profiles in IEEE 802.16e-2005

	Format		Format		Format		Format
0	QPSK CC^a 1/2	14	Reserved	28	64 QAM ZCC 3/4	42	64 QAM LDPC 2/3
1	QPSK CC 3/4	15	QPSK CTC^b 3/4	29	QPSK LDPC 1/2	43	64 QAM LDPC 3/4
2	16 QAM CC 1/2	16	16 QAM CTC 1/2	30	QPSK LDPC 2/3	44^c	QPSK CC 1/2
3	16 QAM CC 3/4	17	16 QAM CTC 3/4	31	QPSK LDPC 3/4	45^c	QPSK CC 3/4
4	64 QAM CC 1/2	18	64 QAM CTC 1/2	32	16 QAM LDPC 1/2	46^c	16 QAM CC 1/2
5	64 QAM CC 2/3	19	64 QAM CTC 2/3	33	16 QAM LDPC 2/3	47^c	16 QAM CC 3/4
6	64 QAM CC 3/4	20	64 QAM CTC 3/4	34	16 QAM LDPC 3/4	48^c	64 QAM CC 2/3
7	QPSK BTC^d 1/2	21	64 QAM CTC 5/6	35	64 QAM LDPC 1/2	49^c	64 QAM CC 3/4
8	QPSK BTC 3/4	22	QPSK ZCC^e 1/2	36	64 QAM LDPC 2/3	50	QPSK LDPC 5/6
9	16 QAM BTC 3/5	23	QPSK ZCC 3/4	37	64 QAM LDPC 3/4	51	16 QAM LDPC 5/6
10	16 QAM BTC 4/5	24	16 QAM ZCC 1/2	38^f	QPSK LDPC 2/3	52	64 QAM LDPC 5/6
11	64 QAM BTC 5/8	25	16 QAM ZCC 3/4	39^f	QPSK LDPC 3/4		> 52 reserved
12	64 QAM BTC 4/5	26	64 QAM ZCC 1/2	40^f	16 QAM LDPC 2/3		
13	QPSK CTC 1/2	27	64 QAM ZCC 2/3	41^f	16 QAM LDPC 3/4		

a. Convolutional code

b. Convolutional turbo code

c.   44–49 use the optional interleaver with the convolutional codes

d. Block turbo codes

e. Zero-terminating convolutional code, which uses a padding byte of 0 x 00 instead of tailbiting

f.   38–43 use the B code for LDPC; other burst profiles with LDPC use A code

then the remainder of the subcarriers are mapped onto the various subchannels, using a permutation scheme [3, 4]. The set of the pilot subcarriers is divided in to two constant sets and two variables sets. The index of the pilot subcarriers belonging to the variable sets changes from one OFDM symbol to the next, whereas the index of the pilot subcarriers belonging to the constant sets remains unchanged. The variable sets allow the receiver to estimate the channel response more accurately across the entire frequency band, which is especially important in channels with

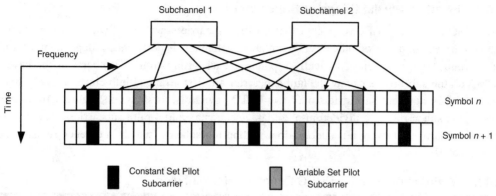

Figure 8.9 FUSC subcarrier permutation scheme

Table 8.5 Parameters of FUSC Subcarrier Permutation

	128	256[a]	512	1,024	2,048
Subcarriers per subchannel	48	N/A	48	48	48
Number of subchannels	2	N/A	8	16	32
Data subcarriers used	96	192	384	768	1,536
Pilot subcarrier in constant set	1	8	6	11	24
Pilot subcarriers in variable set	9	N/A	36	71	142
Left-guard subcarriers	11	28	43	87	173
Right-guard subcarriers	10	27	42	86	172

a. The 256 mode, based on 802.16-2004, does not use FUSC or PUSC but has been listed here for the sake of completeness.

large delay spread (small coherence bandwidth). The various parameters related to the FUSC permutation scheme for different FFT sizes are shown in Table 8.5. When transmit diversity using two antennas is implemented with FUSC, each of the two antennas uses only half of the pilot subcarriers from the variable set and the constant set. This allows the receiver to estimate the channel impulse response from each of the transmitter antennas. Similarly, in the case of transmit diversity with three or four antennas, each antenna is allocated every third or every fourth pilot subcarrier, respectively. The details of space/time coding and how the pilot and data subcarriers are used in that case are explained in more detail in Section 8.8.

8.6.2 Downlink Partial Usage of Subcarriers

DL PUSC is similar to FUSC except that all the subcarriers are first divided into six groups (Table 8.6). Permutation of subcarriers to create subchannels is performed independently within each group, thus, in essence, logically separating each group from the others. In the case of PUSC, all the subcarriers except the null subcarrier are first arranged into clusters. Each cluster consists of 14 adjacent subcarriers over two OFDM symbols, as shown in Figure 8.10. In each cluster, the subcarriers are divided into 24 data subcarriers and 4 pilot subcarriers. The clusters are then renumbered using a pseudorandom numbering scheme, which in essence redistributes the logical identity of the clusters.

Table 8.6 Parameters of DL PUSC Subcarrier Permutation

	128	512	1,024	2,048
Subcarriers per cluster	14	14	14	14
Number of subchannels	3	15	30	60
Data subcarriers used	72	360	720	1,440
Pilot subcarriers	12	60	120	240
Left-guard subcarriers	22	46	92	184
Right-guard subcarriers	21	45	91	183

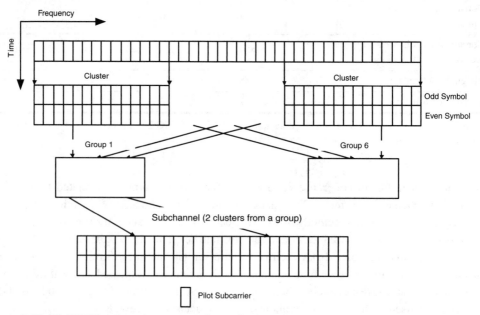

Figure 8.10 DL PUSC subcarrier permutation scheme

After renumbering, the clusters are divided into six groups, with the first one-sixth of the clusters belonging to group 0, and so on. A subchannel is created using two clusters from the same group, as shown in Figure 8.10.

In PUSC, it is possible to allocate all or only a subset of the six groups to a given transmitter. By allocating disjoint subsets of the six available groups to neighboring transmitters, it is possible to separate their signals in the subcarrier space, thus enabling a tighter frequency reuse at the cost of data rate. Such a usage of subcarriers is referred to as *segmentation*. For example, in a BS with three sectors using segmentation, it is possible to allocate two distinct groups to each sector, thus reusing the same RF frequency in all of them. By default, group 0 is always allocated to sector 1, group 2 is always allocated to sector 2, and group 4 is always allocated to sector 3. The distribution of the remaining groups can be done based on demand and can be implementation specific.

By using such a segmentation scheme, all the sectors in a BS can use the same RF channel, while maintaining their orthogonality among subcarriers. This feature of WiMAX systems for OFDMA mode is very useful when the available spectrum is not large enough to permit anything more than a (1,1) frequency reuse. It should be noted that although segmentation can be used with PUSC, PUSC by itself does not demand segmentation.

8.6.3 Uplink Partial Usage of Subcarriers

In UL PUSC, the subcarriers are first divided into various tiles, as shown in Figure 8.11. Each tile consists of four subcarriers over three OFDM symbols. The subcarriers within a tile are divided into eight data subcarriers and four pilot subcarriers. An optional PUSC mode is also allowed in the uplink, whereby each tile consists of three subcarriers over three OFDM symbols as shown in Figure 8.12. In this case, the data subcarriers of a tile are divided into eight data subcarriers and one pilot subcarrier. The optional UL PUSC mode has a lower ratio of pilot subcarriers to data subcarriers, thus providing a higher effective data rate but poorer channel-tracking capability. The two UL PUSC modes allow the system designer a trade-off between higher data rate and more accurate channel tracking depending on the Doppler spread and coherence bandwidth of the channel. The tiles are then renumbered, using a pseudorandom numbering sequence, and divided into six groups. Each subchannel is created using six tiles from a single group. UL PUSC can be used with segmentation in order to allow the system to operate under tighter frequency reuse patterns.

8.6.4 Tile Usage of Subcarriers

The TUSC (tile usage of subcarriers) is a downlink subcarrier permutation mode that is identical to the uplink PUSC. As illustrated in the previous section, the creation of subchannels from the available subcarriers is done differently in the UL PUSC and DL PUSC modes. If closed loop advanced antenna systems (AAS) are to be used with the PUSC mode, explicit feedback of the channel state information (CSI) from the MS to the BS would be required even in the case of TDD, since the UL and DL allocations are not symmetric, and channel reciprocity cannot be used. TUSC allows for a DL allocation that is symmetric to the UL PUSC, thus taking advantage

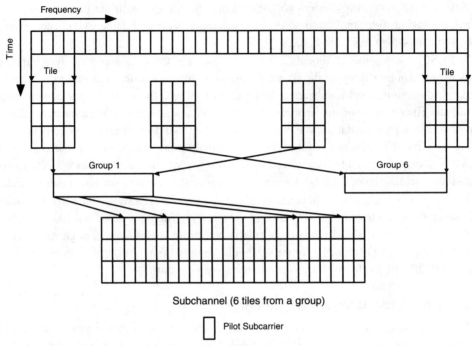

Figure 8.11 UL PUSC subcarrier permutation scheme

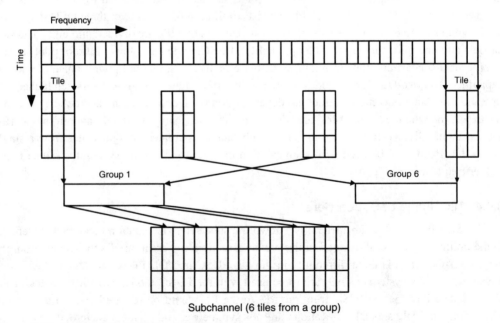

Figure 8.12 Optional UL PUSC subcarrier permutation scheme

of UL and DL allocation symmetry and eliminating the requirement for explicit CSI feedback in the case of closed-loop AAS for TDD systems. The two TUSC modes defined in WiMAX, TUSC1 and TUSC2, correspond to the UL PUSC and the optional UL PUSC modes, respectively.

8.6.5 Band Adaptive Modulation and Coding

Unique to the band AMC permutation mode, all subcarriers constituting a subchannel are adjacent to each other. Although frequency diversity is lost to a large extent with this subcarrier permutation scheme, exploitation of multiuser diversity is easier. Multiuser diversity provides significant improvement in overall system capacity and throughput, since a subchannel at any given time is allocated to the user with the highest SNR/capacity in that subchannel. Overall performance improvement in WiMAX due to multiuser diversity, is shown in Chapters 11 and 12, using link-and-system level simulations. Because of the dynamic nature of the wireless channel, different users get allocated on the subchannel at different instants in time as they go through the crests of their uncorrelated fading waveforms.

In this subcarrier permutation, nine adjacent subcarriers with eight data subcarriers and one pilot subcarrier are used to form a bin, as shown in Figure 8.13. Four adjacent bins in the frequency domain constitute a band. An AMC subchannel consists of six contiguous bins from within the same band. Thus, an AMC subchannel can consist of one bin over six consecutive symbols, two consecutive bins over three consecutive symbols, or three consecutive bins over two consecutive symbols.

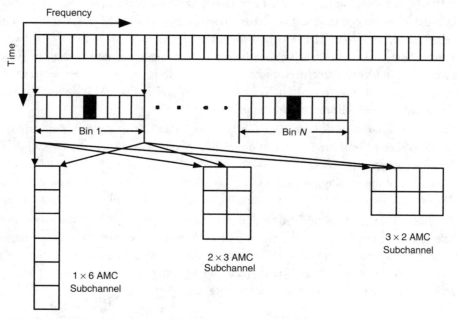

Figure 8.13 Band AMC subcarrier permutation

8.7 Slot and Frame Structure

The MAC layer allocates the time/frequency resources to various users in units of *slots*, which is the smallest quanta of PHY layer resource that can be allocated to a single user in the time/frequency domain. The size of a slot is dependent on the subcarrier permutation mode.

- FUSC: Each slot is 48 subcarriers by one OFDM symbol.
- Downlink PUSC: Each slot is 24 subcarriers by two OFDM symbols.
- Uplink PUSC and TUSC: Each slot is 16 subcarriers by three OFDM symbols.
- Band AMC: Each slot is 8, 16, or 24 subcarriers by 6, 3, or 2 OFDM symbols.

In the time/frequency domain, the contiguous collections of slots that are allocated for a single user from the *data region* of the given user. It should be noted that the scheduling algorithm used for allocating data regions to various users is critical to the overall performance of a WiMAX system. A smart scheduling algorithm should adapt itself to not only the required QoS but also the instantaneous channel and load conditions. Scheduling algorithms and their various advantages and disadvantages are discussed in Chapter 6.

In IEEE 802.16e-2005, both frequency division duplexing and time division duplexing are allowed. In the case of FDD, the uplink and downlink subframes are transmitted simultaneously on different carrier frequencies; in the case of TDD, the uplink and downlink subframes are transmitted on the same carrier frequency at different times. Figure 8.14 shows the frame structure for TDD. The frame structure for the FDD mode is identical except that the UL and DL subframes are multiplexed on different carrier frequencies. For mobile stations, (MS) an additional duplexing mode, known as H-FDD (half-duplex FDD) is defined. H-FDD is a basic FDD duplexing scheme with the restriction that the MS cannot transmit and receive at the same time. From a cost and implementation perspective, an H-FDD MS is cheaper and less complex than its FDD counterpart, but the UL and DL peak data rate of, an H-FDD MS are less, owing to its inability to receive and transmit simultaneously.

Each DL subframe and UL subframe in IEEE 802.16e-2005 is divided into various zones, each using a different subcarrier permutation scheme. Some of the zones, such as DL PUSC, are mandatory; other zones, such as FUSC, AMC, UL PUSC, and TUSC, are optional. The relevant information about the starting position and the duration of the various zones being used in a UL and DL subframe is provided by control messages in the beginning of each DL subframe.

The first OFDM symbol in the downlink subframe is used for transmitting the DL preamble. The preamble can be used for a variety of PHY layer procedures, such as time and frequency synchronization, initial channel estimation, and noise and interference estimation. The subcarriers in the preamble symbol are divided into a group of three carrier sets. The indices of subcarriers associated with a given carrier set are given by

$$Carrier_{n,\,k} = k + 3n, \tag{8.5}$$

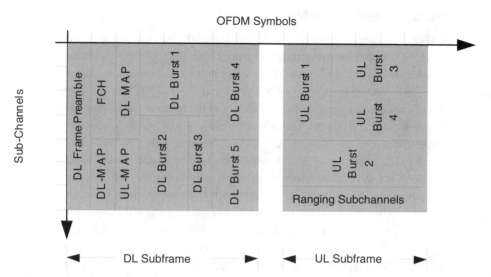

Figure 8.14 TDD frame structure

where the carrier set index, k, runs from 0 to 2, and the subcarrier index runs from 0 to $(N_{used}-3)/3$. Each segment (sector), as defined in the PUSC subcarrier permutation section, uses a preamble composed of only one of the three allowed carrier sets, thus modulating every third subcarrier. A cell-ID-specific PN (pseudonoise) sequence is modulated, using BPSK to create the preamble in the frequency domain. The power of the subcarriers belonging to the carrier set of the preamble is boosted by $2\sqrt{2}$. The frame length, which is defined by the interval between two consecutive DL frame preambles, is variable in WiMAX and can be anywhere between 2msec and 20msec.

In the OFDM symbol following the DL frame preamble, the initial subchannels are allocated for the frame correction header. The FCH is used for carrying system control information, such as the subcarriers used (in case of segmentation), the ranging subchannels, and the length of the DL-MAP message. This information is carried on the DL_Frame_Prefix message contained within the FCH. The FCH is always coded with the BPSK R1/2 mode to ensure maximum robustness and reliable performance, even at the cell edge.

Following the FCH are the DL-MAP and the UL-MAP messages, respectively, which specify the data regions of the various users in the DL and UL subframes of the current frame. By listening to these messages, each MS can identify the subchannels and the OFDM symbols allocated in the DL and UL for its use. Periodically, the BS also transmits the downlink channel descriptor (DCD) and the uplink channel descriptor (UCD) following the UL-MAP message, which contains additional control information pertaining to the description of channel structure and the various burst profiles[4] that are allowed within the given BS. In order to conserve resources, the DCD and the UCD are not transmitted every DL frame.

---

4. As defined previously, a *burst profile* is the combination of modulation constellation, code rate, and the FEC used.

8.8 Transmit Diversity and MIMO

Support for AAS is an integral part of the IEEE 802.16e-2005 and is intended to provide significant improvement in the overall system capacity and spectral efficiency of the network. Expected performance improvements in a WiMAX network owing to multiantenna technology, based on link- and system-level simulations, are presented in Chapter 11 and 12. In IEEE 802.16e-2005, AAS encompasses the use of multiple antennas at the transmitter and the receiver for different purposes, such as diversity, beamforming, and spatial multiplexing (SM). When AAS is used in the open-loop mode—the transmitter does not know the CSI as seen by the specific receiver—the multiple antennas can be used for diversity (space/time block coding), spatial multiplexing, or any combination thereof. When AAS is used in closed-loop mode, the transmitter knows the CSI, either due to channel reciprocity, in case of TDD, or to explicit feedback from the receiver, in the case of FDD, the multiple antennas can be used for either beamforming or closed-loop MIMO, using transmit precoding. In this section, we describe the open- and closed-loop AAS modes of IEEE 802.16e-2005.

8.8.1 Transmit Diversity and Space/Time Coding

Several optional space/time coding schemes with two, three, and four antennas that can be used with both adjacent and diversity subcarrier permutations are defined in IEEE 802.16e-2005. Of these, the most commonly implemented are the two antenna open-loop schemes, for which the following space/time coding matrices are allowed:

$$
B = \begin{bmatrix} S_1 \\ S_2 \end{bmatrix} \qquad A = \begin{bmatrix} S_1 & -S^*_2 \\ S_2 & S^*_1 \end{bmatrix} \quad , \tag{8.6}
$$

where S1 and S2 are two consecutive OFDM symbols, and the space/time encoding matrices are applied on the entire OFDM symbol, as shown in Figure 8.15. The matrix A in Equation (8.6) is the 2×2 Alamouti space/time block codes [1], which are orthogonal in nature and amenable to a linear optimum maximum-likelihood (ML) detector.[5] This provides significant performance benefit by means of diversity in fading channels. On the other hand, the matrix B as provided—see Equation (8.6)—does not provide any diversity but has a space/time coding rate of 2 (spatial multiplexing), which allows for higher data rates. Transmit diversity and spatial multiplexing are discussed in more detail in Chapter 6. Similarly, space/time coding matrices have been defined with three and four antennas. In the case of four antenna transmit diversity, the space/time coding matrix allows for a space/time code rate of 1 (maximum diversity) to a space-time code rate of 4 (maximum capacity), as shown by block coding matrices **A**, **B**, and **C** in Equation (8.7). By using more antennas, the system can perform a finer trade-off between diversity and capacity. For transmit diversity modes with a space/time code rate greater than 1, both horizontal and vertical encoding

---

5. For complex modulation schemes, the full-rate space/time block codes with more than two antennas are no longer orthogonal and do not allow a linear ML detection. More realistic detections schemes involving MRC or MMSE are suboptimal in performance compared to the linear ML detector.

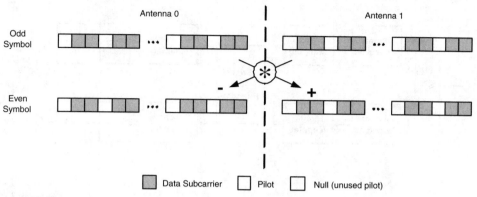

Figure 8.15 Transmit diversity using space/time coding

are allowed, as shown in Figure 8.16. In the case of horizontal encoding, the multiple streams are coded (FEC) and modulated independently before being presented to the space/time encoding block. In the case of vertical encoding, the multiple streams are coded and modulated together before being presented to the space/time encoding block. When multiple antennas are used, the receiver must estimate the channel impulse response from each of the transmit antennas in order to detect the signal. In IEEE 802.16e-2005, this is achieved by the using of MIMO midambles or by distributing the pilot subcarriers among the various transmit antennas.

$$A = \begin{bmatrix} S_1 & -S^*_2 & 0 & 0 \\ S_2 & S^*_1 & 0 & 0 \\ 0 & 0 & S_3 & -S^*_3 \\ 0 & 0 & S_4 & S^*_3 \end{bmatrix} \qquad R = 1 \qquad\qquad (8.7)$$

$$B = \begin{bmatrix} S_1 & -S^*_2 & S_5 & -S^*_7 \\ S_2 & S^*_1 & S_6 & -S^*_8 \\ S_3 & -S^*_4 & S_7 & S^*_5 \\ S_4 & S^*_3 & S_8 & S^*_6 \end{bmatrix} \qquad R = 2$$

$$C = \begin{bmatrix} S_1 \\ S_2 \\ S_3 \\ S_4 \end{bmatrix} \qquad R = 4.$$

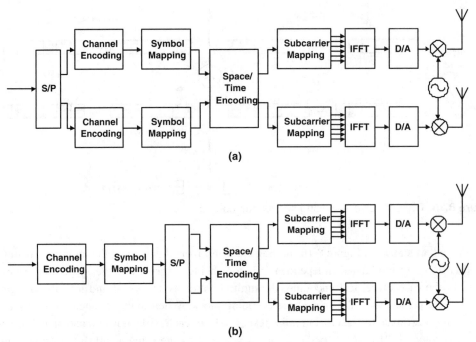

Figure 8.16 (a) Horizontal and (b) vertical encoding for two antennas

When multiple antennas are used with the FUSC subcarrier permutation, the pilot subcarriers in each symbol are divided among antennas. In the case of two antennas, the pilots are divided in the following fashion:

- Symbol 0: Antenna 0 uses variable set 0 and constant set 0, and antenna 1 uses variable set 1 and constant set 1.
- Symbol 1: Antenna 0 uses variable set 1 and constant set 1, and antenna 1 uses variable set 0 and constant set 0.

Similarly when four antennas are used for FUSC subcarrier permutation, the pilots are divided among the antennas in the following fashion.

- Symbol 0: Antenna 0 uses variable set 0 and constant set 0, and antenna 1 uses variable set 1 and constant set 1.
- Symbol 1: Antenna 2 uses variable set 0 and constant set 0, and antenna 3 uses variable set 1 and constant set 1.
- Symbol 2: Antenna 0 uses variable set 1 and constant set 1, and antenna 1 uses variable set 0 and constant set 0.

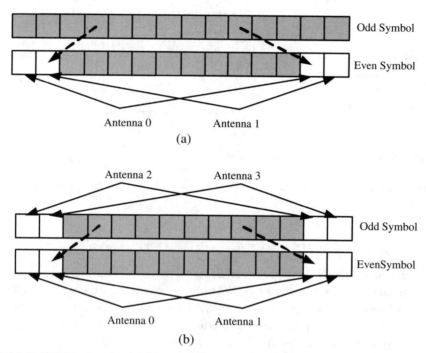

Figure 8.17 PUSC Clusters for (a) two- and four-antenna transmissions

• Symbol 3: Antenna 2 uses variable set 1 and constant set 1, and antenna 3 uses variable set 0 and constant set 0.

For the PUSC subcarrier permutation, a separate cluster structure, as shown in Figure 8.17, is implemented when multiple antennas are used. When three antennas are used for transmission, the pilot pattern distribution is the same as in the case of four antennas, but only the patterns for antennas 50, 1, and 2 are used for transmission.

8.8.2 Frequency-Hopping Diversity Code

In the case of space/time encoding using multiple antennas, the entire OFDM symbol is operated by the space/time encoding matrix, as shown in Figure 8.15. IEEE 802.16e-2005 also defines an optional transmit diversity mode, known as the frequency-hopping diversity code (FHDC), using two antennas in which the encoding is done in the space and frequency domain, as shown in Figure 8.18 rather than the space and time domain. In FHDC, the first antenna transmits the OFDM symbols without any encoding, much like a single-antenna transmission, and the second antenna transmits the OFDM symbol by encoding it over two consecutive subchannels, using the 2×2 Alamouti encoding matrix, as shown in Figure 8.18.

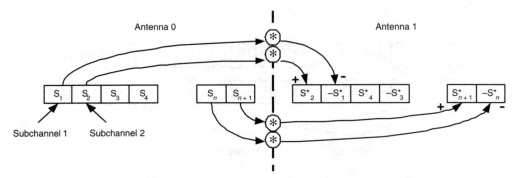

Figure 8.18 Frequency-hopping diversity code

The received signal in the nth and $(n + 1)$th subchannel can then be written as

$$
\begin{bmatrix} r_n \\ r^*_{n+1} \end{bmatrix} = \begin{bmatrix} h_{1,\,n} & h_{2,\,n} \\ -h^*_{2,\,n+1} & h^*_{1,\,n+1} \end{bmatrix} \begin{bmatrix} s_n \\ s^*_{n+1} \end{bmatrix} + \begin{bmatrix} z_n \\ z^*_{n+1} \end{bmatrix}. \tag{8.8}
$$

Although equation (8.8) shows the received signal in the nth and $(n + 1)$th subchannel, the reception is done on a per subcarrier basis. When the subcarriers corresponding to the nth and $(n + 1)$th subchannel are far apart relative to the coherence bandwidth of the channel, the space/time coding is not orthogonal, and the maximum-likelihood detector is not linear. In such a case, an MMSE or BLAST space/time detection scheme is required.

8.9 Closed-Loop MIMO

The various transmit diversity and spatial-multiplexing schemes of IEEE 802.16e-2005 described in the previous section do not require the transmitter to know the CSI for the receiver of interest. As discussed in Chapter 5, MIMO and diversity schemes can benefit significantly if the CSI is known at the transmitter. CSI information at the transmitter can be used to select the appropriate MIMO mode—number of transmit antennas, number of simultaneous streams, and space/time encoding matrix—as well as to calculate an optimum precoding matrix that maximizes system capacity. The CSI can be known at the transmitter due to channel reciprocity, in the case of TDD, or by having a feedback channel, in the case of FDD. The uplink bandwidth required to provide the full CSI to the transmitter—the MIMO channel matrix for each subcarrier in a multiuser FDD MIMO-OFDM system—is too large and thus impractical for a closed-loop FDD MIMO system. For practical systems, it is possible only to send some form of quantized information in the uplink. The framework for closed-loop MIMO in IEEE 802.16e-2005, as shown in Figure 8.19, consists of a space/time encoding stage identical to an open-loop system and a MIMO precoding stage. The MIMO precoding matrix in general is a complex matrix, with the number of rows equal to the number of transmit antennas and the number of columns equal to the output of the space/time encoding block. The linear precoding matrix spatially mixes the various parallel streams among the various antennas, with appropriate amplitude and phase adjustment.

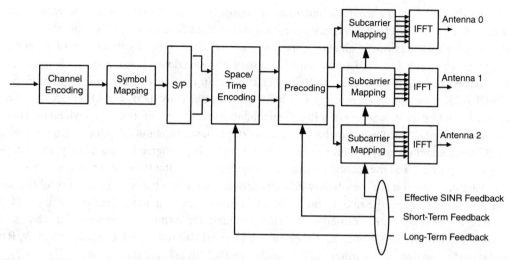

Figure 8.19 Closed-loop MIMO framework in IEEE 802.16e-2005

In order to determine the appropriate amplitude and phases of the various weights, the transmitter requires some feedback from the MS. In the case of closed-loop MIMO, the feedback falls broadly into two categories: long-term feedback and short-term feedback. The long-term feedback provides information related to the maximum number of parallel streams: the rank of the precoding matrix to be used for DL transmissions. The short-term feedback provides information about the precoding matrix weights to be used. The IEEE 802.16e-2005 standard defines the following five mechanisms so that the BS can estimate the optimum precoding matrix for closed-loop MIMO operations:

1. **Antenna selection.** The MS indicates to the BS which transmit antenna(s) should be used for transmission in order to maximize the channel capacity and/or improve the link reliability.
2. **Antenna grouping.** The MS indicates to the BS the optimum permutation of the order of the various antennas to be used with the current space/time encoding matrix.
3. **Codebook based feedback.** The MS indicates to the BS the optimum precoding matrix to be used, based on the entries of a predefined codebook.
4. **Quantized channel feedback.** The MS quantizes the MIMO channel and sends this information to the BS, using the MIMO_FEEDBACK message. The BS can use the quantized MIMO channel to calculate an optimum precoding matrix.
5. **Channel sounding.** The BS obtains exact information about the CSI of the MS by using a dedicated and predetermined signal intended for channel sounding.

8.9.1 Antenna Selection

When the number of the transmit antennas N_t is larger than the number of parallel streams N_s—rank of the precoding matrix based on the long-term feedback—the antenna-selection feedback tells the BS which of the available antennas are optimal for DL transmission. The MS usually calculates the MIMO channel capacity for each possible antenna combination and chooses the combination that maximizes channel capacity. The MS then indicates its choice of antennas, using the secondary fast-feedback channel. Primary and secondary fast-feedback channels can be allocated to individual MSs, which the MS can use in a unicast manner to send the FAST-FEEDBACK message. Each primary fast-feedback channel consists of one OFDMA slot. The MS uses the 48 data subcarriers of a PUSC subchannel to carry an information payload of 6 bits. The secondary fast-feedback subchannel, on the other hand, uses the 24 pilot subcarriers of a PUSC subchannel to carry a 4-bit payload. Due to such a high degree of redundancy, the reception of the primary and the secondary fast-feedback message at the BS is less prone to errors.

Antenna selection is a very bandwidth-efficient feedback mechanism and is a useful feature at higher speeds, when the rate of the feedback is quite high. Antenna selection has the added advantage that unlike other closed-loop MIMO modes, the number of required RF chains is equal to the number of streams N_s. Other closed-loop MIMO schemes require a total of N_t RF chains at the transmitter, regardless of how many parallel streams are transmitted.

8.9.2 Antenna Grouping

Antenna grouping is a concept that allows the BS to permutate the logical order of the transmit antennas. As shown in Equation (8.9), if A1 is considered the natural order, A2 implies that the logical order of the transmit antennas 2 and 3 is switched. Similarly, A3 implies that first, the logical order of the antennas 2 and 4 is switched, and then the logical order of antennas 3 and 4 is switched. The MS indicates the exact permutation and the number of transmit antennas to be used by the primary fast-feedback channel. Antenna grouping can also be performed with all the space/time encoding matrices, as described in the previous section for two, three, and four antennas.

$$
A1 = \begin{bmatrix}
S_1 & -S^*_2 & 0 & 0 \\
S_2 & S^*_1 & 0 & 0 \\
0 & 0 & S_3 & -S^*_3 \\
0 & 0 & S_4 & S^*_3
\end{bmatrix}
$$

$$
A2 = \begin{bmatrix}
S_1 & -S^*_2 & 0 & 0 \\
0 & 0 & S_3 & -S^*_3 \\
S_2 & S^*_1 & 0 & 0 \\
0 & 0 & S_4 & S^*_3
\end{bmatrix}
$$

$$
A3 = \begin{bmatrix}
S_1 & -S^*_2 & 0 & 0 \\
0 & 0 & S_3 & -S^*_3 \\
0 & 0 & S_4 & S^*_3 \\
S_2 & S^*_1 & 0 & 0
\end{bmatrix}
$$

(8.9)

8.9.3 Codebook Based Feedback

Codebook based feedback allows the MS to explicitly identify a precoding matrix based on a codebook that should be used for DL transmissions. Separate codebooks are defined in the standard for various combinations of number of streams N_s and number of transmit antennas N_t. For each combination of N_s and N_t, the standard defines two codebooks: the first with 8 entries and the second with 64 entries. If it chooses a precoding matrix from the codebook with 8 entries, the MS can signal this to the BS by using a 3-bit feedback channel. On the other hand, if it chooses a precoding matrix from the codebook with 64 entries, the MS can indicate its choice to the BS by using a 6-bit feedback channel. This choice of two codebooks allows the system to perform a controlled trade-off between performance and feedback efficiency. For band AMC operation, the BS can instruct the MS to provide either a single precoder for all the bands of the preferred subchannels or different precoders for the N best bands.

The IEEE 802.16e-2005 standard does not specify what criteria the MS should use to calculate the optimum precoding matrix. However, two of the more popular criteria are maximization of sum capacity and minimization of mean square error (MSE). Link performances of the various codebook selection criteria and their comparison to more optimal closed-loop precoding techniques, such as based on singular-value decomposition precoding, are provided in Chapter 11.

8.9.4 Quantized Channel Feedback

Quantized MIMO feedback allows the MS to explicitly inform the BS of its MIMO channel state information. The MS quantizes the real and imaginary components of the $N_t \times N_r$ MIMO channel to a 6-bit binary number and then sends this information to the BS, using the fast-feedback channel. Clearly, the quantized channel feedback requires much more feedback bandwidth in the UL compared to the codebook-based method. For example, in the case of a IEEE 802.16e-2005 system with four antennas at the transmitter and two antennas at the receiver, a quantized channel feedback would require 16×6 bits to send the feedback as opposed to the codebook based method, which would require only 6 bits. Owing to the high-bandwidth requirement of the quantized channel feedback mode, we envision this mode to be useful only in pedestrian and stationary conditions. In such slow-varying channel conditions, the rate at which the MS needs to provide this feedback is greatly reduced, thus still maintaining a reasonable bandwidth efficiency.

Again, the IEEE 802.16e-2005 standard does not specify what criteria the BS needs to use in order to calculate an optimum precoder, but two of the most popular criteria are maximization of sum capacity and minimization of MSE. Link performances of various optimization criteria and their performance relative to other techniques are provided in Chapter 11.

8.9.5 Channel Sounding

As defined in the standard for TDD operations, the channel-sounding mechanism involves the MS's transmitting a deterministic signal that can be used by the BS to estimate the UL channel from the MS. If the UL and DL channels are properly calibrated, the BS can then use the UL channel as an estimate of the DL channel, due to channel reciprocity.

The BS indicates to the MS, using the UL_MAP message if a UL sounding zone has been allocated for a user in a given frame. On the receipt of such instructions, the MS sends a UL channel-sounding signal in the allocated sounding zone. The subcarriers within the sounding zone are divided into nonoverlapping sounding frequency bands, with each band consisting of 18 consecutive subcarriers. The BS can instruct the MS to perform channel sounding over all the allowed subcarriers or a subset thereof. For example, when 2,048 subcarriers are used, the maximum number of usable subcarriers is 1,728. Thus, the entire channel bandwidth can be divided into $1,728/18 = 96$ sounding frequency bands. In order to enable DL channel estimation at the BS in mobile environments, the BS can also instruct the MS to perform periodic UL channel sounding.

The channel-sounding option for closed-loop MIMO operation is the most bandwidth-intensive MIMO channel-feedback mechanism, but it provides the BS with the most accurate estimate of the DL channel, thus providing maximum capacity gain over open-loop modes.

8.10 Ranging

In IEEE 802.16e-2005, ranging is an uplink physical layer procedure that maintains the quality and reliability of the radio-link communication between the BS and the MS. When it receives the ranging transmission from a MS, the BS processes the received signal to estimate various radio-link parameters, such as channel impulse response, SINR, and time of arrival, which allows the BS to indicate to the MS any adjustments in the transmit power level or the timing offset that it might need relative to the BS. Initial and periodic ranging processes that allow the BS and the MS to perform time and power synchronization with respect to each other during the initial network reentry and periodically, respectively are supported.

The ranging procedure involves the transmission of a predermined sequence, known as the ranging code, repeated over two OFDM symbols using the ranging channel, as shown in Figure 8.20. For the purposes of ranging, it is critical that no phase discontinuity[6] occur at the OFDM symbol boundaries, even without windowing, which is guaranteed by constructing the OFDM symbols in the manner shown in Figure 8.20. The first OFDM symbol of the ranging subchannels is created like any normal OFDM symbol: performing an IFFT operation on the ranging code and then appending, at the begining, a segment of length T_g from the end. The second OFDM symbol is created by performing an IFFT on the same ranging code and by then appending, at the end, a segment of length T_g from the beginning of the symbol.

Creating the second OFDM symbol of the ranging subchannels in this manner guarantees that there is no phase discontinuity at the boundary between the two consecutive symbols. Such a construction of the ranging code allows the BS to properly receive the requests from an un-

---

6. During a ranging process, the BS determines the parameters of ranging by correlating the received signal with an expected copy of the signal, which is known by the BS a priori. In order for the correlation process to work over the entire ranging signal, which spans multiple OFDM symbols, there must be no discontinuity of the signal across OFDM symbols.

OFDM Symbol Construction

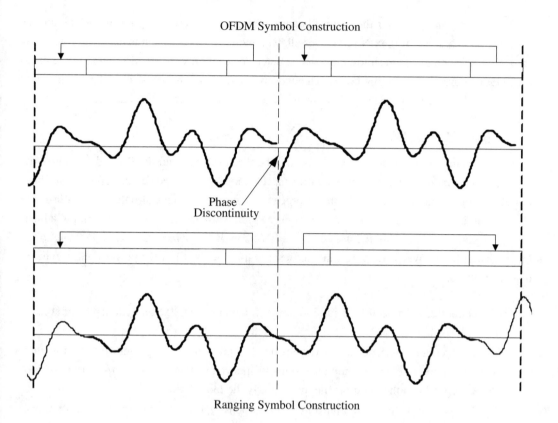

Ranging Symbol Construction

Figure 8.20 Ranging Symbol Construction

ranged MS with a time/synchronization mismatch much larger than the cyclic prefix, which is likely during initial network acquisitions. The MS can optionally use two consecutive ranging codes transmitted over four OFDM symbol periods. This option decreases the probability of failure and increases the ranging capacity to support larger numbers of simultaneously ranging MSs. The four-symbol ranging also allows for a larger timing mismatch between the BS and the SS, which might be useful when cell radii are very large. Typically, the ranging channel comprises of six subchannels and up to five consecutive OFDM symbols, the indices of which in the time and frequency domain are provided in the FCH message. The ranging channel may not be allocated in all uplink subframes and is accordingly indicated in the FCH message.

To process an initial ranging request, a ranging code is repeated twice and transmitted in two consecutive OFDM symbols with no phase discontinuity between them. The ranging codes in IEEE 802.16e-2005 are PN sequences of length 144 chosen from a set of 256 codes. Of the available codes the first N are for initial ranging, the next M are for periodic ranging, the next O are for bandwidth request, and the last S are for handover ranging. The values N, M, O, and S are decided by the BS and conveyed over the control channels. During a specific ranging procedure, an MS randomly chooses one of the PN sequences allowed by the BS. This ensures that even if two SSs collide during a ranging procedure they can be detected separately by the MS owing to the pseudorandom nature of the ranging codes. The chosen PN sequence is BPSK modulated and transmitted over the subchannels and OFDM symbols allocated for the ranging channel.

8.11 Power Control

In order to maintain the quality of the radio link between the MS and the BS and to control the overall system interference, a power-control mechanism is supported for the uplink with both initial calibration and periodic adjustment procedure, without the loss of data. The BS uses the UL ranging channel transmissions from various MSs to estimate the initial and periodic adjustments for power control. The BS uses dedicated *MAC managements messages* to indicate to the MS the necessary power-level adjustments. Basic requirements [6] of the power-control mechanism as follows.

  • Power control must be able to support power fluctuations at 30dB/s with depths of at least 10dB.
  • The BS accounts for the effect of various burst profiles on the amplifier saturation while issuing the power-control commands. This is important, since the *peak-to-average ratio* (PAR) depends on the burst profile, particularly the modulation.
  • The MS maintains the same transmitted power density, regardless of the number of active subchannels assigned. Thus, when the number of allocated subchannels to a given MS is decreased or increased, the transmit power level is proportionally decreased or increased without additional power-control messages.

In order to maintain a power-spectral density and SINR at the receiver consistent with the modulation and code rate in use, the BS can adjust the power level and/or the modulation and code rate of the transmissions. In some situations, however, the MS can temporarily adjust its power level and modulation and code rate without being instructed by the BS.

The MS reports to the BS the maximum available power and the transmitted power that may be used by the BS for optimal assignment of the burst profile and the subchannels for UL transmissions. The maximum available power reported for QPSK, 16 QAM and 64 QAM constellations must account for any required backoff owing to the PAR of these modulation constellations.

On the downlink, there is no explicit support provided for a closed-loop power control, and it is left up to the manufacturer to implement a power-control mechanism, if so desired, based on the DL channel-quality feedback provided by the SS.

8.12 Channel-Quality Measurements

The downlink power-control process and modulation and code rate adaptation are based on such channel-quality measurements as RSSI (received signal strength indicator) and SINR (signal-to-interference-plus-noise ratio) that the MS is required to provide to the BS on request. The MS uses the channel quality feedback (CQI) to provide the BS with this information. Based on the CQI, the BS can either and/or:

- Change modulation and/or coding rate for the transmissions: change the burst profile
- Change the power level of the associated DL transmissions

Owing to the dynamic nature of the wireless channel, both the mean and the standard deviation of the RSSI and SINR are included in the definition of CQI. The RSSI measurement as defined by the IEEE 802.16e-2005 standard does not require the receiver to actively demodulate the signal, thus reducing the amount of processing power required. When requested by the BS, the MS measures the instantaneous RSSI. A series of measured instantaneous RSSI values are used to derive the mean and standard deviation of the RSSI. The mean $\mu_{RSSI}[k]$ and standard deviation $\sigma_{RSSI}[k]$ of the RSSI during the kth measurement report are given by equation (8.10):

$$\mu_{RSSI}[k] = (1-\alpha)\mu_{RSSI}[k-1] + \alpha RSSI[k] \tag{8.10}$$

$$\chi^2_{RSSI}[k] = (1-\alpha)\chi^2_{RSSI}[k] + \alpha|RSSI[k]|^2$$

$$\sigma_{RSSI}[k] = \sqrt{\chi^2_{RSSI}[k] - \mu^2_{RSSI}[k]},$$

where $RSSI[k]$ is the kth measured values of RSSI, and α is an averaging parameter whose value is implementation specific and can in principle be adapted, depending on the coherence time of the channel.[7] In equation (8.10), the instantaneous value, mean, and standard deviation of the RSSI are all expressed in the linear scale. The mean and the standard deviation of the RSSI are then converted to the dB scale before being reported to the BS.

The SINR measurements, unlike the RSSI measurement, require active demodulation of the signal and are usually a better indicator of the true channel quality. Similar to the RSSI measurement the mean and the standard deviation of the SINR during the kth measurement report are given by equation (8.11):

$$\mu_{SINR}[k] = (1-\alpha)\mu_{SINR}[k-1] + \alpha SINR[k] \tag{8.11}$$

$$\chi^2_{SINR}[k] = (1-\alpha)\chi^2_{AINR}[k] + \alpha|SINR[k]|^2$$

$$\sigma_{SINR}[k] = \sqrt{\chi^2_{SINR}[k] - \mu^2_{SINR}[k]}.$$

The mean and the standard deviation of the SINR are converted to the dB scale before being reported to the BS.

---

7. Depends on the Doppler spread of the channel.

8.13 Summary and Conclusions

This chapter described the WiMAX PHY layer, based on the IEEE 802.16-2004 and IEEE 802.16e-2005 standards. The level of detail provided should be sufficient to fully comprehend the nature of the WiMAX physical layer and understand the various benefits and trade-offs associated with the various options/modes of the WiMAX PHY layer.

- The PHY layer of WiMAX can adapt seamlessly, depending on the channel, available spectrum, and the application of the technology. Although the standard provides some guidance, the overall choice of various PHY-level parameters is left to the discretion of the system designer. It is very important for an equipment manufacturer and the service provider to understand the basic trade-off associated with the choice of these parameters.
- A unique feature of the WiMAX PHY layer is the choice of various subcarrier permutation schemes which are summarized in Table 8.7. The system allows for both distributed and adjacent subcarrier permutations for creating a subchannel. The distributed subcarrier mode provides frequency diversity; the adjacent subcarrier mode provides multiuser diversity and is better suited for beamforming.
- The WiMAX PHY layer has been designed from the ground up for multiantenna support. The multiple antennas can be used for diversity, beamforming, spatial multiplexing and various combinations thereof. This key feature can enable WiMAX-based networks to have very high capacity and high degree of reliability, both of which are shortcoming of current generations of cellular wireless networks.

8.14 Bibliography

[1] S. Alamouti, A simple transmit diversity technique for wireless communications, *IEEE Journal on Selected Areas in Communications*, October 1998.

[2] C. Berrou and M. Jezequel, Non binary convolutional codes and turbo coding, *Electronics Letters*, 35 (1): January 1999.

[3] IEEE. Standard 802.16-2004, Part 16: Air interface for fixed broadband wireless access systems, June 2004.

[4] IEEE. Standard 802.16-2005, Part 16: Air interface for fixed and mobile broadband wireless access systems, December 2005.

Table 8.7 Summary of Subcarrier Permutation Schemes[a]

Name	Basic Unit	Subcarrier Groups	Subchannel
FUSC	not applicable	not applicable	48 distributed subcarriers
DL PUSC	*Cluster:* 14 adjacent subcarriers over 2 symbols with 4 embedded pilot subcarriers	Clusters divided into 6 groups (0–5)	2 clusters from the same group
UL PUSC	*Tile:* 4 adjacent subcarriers over 3 symbols with 4 embedded pilot subcarriers	Tiles divided into 6 groups (0-5)	6 tiles from the same group
Optional UL PUSC	*Tile:* 3 adjacent subcarriers over 3 symbols with 1 embedded pilot subcarriers	Tiles divided into 6 groups (0–5)	6 tiles from the same group
TUSC 1	*Tile:* 4 adjacent subcarriers over 3 symbols with 4 embedded pilot subcarriers	Tiles divided into 6 groups (0–5)	6 tiles from the same group
TUSC 2	*Tile:* 3 adjacent subcarriers over 3 symbols with 1 embedded pilot subcarriers	Tiles divided into 6 groups (0–5)	6 tiles from the same group
Band AMC	*Bin:* 9 adjacent subcarriers over 1 symbol with 1 embedded pilot	not applicable	6 adjacent bins over 6 consecutive OFDM symbol (or 2 bins over 3 OFDM symbols or 3 bins ×2 OFDM symbols)

a. Only the DL PUSC, UL PUSC, and band AMC are a part of the initial WiMAX profile.

CHAPTER 9

MAC Layer of WiMAX

C hapter 8 described theWiMAX physical (PHY) layer, also referred to as layer 1 of the open systems interconnect (OSI) stack. In a network, the purpose of the PHY layer is to reliably deliver information bits from the transmitter to the receiver, using the physical medium, such as radio frequency, light waves, or copper wires. Usually, the PHY layer is not informed of quality of service (QoS) requirements and is not aware of the nature of the application, such as VoIP, HTTP, or FTP. The PHY layer can be viewed as a pipe responsible for information exchange over a single link between a transmitter and a receiver. The Media Access Control (MAC) layer, which resides above the PHY layer, is responsible for controlling and multiplexing various such links over the same physical medium. Some of the important functions of the MAC layer in WiMAX are to

- Segment or concatenate the service data units (SDUs) received from higher layers into the MAC PDU (protocol data units), the basic building block of MAC-layer payload
- Select the appropriate burst profile and power level to be used for the transmission of MAC PDUs
- Retransmission of MAC PDUs that were received erroneously by the receiver when automated repeat request (ARQ) is used
- Provide QoS control and priority handling of MAC PDUs belonging to different data and signaling bearers
- Schedule MAC PDUs over the PHY resources
- Provide support to the higher layers for mobility management
- Provide security and key management
- Provide power-saving mode and idle-mode operation

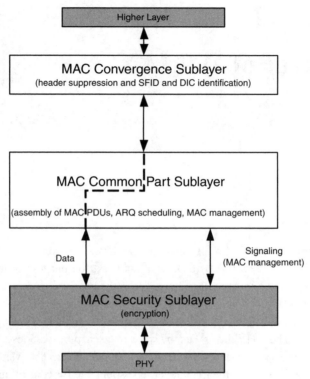

Figure 9.1 The WiMAX MAC layer

The MAC layer of WiMAX, as shown in Figure 9.1, is divided into three distinct components: the service-specific convergence sublayer (CS), the common-part sublayer, and the security sublayer. The CS, which is the interface between the MAC layer and layer 3 of the network, receives data packets from the higher layer. These higher-layer packets are also known as MAC service data units (SDU). The CS is responsible for performing all operations that are dependent on the nature of the higher-layer protocol, such as header compression and address mapping. The CS can be viewed as an adaptation layer that masks the higher-layer protocol and its requirements from the rest of the MAC and PHY layers of a WiMAX network.

The common-part sublayer of the MAC layer performs all the packet operations that are independent of the higher layers, such as fragmentation and concatenation of SDUs into MAC PDUs, transmission of MAC PDUs, QoS control, and ARQ. The security sublayer is responsible for encryption, authorization, and proper exchange of encryption keys between the BS and the MS.

In this chapter, we first describe the CS and its various functions. Next, we describe the MAC common-part sublayer, the construction of MAC PDUs, bandwidth allocation process, QoS control, and network-entry procedures. We then turn to the mobility-management and power-saving features of the WiMAX MAC layer.

9.1 Convergence Sublayer

Table 9.1 shows the various higher-layer protocol convergence sublayers—or combinations—that are supported in WiMAX. Apart from header compression, the CS is also responsible for mapping higher-layer addresses, such as IP addresses, of the SDUs onto the identity of the PHY and MAC connections to be used for its transmission. This functionality is required because there is no visibility of higher-layer addresses at the MAC and PHY layers.

The WiMAX MAC layer is connection oriented and identifies a logical connection between the BS and the MS by a unidirectional connection indentifier (CID). The CIDs for UL and DL connections are different. The CID can be viewed as a temporary and dynamic layer 2 address assigned by the BS to identify a unidirectional connection between the peer MAC/PHY entities and is used for carrying data and control plane traffic. In order to map the higher-layer address to the CID, the CS needs to keep track of the mapping between the destination address and the respective CID. It is quite likely that SDUs belonging to a specific destination address might be carried over different connections, depending on their QoS requirements, in which case the CS determines the appropriate CID, based on not only the destination address but also various other factors, such as *service flow*[1] ID (SFID) and source address. As shown in Table 9.1 the IEEE 802.16 suite of standards defines a CS for ATM (asynchronous transfer mode) services and packet service. However, the WiMAX Forum has decided to implement only IP and Ethernet (802.3) CS.

9.1.1 Packet Header Suppression

One of the key tasks of the CS is to perform packet header suppression (PHS). At the transmitter, this involves removing the repetitive part of the header of each SDU. For example, if the SDUs delivered to the CS are IP packets, the source and destination IP addresses contained in the header of each IP packet do not change from one packet to the next and thus can be removed before being transmitted over the air. Similarly at the receiver: The repetitive part of the header can be reinserted into the SDU before being delivered to the higher layers. The PHS protocol establishes and maintains the required degree of synchronization between the CSs at the transmitter and the receiver.

In WiMAX, PHS implementation is optional; however, most systems are likely to implement this feature, since it improves the efficiency of the network to deliver such services as VoIP. The PHS operation is based on the *PHS rule*, which provides all the parameters related to header suppression of the SDU. When a SDU arrives, the CS determines the PHS rule to be used, based on such parameters as destination and source addresses. Once a matching rule is found, it provides a SFID, a CID and PHS-related parameters to be used for the SDU. The PHS rule can be dependent on the type of service, such as VoIP, HTTP, or FTP, since the number of bytes that can be suppressed in the header is dependent on the nature of the service. In of VoIP, for example, the repetitive part of the header includes not only the source and destination IP addresses but

---

1. The concept of service flow is discussed in Section 9.2.

Table 9.1 Convergence Sublayers of WiMAX

Value	Convergence Sublayer
0	ATM CS
1	Packet CS IPv4
2	Packet CS IPv6
3	Packet CS 802.3 (Ethernet)
4	Packet CS 802.1/Q VLAN
5	Packet CS IPv4 over 802.3
6	Packet CS IPv6 over 802.3
7	Packet CS IPv4 over 802.1/Q VLAN
8	Packet CS IPv6 over 802.1/Q VLAN
9	Packet CS 802.3 with optional VLAN tags and ROHC header compression
10	Packet CS 802.3 with optional VLAN tags and ERTCP header compression
11	Packet IPv4 with ROHC header compression
12	Packet IPv6 with ROHC header compression
13–31	Reserved

also the length indicator; for HTTP, the length indicator can change from one SDU to the next. Although it standard instructs the CS to use PHS rules, the WiMAX does not specify how and where such rules are created. It is left to a higher-layer entity to create the PHS rules.

Figure 9.2 shows the basic operation of header suppression in WIMAX. When an SDU arrives, the CS first determines whether an associated PHS rule exists for the SDU . If a matching rule is found, the CS determines the part of the header that is not to be suppressed, using a PHS mask (PHSM) associated with the SDU. The portion of the header to be suppressed is referred to as the PHS field (PHSF). If PHS verify (PHSV) is used, the CS first compares the bits in the PHSF with what they are expected, to be based on the PHS rule. If the PHSF of the SDU matches the cached PHSF, the bytes corresponding to the PHSF are removed, and the SDU is appended by the PHS index (PHSI) as provided by the matching rule. The PHSI is an 8-bit field that refers to the cached PHSF for the matched PHS rule. Similarly, if the PHSF of the SDU does not match the PHSF of the associated rule, the PHSFs are not suppressed, and the SDU is appended with a PHSI of 0.

If PHSV is not used, the CS does not compare the PHSF of the SDU with the cached PHSF, and header suppression is performed on all SDUs. PHSV operation guarantees that the regenerated.

SDU header at the receiver matches the original SDU header. Figure 9.3 shows the various steps involved in a typical PHS operation in WiMAX.

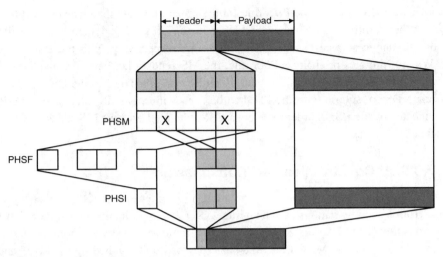

Figure 9.2 Header suppression in WiMAX

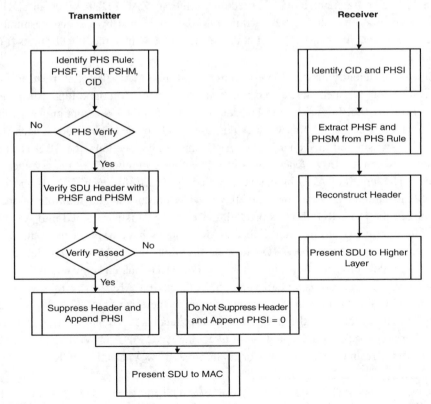

Figure 9.3 PHS operation in WiMAX

In order for PHS to work, the PHS rules at the transmitter and the receiver need to be synchronized. When it initiates or changes the PHS rule, the BS sends a dynamic service allocation (DSA) or a dynamic service change (DSC) message, respectively, with the PSHF, PHSI, or PHSM. When it initiates or changes PHS rule, the MS sends a DSA or DSC message, respectively, with all the elements except the PHSI. The BS, as a response to the DSA or DSC message, sends a DSC response with the PHSI to be used with this PHS rule. Like CID, PHSI is always allocated by the BS. In order to delete a PHS rule, the BS or the MS sends a dynamic service delete (DSD) message.

9.2 MAC PDU Construction and Transmission

As the name suggests, the MAC common-part sublayer is independent of the higher-layer protocol and performs such operations as scheduling, ARQ, bandwidth allocations, modulation, and code rate selection. The SDUs arriving at the MAC common-part sublayer from the higher layer are assembled to create the MAC PDU, the basic payload unit handled by the MAC and PHY layers. Based on the size of the payload, multiple SDUs can be carried on a single MAC PDU, or a single SDU can be fragmented to be carried over multiple MAC PDUs. When an SDU is fragmented, the position of each fragment within the SDU is tagged by a sequence number. The sequence number enables the MAC layer at the receiver to assemble the SDU from its fragments in the correct order.

In order to efficiently use the PHY resources, multiple MAC PDUs destined to the same receiver can be concatenated and carried over a single transmission opportunity or data region, as shown in Figure 9.4. In the UL and DL data regions of an MS is a contiguous set of slots[2] reserved for its transmission opportunities. For non-ARQ-enabled connections, each fragment of the SDU is transmitted in sequence. For ARQ-enabled connections, the SDU is first partitioned into fixed-length *ARQ blocks,* and a block sequence number (BSN) is assigned to each ARQ block. The length of ARQ blocks is specified by the BS for each CID, using the ARQ-BLOCK-SIZE parameter. If the length of the SDU is not an integral multiple of the ARQ-BLOCK-SIZE, the last ARQ block is padded. Once the SDU is partitioned into ARQ blocks, the partitioning remains in effect until all the ARQ blocks have been received and acknowledged by the receiver. After the ARQ block partitioning, the SDU is assembled into MAC PDUs in a normal fashion, as shown in Figure 9.4. For ARQ-enabled connections, the *fragmentation and packing subheader* contains the BSN of the first ARQ block following the subheader. The ARQ feedback from the receiver comes in the form of ACK (acknowledgment), indicating proper reception of the ARQ blocks. This feedback is sent either as a stand-alone MAC PDU or piggybacked on the payload of a regular MAC PDU. In WiMAX, the ARQ feedback can be in the form of *selective ACK* or *cumulative ACK.* A selective ACK for a given BSN

---

2. A *slot,* the basic unit of PHY-layer resources, can be used for allocation and consists of one subchannel over one, two, or three OFDM symbols, depending on the subcarrier permutation. This is discussed more detail in Chapter 8.

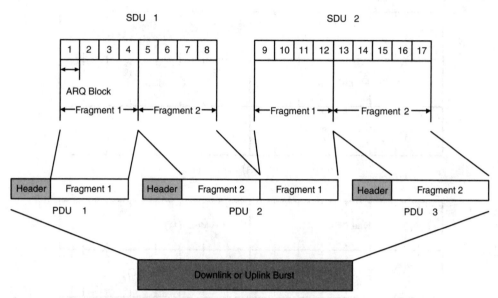

Figure 9.4 Segmentation and concatenation of SDUs in MAC PDUs

indicates that the ARQ block has been received without errors. A cumulative ACK for a given BSN, on the other hand, indicates that all blocks with sequence numbers less than or equal to the BSN have been received without error.

Each MAC PDU consists of a header followed by a payload and a cyclic redundancy check (CRC).[3] The CRC is based on IEEE 802.3 and is calculated on the entire MAC PDU; the header and the payload. WiMAX has two types of PDUs, each with a very different header structure, as shown in Figure 9.5.

1. The *generic* MAC PDU is used for carrying data and MAC-layer signaling messages. A generic MAC PDU starts with a generic header whose structure is shown in Figure 9.5 as followed by a payload and a CRC. The various information elements in the header of a generic MAC PDU are shown in Table 9.2.

2. The *bandwidth* request PDU is used by the MS to indicate to the BS that more bandwidth is required in the UL, due to pending data transmission. A bandwidth request PDU consists only of a bandwidth-request header, with no payload or CRC. The various information elements of a bandwidth request header are provided in Table 9.3.

---

3. The CRC is mandatory for the SCa, OFDM, and OFDMA PHY. In the case of SC PHY, the CRC is optional.

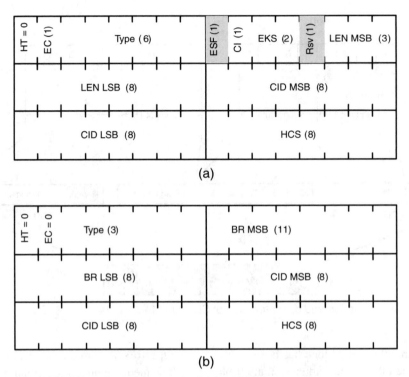

Figure 9.5 WiMAX PDU headers: (a) generic; (b) bandwidth request

Table 9.2 Generic MAC Header Fields

Field	Length (bits)	Description
HT	1	Header type (set to 0 for such header)
EC	1	Encryption control (0 = payload not encrypted; 1 = payload encrypted)
Type	6	Type
ESF	1	Extended subheader field (1 = ES present; 0 = ES not present)
CI	1	CRC indicator (1 = CRC included; 0 = CRC not included)
EKS	2	Encryption key sequence (index of the traffic encryption key and the initialization vector used to encrypt the payload)
Rsv	1	Reserved
LEN	11	Length of MAC PDU in bytes, including the header
CID	16	Connection identifier on which the payload is to be sent
HCS	8	Header check sequence; generating polynomial $D^8 + D^2 + D + 1$

Table 9.3 Bandwidth Request MAC Header Fields

Field	Length (bits)	Description
HT	1	Header type (set to 1 for such header)
EC	1	Encryption control (set to 0 for such header)
Type	3	Type
BR	19	Bandwidth request (the number of bytes of uplink bandwidth requested by the SS for the given CID)
CID	16	Connection indentifier
HCS	8	Header check sequence

Apart from the generic and bandwidth request headers, WiMAX also defines five subheaders that can be used in a generic MAC PDU:

1. **Mesh subheader.** Follows generic header when mesh networking is used.
2. **Fragmentation subheader.** Follows the generic MAC header and indicates that the SDU is fragmented over multiple MAC PDUs.
3. **Packing subheader.** Indicates that multiple SDUs or SDU fragments are packed into a single MAC PDU and are placed at the beginning of each SDU or SDU fragment.
4. **Fast-feedback allocation subheader.** Indicates that the PDU contains feedback from the MS about the DL channel state information. This subheader provides the functionality for channel state information feedback for MIMO and non-MIMO implementations.
5. **Grant-management subheader.** Used by the MS, conveys various messages related to bandwidth management, such as polling request and additional-bandwidth request. Using this subheader is more efficient than the bandwidth-request PDU for additional bandwidth during an ongoing session, since it is more compact and does not require the transmission of a new PDU. The bandwidth-request PDU is generally used for the initial bandwidth request.

Once a MAC PDU is constructed, it is handed over to the scheduler, which schedules the MAC PDU over the PHY resources available. The scheduler checks the service flow ID and the CID of the MAC PDU, which allows it to gauge its QoS requirements. Based on the QoS requirements of the MAC PDUs belonging to different CIDs and service flow IDs, the scheduler determines the optimum PHY resource allocation for all the MAC PDUs, on a frame-by-frame basis. (Scheduling algorithms and their various pros and cons in the context of an OFDMA system are discussed in Chapter 6). The scheduling procedure is outside the scope of the WiMAX standard and has been left to the equipment manufacturers to implement. Since the scheduling algorithm has a profound impact on the overall capacity and performance of the system, it can be a key feature distinguishing among implementations of various equipment manufacturers.

9.3 Bandwidth Request and Allocation

In the downlink, all decisions related to the allocation of bandwidth to various MSs are made by the BS on a per CID basis, which does not require the involvement of the MS. As MAC PDUs arrive for each CID, the BS schedules them for the PHY resources, based on their QoS requirements. Once dedicated PHY resources have been allocated for the transmission of the MAC PDU, the BS indicates this allocation to the MS, using the DL-MAP message.

In the uplink, the MS requests resources by either using a stand-alone bandwidth-request MAC PDU or piggybacking bandwidth requests on a generic MAC PDU, in which case it uses a grant-management subheader. Since the burst profile associated with a CID can change dynamically, all resource requests are made in terms of bytes of information, rather than PHY-layer resources, such as number of subchannels and/or number of OFDM symbols.

Bandwidth requests in the UL can be incremental or aggregate requests. When it receives an incremental bandwidth request for a particular CID, the BS adds the quantity of bandwidth requested to its current perception of the bandwidth need. Similarly, when it receives an aggregate bandwidth request for a particular CID, the BS replaces its perception of the bandwidth needs of the connection with the amount of bandwidth requested. The *Type* field in the bandwidth-request header indicates whether the request is incremental or aggregate. Bandwidth requested by piggybacking on a MAC PDU can be only incremental.

When multiple CIDs are associated with a particular MS, the BS-allocated UL aggregate resources for the MS rather than individual CIDs. When the resource granted by the BS is less than the aggregate resources requested by the MS, the UL scheduler at the MS determines that allocation and distribution of the granted resource among the various CIDs, based on the amount of pending traffic and their QoS requirements.

In WiMAX, *polling* refers to the process whereby dedicated or shared UL resources are provided to the MS to make bandwidth requests. These allocations can be for an individual MS or a group of MSs. When an MS is polled individually, the polling is called *unicast,* and the dedicated resources in the UL are allocated for the MS to send a bandwidth-request PDU. The BS indicates to the MS the UL allocations for unicast polling opportunities by the UL MAP[4] message of the DL subframe. Since the resources are allocated on a per MS basis, the UL MAP uses the *primary* CID of the MS to indicate the allocation. The primary CID is allocated to the MS during the network entry and initialization stage and is used to transport all MAC-level signaling messages. An MS can also dynamically request additional CIDs, known as *secondary CIDs,* which it can use only for transporting data. MSs that have an active unsolicited grant service[5] (UGS) connection are not polled, since the bandwidth request can be sent on the UGS allocation

---

4. The UL MAP and DL MAP, control messages in the initial part of the DL subframe, signal the UL and DL allocations, respectively, for all the MSs.
5. UGS is the type of connection in which the MS is given UL and DL resources on a periodic basis without explicit request for such allocations. This feature of WiMAX is optimal for transporting constant bit rate services with very low latency requirements, such as VoIP.

either in the form of a bandwidth request PDU or by piggybacking on generic MAC PDUs. If the MS does not have additional bandwidth requirements, it sends a dummy MAC PDU during the unicast poll, and the *Type* field of the header indicates that it is a dummy MAC PDU. Note that the MS is not allowed to remain silent during the unicast poll.

If sufficient bandwidth is not available to poll each MS individually, multicast or broadcast polling is used to poll a group of users or all the users at a time. All MSs belonging to the polled group can request bandwidth during the multicast/broadcast polling opportunity. In order to reduce the likelihood of collision, only MSs with bandwidth requirements respond. WiMAX uses a truncated binary exponential backoff algorithm for contention-resolution during a multicast/broadcast poll. When it needs to send a bandwidth request over a multicast/broadcast poll, the MS first enters a contention resolution phase, if selecting a uniformly distributed random number between 0 and *BACKOFF WINDOW*. This random value indicates the number of transmission opportunities—allocated resources for multicast/broadcast poll—the MS will wait before sending its bandwidth request. BACKOFF WINDOW is the maximum number of transmission opportunities an MS can wait before sending the pending bandwidth request. If it does not receive a bandwidth allocation based on the UL MAP message within a time window specified by the T16 timer, the MS assumes that its bandwidth request message was lost, owing to collision with another MS, in which case MS increases is backoff window by a factor of 2—as long as it is less than a *maximum backoff window*—and repeats the process. If bandwidth is still not allocated after a maximum number of retries, the MAC PDU is discarded. The maximum number of retries for the bandwidth request is a tunable parameter and can be adjusted by either the service provider or the equipment manufacturer, as needed.

9.4 Quality of Service

One of the key functions of the WiMAX MAC layer is to ensure that QoS requirements for MAC PDUs belonging to different service flows are met as reliably as possible given the loading conditions of the system. This implies that various negotiated performance indicators that are tied to the overall QoS, such as latency, jitter, data rate, packet error rate, and system availability, must be met for each connection. Since the QoS requirements of different data services can vary greatly, WiMAX has various handling and transporting mechanisms to meet that variety.

9.4.1 Scheduling Services

The WiMAX MAC layer uses a *scheduling service* to deliver and handle SDUs and MAC PDUs with different QoS requirements. A scheduling service uniquely determines the mechanism the network uses to allocate UL and DL transmission opportunities for the PDUs. WiMAX defines five scheduling services.

1. *The unsolicited grant service (UGS)* is designed to support real-time service flows that generate fixed-size data packets on a periodic basis, such as T1/E1 and VoIP. UGS offers fixed-size grants on a real-time periodic basis and does not need the SS to explicitly

request bandwidth, thus eliminating the overhead and latency associated with bandwidth request.

2. *The real-time polling services (rtPS)* is designed to support real-time services that generate variable-size data packets on a periodic basis, such as MPEG (Motion Pictures Experts Group) video. In this service class, the BS provides unicast polling opportunities for the MS to request bandwidth. The unicast polling opportunities are frequent enough to ensure that latency requirements of real-time services are met. This service requires more request overhead than UGS does but is more efficient for service that generates variable-size data packets or has a duty cycle less than 100 percent.

3. *The non-real-time polling services (nrtPS)* is very similar to rtPS except that the MS can also use contention-based polling in the uplink to request bandwidth. In nrtPS, it is allowable to have unicast polling opportunities, but the average duration between two such opportunities is in the order of few seconds, which is large compared to rtPS. All the MSs belonging to the group can also request resources during the contention-based polling opportunity, which can often result in collisions and additional attempts.

4. *The best-effort service (BE)* provides very little QoS support and is applicable only for services that do not have strict QoS requirements. Data is sent whenever resources are available and not required by any other scheduling-service classes. The MS uses only the contention-based polling opportunity to request bandwidth.

5. *The extended real-time polling service (ertPS),* a new scheduling service introduced with the IEEE 802.16e standard, builds on the efficiencies of UGS and rtPS. In this case, periodic UL allocations provided for a particular MS can be used either for data transmission or for requesting additional bandwidth. This features allows ertPS to accommodate data services whose bandwidth requirements change with time. Note that in the case of UGS, unlike ertPS, the MS is allowed to request additional bandwidth during the UL allocation for only non-UGS-related connections.

9.4.2 Service Flow and QoS Operations

In WiMAX, a *service flow* is a MAC transport service provided for transmission of uplink and downlink traffic and is a key concept of the QoS architecture. Each service flow is associated with a unique set of QoS parameters, such as latency, jitter throughput, and packet error rate, that the system strives to offer. A service flow has the following components:

- *Service flow ID,* a 32-bit identifier for the service flow.
- *Connection ID*, a 16-bit identifier of the logical connection to be used for carrying the service flow. The CID is analogous to the identity of an MS at the PHY layer. As mentioned previously, an MS can have more that one CID at a time, that is, a primary CID and multiple secondary CIDs. The MAC management and signaling messages are carried over the primary CID.

- *Provisioned QoS parameter set,* the recommended QoS parameters to be used for the service flow, usually provided by a higher-layer entity.
- *Admitted QoS parameter set,* the QoS parameters actually allocated for the service flow and for which the BS and the MS reserve their PHY and MAC resources. The admitted QoS parameter set can be a subset of the provisioned QoS parameter set when the BS is not able, for a variety of reasons, to admit the service with the provisioned QoS parameter set.
- *Active QoS parameter set,* the QoS parameters being provided for the service flow at any given time.
- *Authorization module,* logical BS function that approves or denies every change to QoS parameters and classifiers associated with a service flow.

The various service flows admitted in a WiMAX network are usually grouped into *service flow classes,* each identified by a unique set of QoS requirements. This concept of service flow classes allows higher-layer entities at the MS and the BS to request QoS parameters in globally consistent ways. WiMAX does not explicitly specify what the service flow classes are, leaving it to the service provider or the equipment manufacturer to define. As a general practice, services with very different QoS requirements, such as VoIP, Web browsing, e-mail, and interactive gaming, are usually associated with different service flow classes. The overall concept of service flow and service flow classes is flexible and powerful and allows the service provider full control over multiple degrees of freedom for managing QoS across all applications.

9.5 Network Entry and Initialization

When an MS acquires the network after being powered up a WiMAX network undergoes various steps. An overview of this process, also referred to as *network entry,* is shown in Figure 9.9.

9.5.1 Scan and Synchronize Downlink Channel

When an MS is powered up, it first scans the allowed downlink frequencies to determine whether it is presently within the coverage of a suitable WiMAX network. Each MS stores a nonvolatile list of all operational parameters, such as the DL frequency used during the previous operational instance. The MS first attempts to synchronize with the stored DL frequency. If this fails, the MS it scans other frequencies in an attempt to synchronize with the DL of the most suitable BS. Each MS also maintains a list of preferred DL frequencies, which can be modified to suit a service provider's network.

During the DL synchronization, the MS listens for the DL frame preambles. When one is detected, the MS can synchronize[6] itself with respect to the DL transmission of the BS. Once it obtains DL synchronization, the MS listens to the various control messages, such as FCH, DCD,

---

6. The initial synchronization involves both frequency and timing synchronization, obtained by listening to the DL frame preamble.

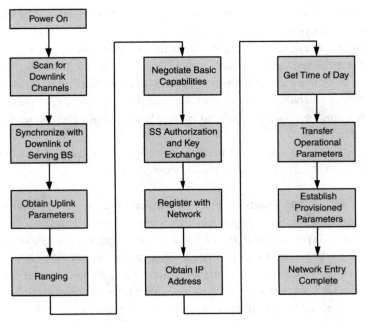

Figure 9.6 Process of network entry

UCD, DL-MAP, and UL-MAP, that follow the preamble to obtain the various PHY- and MAC-related parameters corresponding to the DL and UL transmissions.

9.5.2 Obtain Uplink Parameters

Based on the UL parameters decoded from the control messages, the MS decides whether the channel is suitable for its purpose. If the channel is not suitable, the MS goes back to scanning new channels until it finds one that is. If the channel is deemed usable, the MS listens to the UL MAP message to collect information about the ranging opportunities.

9.5.3 Perform Ranging

At this stage, the MS performs *initial ranging* with the BS to obtain the relative timing and power-level adjustment required to maintain the UL connection with the BS. Once the a UL connection has been established, the MS should do *periodic ranging* to track timing and power-level fluctuations. These fluctuations can arise because of mobility, fast fading, shadow fading, or any combinations thereof. Since the MS does not have a connection established at this point, the initial ranging opportunity is contention based. Based on the UL and DL channel parameters, the MS uses the following formula to calculate the transmit-power level for the initial ranging:

$$P_{TX} = EIRxP_{IR, \, MAX} + BSEIRP - RSSI, \tag{9.1}$$

where the parameters $EIRxP_{IR,MAX}$ and $BSEIRP$ are provided by the BS over the DCD[7] message, and $RSSI$ is the received signal strength at the MS. For the OFDMA PHY, the MS sends a CDMA ranging code[8] with the power level, as shown in equation (9.1), at the first determined ranging slot. Similarly, in the OFDM PHY, the MS sends a RNG-REQ message with the CID set to *initial ranging CID*. If it does not receive any response from the BS within a certain time window, then the MS considers the previous ranging attempt to be unsuccessful and enters the contention-resolution stage. Therein, the MS sends a new CDMA ranging code at the next ranging opportunity, after an appropriate back-off delay, and with a one-step-higher power level. Figure 9.7 shows the ranging and automatic parameter-adjustment procedure in WiMAX.

If in the DL the MS receives a RNG-RSP message containing the parameters of the CDMA code used—or the initial ranging CID for OFDM PHY—the MS considers the ranging to be successful with *status continue* and implements the parameter adjustment as indicated in the RNG-RSP message, which also contains the basic and primary CIDs allocated to the MS. The BS uses the initial ranging CID or the CDMA code parameter in the DL-MAP message to signal the DL allocations for the MS containing the RNG-RSP message. On the other hand, if it receives an allocation in the UL MAP with the parameters of the CDMA code—or initial ranging CID for OFDM PHY—the MS considers ranging to be unsuccessful with *status continue*. This UL allocation provides a unicast ranging opportunity during which the SS can send another RNG-REQ message, using the initial ranging CID in the header.

On receipt of this RNG-REQ message, the BS sends an RNG-RSP message using the initial ranging CID.[9] This message contains the basic and primary management CID allocated to the MS. From here on, the basic and primary management CID is used by the MS and the BS to send most of the MAC management messages. Additional CIDs may be allocated to the MS in the future, as needed. Apart from the primary CID, the RNG-RSP message can also contain additional power and timing offset adjustments. If this message is received with *status continue,* the MS assumes that further ranging is required. The SS waits for a unicast ranging opportunity allocated to the primary CID of the MS and at the first such opportunity sends another RNG-REQ message, on receipt of which the BS sends a RNG-RSP message with additional adjustment to the power level and timing offset. This process continues until the MS receives a RNG-RSP message with *status complete.* At this point, the initial ranging process is assumed to be complete, and the MS may start its UL transmission.

---

7. The $EIRxP_{IR,max}$ is the maximum equivalent isotropic received power computed for a simple single-antenna receiver as $RSS_{IR,max} - GANT\_BS\_Rx$, where the $RSS_{IR,max}$ is the received signal strength at antenna output and $GANT\_BS\_Rx$ is the receive antenna gain. The $BSEIRP$ is the equivalent isotropic radiated power of the base station, which is computed for a simple single-antenna transmitter as P_{Tx} + $GANT\_BS\_Tx$, where P_{Tx} is the transmit power, and $GANT\_BS\_Tx$ is the transmit-antenna gain.
8. The PHY-layer ranging procedure and CDMA ranging codes are discussed in Chapter 8.
9. The initial ranging CID is a BS-specific parameter and is specified in the broadcast control messages.

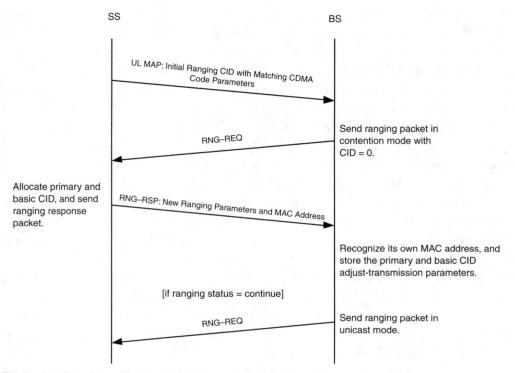

Figure 9.7 Ranging and parameter-adjustment procedure

9.5.4 Negotiate Basic Capabilities

After initial ranging, the MS sends an SBC-REQ message informing the BS about its basic capability set, which includes various PHY and bandwidth-allocation-related parameters, as shown in Table 9.4. On the of this message, the BS responds with an SBC-RSP, providing the PHY and bandwidth-allocation parameters to be used for UL and DL transmissions. The operational PHY and bandwidth-allocation parameters can be the same as the basic capability set of the SS or a subset of it.

9.5.5 Register and Establish IP Connectivity

After negotiating the basic capabilities and exchanging the encryption key, the MS registers itself with the network. In WiMAX, registration is the process by which the MS is allowed to enter the network and can receive secondary CIDs. The registration process starts when the MS sends a REG-REQ message to the BS. The message contains a *hashed message authentication code* (HMAC), which the BS uses to validate the authenticity of this message. Once it determines that the request for registration is valid, the BS sends to the MS a REG-RSP message in which it provides the secondary management CID. In the REG-REQ message, the MS also indicates to the BS its secondary capabilities not covered under the *basic capabilities,* such as IP

Table 9.4 Parameters in Basic Capability Set of BS and MS

PHY Related Parameters	Meaning
Transmission gap	The transmission gap between the UL and DL subframe supported by the SS for TDD and HF-FDD
Maximum transmit power	Maximum transmit power available for BPSK, QPSK, 16 QAM, and 64 QAM modulation
Current transmit power	The transmit power used for the current MAC PDU (containing the SBC-REQ message)
FFT size	The supported FFT sizes (128, 512, 1,024, and 2,048 for OFDMA mode; 256 for OFDM mode)
64 QAM support	Support for 64 QAM by the modulator and demodulator
FEC support	Which optional FEC modes are supported: CTC, LDPC, and so on
HARQ support	Support for HARQ
STC and MIMO support	The various space/time coding and MIMO modes
AAS private MAP support	Support for various AAS private MAP
Uplink power–control support	Uplink power-control options (open loop, closed loop, and AAS preamble power control)
Subcarrier permutation support	Support for various optional PUSC, FUCSC, AMC, and TUSC modes
Bandwidth-Allocation-Related Parameters	
Half-duplex/full-duplex FDD support	Support for half-duplex and full-duplex FDD modes in case of FDD implementation

version supported, convergence sublayer supported, and ARQ support. The MS may indicate the supported IP versions to the BS in the REG-REQ message, in which case the BS indicates the IP version to be used in the REG-RSP message. The BS allows the use of exactly one of the IP versions supported by the MS. If the information about the supported IP version is omitted in the REG-REQ message, the BS assumes that the MS can support only IPv4. After receiving the REG-RSP message from the BS, the SS can use DHCP to obtain an IP address.

9.5.6 Establish Service Flow

The creation of service flows can be initiated by either the MS or the BS, based on whether initial traffic arrives in the uplink or the downlink. When it an MS chooses to initiate the creation of a service flow, an MS sends a DSA-REQ message containing the required QoS set of the service flow (Figure 9.8). On receipt of the DSA-REQ message, the BS first checks the integrity of the message and sends a DSX-RVD message indicating whether the request for a new service flow was received with its integrity preserved. Then the BS checks whether the requested QoS set can

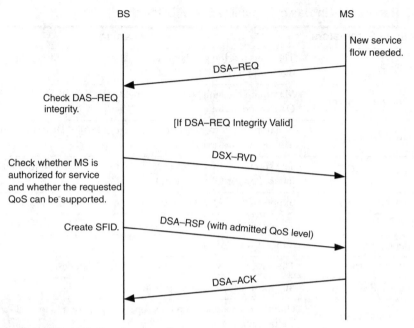

Figure 9.8 MS-initiated service flow creation

be supported, creates a new SFID and sends an appropriate DSA-RSP indicating the admitted
QoS set. TheMS completes the process by sending a DSA-ACK message.

   If it needs to initiate the creation of a service flow, the BS first checks whether the MS is
authorized for such service and whether the requested level of QoS can be supported. The
request for such service usually comes from a higher-layer entity and is outside the scope of the
IEEE 802.16e.2005/802.16-2004 standard. If the MS is authorized for service, the BS creates a
new SFID and sends a DSA-REQ message with the admitted QoS set and the CID to be used, as
shown in Figure 9.9. On receipt of this request, the MS sends a DSA-RSP message indicating its
acceptance. The BS completes this process by sending a DSA-ACK message. After the creation
of the requested service flow, the MS and the BS are ready to exchange data and management
messages over the specified CID.

9.6 Power-Saving Operations

The mobile WiMAX standard (IEEE 802.16e) introduces several new concepts related to mobility management and power management, two of the most fundamental requirements of a mobile
wireless network. Although mobility and power management are often referred to together, they
are conceptually different. Power management enables the MS to conserve its battery resources, a
critical feature required for handheld devices. Mobility management, on the other hand, enables
the MS to retain its connectivity to the network while moving from the coverage area of one BS to
the next. In this section, we describe the power-management features of a WiMAX network.

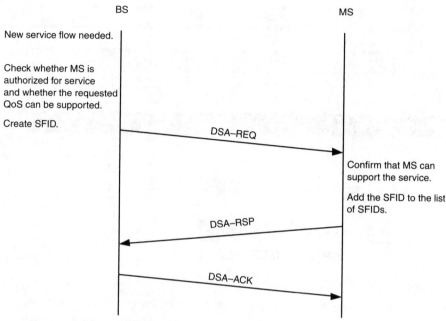

Figure 9.9 BS-initiated service flow creation

9.6.1 Sleep Mode

Sleep mode is an optional mode of operation in WiMAX. An MS with active connections—one or more CIDs—negotiates with the BS to temporarily disrupt its connection over the air interface for a predetermined amount of time, called the *sleep window*. Each sleep window is followed by a *listen window*, during which the MS restores its connection. As shown in Figure 9.10, the MS goes through alternating sleep and listen windows for each connection. The length of each sleep and listen window is negotiated between the MS and the BS and is dependent on the *power-saving class* of the sleep-mode operation. The period of time when all the MS connections are in their sleep windows is referred to as the *unavailability interval*, during which the MS cannot receive any DL transmission or send any UL transmission. Similarly, during the *availability interval*, when one or more MS connections are not in sleep mode, the MS receives all DL transmissions and sends UL transmissions in a normal fashion on the CIDs that are in their listen windows. During the unavailability interval, the BS does not schedule any DL transmissions to the MS, so that it can power down one or more hardware components required for communication. The BS may buffer or drop all arriving SDUs associated with a *unicast* transmission to the MS. For *multicast* transmissions, the BS delays all SDUs until the availability interval common to all MSs in the multicast group.

Based on their respective parameters, sleep-mode operation takes place in one of three power-saving classes.

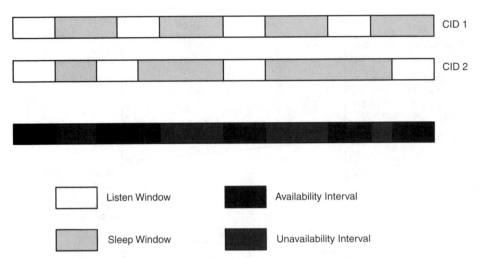

CID 1

CID 2

☐ Listen Window   ■ Availability Interval

▨ Sleep Window   ■ Unavailability Interval

Figure 9.10 Sleep-mode operation in IEEE 802.16e-2005

1. In *power-saving class 1,* each listen window of fixed length is followed by a sleep window such that its length is twice the length of the previous sleep window but not greater than a *final sleep window size.* Before entering power-saving class 1, the BS indicates to the MS the *initial sleep window size* and the *final sleep window size.* Once the final sleep window size is reached, all the subsequent sleep windows are of the same length. At any time during the sleep-mode operation, the BS can reset the window size to the initial sleep window size, and the process of doubling sleep window sizes is repeated. For DL allocations, the reset happen when the BS feels that the amount of listen window is not sufficient to send all the traffic. Similarly for UL allocations: The reset happens based on a request from the MS. Power-saving class 1 is recommended for best-effort or non-real-time traffic.

2. In *power-saving class 2,* all the sleep windows are of fixed length and are followed by a listen window of fixed length. Before entering power-saving class 2 mode, the BS indicates to the MS the sleep and listen window sizes. Power-saving class 2 is the recommended sleep-mode operation for UGS connections.

3. *Power-saving class 3* operation, unlike the other classes, consists of a single sleep window. The start time and the length of the sleep window are indicated by the BS before entering this mode. At the end of the sleep window, the power-saving operation becomes inactive. This power-saving class operation is recommended for multicast traffic or for MAC management traffic. For multicast service, the BS may guess when the next portion of data will appear. Then the BS allocates a sleep window for all times when it does not expect the multicast traffic to arrive. After expiration of the sleep window multicast data may be transmitted to relevant SSs. After that, the BS may decide to reinitiate power-saving operation.

9.6.2 Idle Mode

In mobile WiMAX, idle mode is a mechanism that allows the MS to receive broadcast DL transmission from the BS without registering itself with the network. Support for idle mode is optional in WiMAX and helps mobile MS by eliminating the need for handoff when it is not involved in any active data session. Idle mode also helps the BS to conserve its PHY and MAC resources, since it does not need to perform any of the handoff-related procedures or signaling for MSs that are in idle mode.

For idle-mode operation, groups of BSs are assigned to a *paging group,* as shown in Figure 9.11. An MS in idle mode periodically monitors the DL transmission of the network to determine the paging group of its current location. On detecting that it has moved to a new paging group, an MS performs a *paging group update*, during which it informs the network of the current paging group in which it is present. When, due to pending downlink traffic, the network needs to establish a connection with an MS in idle mode, the network needs to page the MS only in all the BSs belonging to the current paging group of the MS. Without the concept of the paging area, the network would need to page the MSs in all the BSs within the entire network. Each paging area should be large enough so that the MS is not required to perform a paging area update too often and should be small enough so that the paging overhead associated with sending the page on multiple BSs is low enough.

During idle-mode operation, the MS can be in either *MS paging-unavailable interval* or in *MS paging-listen interval.* During the MS paging-unavailable interval, the MS is not available for paging and can power down, conduct ranging with a neighboring BS, or scan the neighboring BS for the received signal strength and/or signal-to-noise ratio. During the MS paging-listen interval, the MS listens to the DCD and DL MAP message of the serving BS to determine when the broadcast paging message is scheduled. If the MS is paged in the broadcast paging message, the MS responds to the page and terminates its idle-mode operation. If the MS is not paged in the broadcast paging message, the MS enters the next MS paging-unavailable interval.

9.7 Mobility Management

In WiMAX, as in any other cellular network, the handoff procedure requires support from layers 1, 2, and 3 of the network. Although the ultimate decision for the handoff is determined by layer 3, the MAC and PHY layers play a crucial role by providing information and triggers required by layer 3 to execute the handoff. In this section, we discuss the mobility-management-related features of the WiMAX MAC layer.

In order to be aware of its dynamic radio frequency environment, the BS allocates time for each MS to monitor and measure the radio condition of the neighboring BSs. This process is called *scanning,* and the time allocated to each MS is called the *scanning interval.* Each scanning interval is followed by an interval of normal operation, referred to as the *interleaving interval.* In order to start the scanning process, the BS issues a MOB_SCN-REQ message that specifies to the MS the length of each scanning interval, the length the of interleaving interval, and the number of scanning events the MS is required to execute. In order to reduce the number of times such messages as

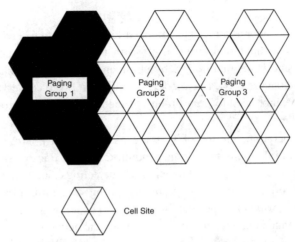

Figure 9.11 Paging area example

MOB_SCN-REQ and MOB_SCN-RES—the response from the MS after a MOB_SCN-REQ is issued to it—that are sent over the air, the BS can direct the MS to perform multiple scanning events. The identity of neighboring BSs and the frequencies that a MS is required to scan are provided in the MOB_NBR-ADV message sent over the broadcast channel.

During a scanning interval, the MS measures the received signal strength indicator (RSSI) and the signal-to-noise-plus noise ratio (SINR) of the neighboring BS and can optionally *associate* with some or all the BSs in the neighbor list, which requires the MS to perform some level of initial ranging with the neighboring BS. Three levels of association are possible during the scanning process.

1. During *association level 0 (scan/association without coordination),* the MS performs ranging without coordination from the network. As a result, the only ranging interval available to the MS is contention-based scanning. When the ranging with the neighboring BS is successful the MS receives an RNG-RSP message indicating success.

2. During *association level 1 (scan/association with coordination),* the serving BS coordinates the association procedure with the neighboring BS. The network provides the MS with a ranging code and a transmission interval for each of the neighboring BSs. A neighboring BS may assign the same code or transmission opportunity to more than one MS but not both. This gives the MS an opportunity for unicast ranging, and collision among various MSs can be avoided. When the ranging with the neighboring BS successful, the MS receives an RNG-RSP message with ranging-status success.

3. *Association level 2 (network assisted association reporting)* is similar to association level 1 except that after the unicast ranging transmission, the MS does not need to wait for the RNG-RSP message from the neighboring BS. Instead, the RNG-RSP information on PHY offsets will be sent by each neighbor BS (over the backbone) to the serving BS . The serving BS may aggregate all ranging related information into a single MOB_ASC_REPORT message.

9.7.1 Handoff Process and Cell Reselection

In WiMAX, the handoff process is defined as the set of procedures and decisions that enable an MS to migrate from the air interface of one BS to the air interface of another and consists of the following stages.

1. **Cell reselection:** During this stage, the MS performs scanning and association with one or more neighboring BSs to determine their suitability as a handoff target. After performing cell reselection, the MS resumes normal operation with the serving BS.

2. **Handoff decision and initiation:** The handoff process begins with the decision for the MS to migrate its connections from the serving BS to a new target BS. This decision can be taken by the MS, the BS, or some other external entity in the WiMAX network and is dependent on the implementation. When the handoff decision is taken by the MS, it sends a MOB_MSHO-REQ message to the BS, indicating one or more BSs as handoff targets. The BS then sends a MOB_BSHO-RSP message indicating the target BSs to be used for this handoff process. The MS sends a MOB_MSHO-IND indicating which of the BSs indicated in MOB_BSHO-RSP will be used for handoff. When the handoff decision is taken by the BS, it sends a MOB_BSHO-REQ message to the MS, indicating one or more BSs for the handoff target. The MS in this case sends a MOB_MSHO-IND message indicating receipt of the handoff decision and its choice of target BS. After the handoff process has been initiated, the MS can cancel it at any time.

3. **Synchronization to the target BS:** Once the target BS is determined, the MS synchronizes with its DL transmission. The MS begins by processing the DL frame preamble of the target BS. The DL frame preamble provides the MS with time and frequency synchronization with the target BS. The MS then decodes the DL-MAP, UL-MAP, DCD, and UCD messages to get information about the ranging channel. This stage can be shortened if the target BS was notified about the impending handoff procedure and had allocated unicast ranging resources for the MS.

4. **Ranging with target BS:** The MS uses the ranging channel to perform the initial ranging process to synchronize its UL transmission with the BS and get information about initial timing advance and power level. This initial ranging process is similar to the one used during network entry. The MS can skip or shorten this stage if it performed association with the target BS during the cell reselection/scanning stage.

5. **Termination of context with previous BS:** After establishing connection with the target BS, the MS may decide to terminate its connection with the serving BS, sending a MOB_HO-IND message to the BS. On receipt of this message, the BS starts the *resource-retain timer* and keeps all the MAC state machines and buffered MAC PDUs associated with the MS until the expiry of this timer. Once the resource retain timer expires, the BS discards all the MAC state machines and MAC PDUs belonging to the MS, and the handoff process is assumed to be complete.

A *call drop* during the handoff process is defined as the situation in which an MS has stopped communication with its serving BS in either the downlink or the uplink before the normal handoff sequence has been completed. When the MS detects a call drop, it attempts a network reentry procedure with the target BS to reestablish its connection with the network.

9.7.2 Macro Diversity Handover and Fast BS Switching

Apart from the conventional handoff process, WiMAX also defines two optional handoff procedures: macro diversity handover (MDHO) and fast base station switching (FBSS). In the case of MDHO, the MS is allowed to simultaneously communicate using the air interface of more than one BS. All the BSs involved in the MDHO with a given MS are referred to as the *diversity set.*

The normal mode of operation, no MDHO, can be viewed as a special case of MDHO in which the diversity set consists of a single BS.

When the diversity set of an MS consists of multiple BSs, one of them is considered the *anchor BS,* which often acts as the controlling entity for DL and UL allocations. In WiMAX, there are two modes by which an MS involved in MDHO can monitor its DL and UL allocation. In the first mode, the MS monitors only the DL MAP and UL MAP of the anchor BS, which provides the DL and UL allocations of the MS for the anchor BS and the all the nonanchor BSs. In the second mode, the MS monitors the DL MAP and the UL MAP of all the BSs in the diversity set separately for the DL and UL allocations, respectively. As shown in Figure 9.12, the DL signals from all the BSs in the diversity set are combined before being decoded by the FEC stage. The standard does not specify how the signals from all the BSs in the diversity set should be combined. In principle, this task can be performed in two ways. The more optimum way to combine the signals from different BSs would require the MS to demodulate these signals independently and combine them at the baseband level before the FEC decoder stage.

In such an implementation, it is possible to combine the signals optimally to achieve a certain objective, such as SNR maximization or mean square error (MSE) minimization. This method of combining is similar in concept to the maximum ratio combining (MRC) used in most CDMA handsets when it is in soft handoff with two or more BSs. In an OFDM system, in order to demodulate the signals from the different BSs that are on the same carrier frequency, the MS would require multiple antennas, with the number of antennas being equal to or greater than the number of BSs in the diversity set. If the BSs of the diversity set are on different carrier frequencies, multiple RF chains will be required. In either case, it is not possible to cost-effectively demodulate the OFDM signals from multiple BSs at baseband.

A suboptimal but more practical way to combine the signals from various BSs in WiMAX is to combine them at the RF level, which implies that all the BSs in the diversity set should not only be synchronized in time and frequency but also use the same CID, encryption mask, modulation format, FEC code rate, H-ARQ redundancy version, subcarrier permutation, and subchannels for the target MS. In this case, since the signals from various BSs are simply added in the front end of the receiver, and the achievable link gains are expected to be less than what is experienced in CDMA systems from soft handoff. This, however, requires a significant amount of

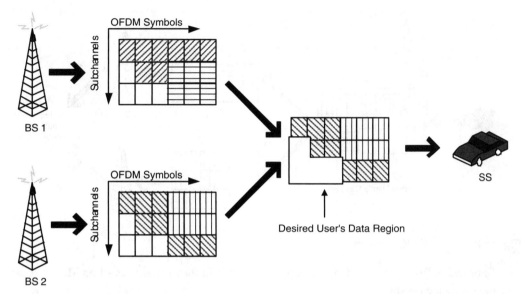

Figure 9.12 DL MOHO: combining

signaling over the backbone network to coordinate the transmission of all the BSs in the diversity set.

In the UL shown in Figure 9.13, each BS separately decodes the FEC blocks form the MS and forwards the decoded packet to a central entity—typically, the anchor BS—which selects the copy that was received without errors. In principle, this is very similar to the soft-handoff implementation in current CDMA systems.

FBBS is similar to MDHO that each MS maintains a diversity set that consists of all the BSs with which the MS has an active connection; that is the MS has established one or more CIDs and conducts periodic ranging with theses BSs. However, unlike MDHO, the MS communicates in the uplink and downlink with only one BS at a time, also referred to as the *anchor BS*.

When it needs to add a new BS to its diversity set or remove an existing one owing to variations in the channel, the MS sends a MS_MSHO-REQ message indicating a request to update its diversity set. Each FBSS-capable BS broadcasts its H_Add and H_Delete thresholds, which indicate the mean SINR, as observed by the MS, required to add or delete the BS from the diversity set. The anchor BS, when it receives a request from the MS to update its diversity set, responds with a MS_BSHO-RSP message indicating the updated diversity set.

The MS or the BS can change the anchor BS by sending a MS_MSHO-REQ (MOB_BSHO-REQ) message with such a request. If the MS is allowed to switch its anchor BS, a regular handoff procedure is not required, since the MS already has CIDs established with all the BSs in the diversity set. FBSS eliminates the various steps involved in a typical handoff and their associated message exchange and is significantly faster than the conventional handoff mechanism.

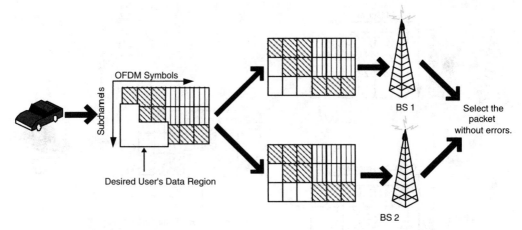

Figure 9.13 UL MDHO: Selection

In order for FBSS or MDHO to be feasible, the BSs in the diversity set of an MS must satisfy the following conditions.

- BSs involved in FBSS are synchronized, based on a common timing source.
- The DL frames sent from the BSs arrive at the MS within the cyclic prefix interval.
- BSs involved in FBSS must be on the same carrier frequency.
- BSs involved in FBSS must have synchronized frames in the DL and the UL.
- BSs involved in FBSS are also required to share all information that MS and BS normally exchange during network entry.
- BSs involved in FBSS must share all informations, such as SFID, CID, encryption, and authentication keys.

9.8 Summary and Conclusions

This chapter descrbed the MAC layer of WiMAX. Various features and functions of the MAC layer, such as construction of PDUs, ARQ, the bandwidth-request mechanism, QoS control, mobility management, and power saving were described.

- The WiMAX MAC layer has been designed from ground up to provide a flexible and powerful architecture that can efficiently support a variety of QoS requirements. WiMAX defines several scheduling services, which handle and schedule data packets differently, based on their QoS needs.
- WiMAX has several optional features, such as sleep mode, idle mode, and a handover mechanism, that can support mobility. The mobility-related features can be used to provide various levels of handoff capability, starting from simple network-reentry-based

handoff to full mobility as supported by current cellular networks. These features can also be turned off or not implemented if the network is to be optimized for fixed applications.
- WiMAX also defines several powerful encryption and authentication schemes that allow for a level of security comparable with that of wireline networks.

9.9 Bibliography

[1] R. Droms, Dynamic host configuration protocol, *IETF RFC 2131*, March 1997.
[2] IEEE. Standard 802.16-2004. Part 16: Air interface for fixed broadband wireless access systems, June 2004.
[3] IEEE. Standard 802.16-2005. Part 16: Air interface for fixed and mobile broadband wireless access systems, December 2005.

WiMAX Network Architecture

C hapters 8 and 9 covered the details of the IEEE 802.16-2004 and IEEE 802.16e-2005 PHY and MAC layers. In this chapter, we focus on the end-to-end network system architecture of WiMAX.[1] Simply specifying the PHY and MAC of the radio link alone is not sufficient to build an interoperable broadband wireless network. Rather, an interoperable network architecture framework that deals with the end-to-end service aspects such as IP connectivity and session management, security, QoS, and mobility management is needed. The WiMAX Forum's Network Working Group (NWG) has developed and standardized these end-to-end networking aspects that are beyond the scope of the IEEE 802.16e-2005 standard.

This chapter looks at the end-to-end network systems architecture developed by the WiMAX NWG. The WiMAX Forum has adopted a three-stage standards development process similar to that followed by 3GPP. In stage 1, the use case scenarios and service requirements are listed;[2] in stage 2, the architecture that meets the service requirements is developed; and in stage 3, the details of the protocols associated with the architecture are specified. At the time of this writing, the WiMAX NWG is close to completing all three stages of its first version, referred to as Release 1, with ongoing work on the next version, referred to as Release 1.5. This chapter will focuses mostly on the stage 2 specifications for Release 1, which specifies the end-to-end network systems architecture. [3]. *Disclaimer*: Although care has been taken to keep the contents of this chapter current and accurate, it should be noted that owing to ongoing developments, the final published specifications may differ in minor ways from what is summarized in this chapter.

---

1. Most of the material in this chapter has been adapted with permission from WiMAX Forum documents [1, 2, 3, 4].
2. Stage 1 requirements are developed within the WiMAX Service Provider Working Group (SPWG).

We begin this chapter with an outline of the design tenets followed by the WiMAX Forum NWG. We then introduce the WiMAX network reference model and define the various functional entities and their interconnections. Next, we discuss the end-to-end protocol layering in a WiMAX network, network selection and discovery, and IP address allocation. Then, we describe in more detail the functional architecture and processes associated with security, QoS, and mobility management.

10.1 General Design Principles of the Architecture

Development of the WiMAX architecture followed several design tenets, most of which are akin to the general design principles of IP networks. The NWG was looking for greater architectural alignment with the wireline broadband access networks, such as DSL and cable, while at the same time supporting high-speed mobility. Some of the important design principles that guided the development of the WiMAX network systems architecture include the following:

Functional decomposition

The architecture shall be based on functional decomposition principles, where required features are decomposed into functional entities without specific implementation assumptions about physical network entities. The architecture shall specify open and well-defined *reference points* between various groups of network functional entities to ensure multivendor interoperability. The architecture does not preclude different vendor implementations based on different decompositions or combinations of functional entities as long as the exposed interfaces comply with the procedures and protocols applicable for the relevant reference point.

Deployment modularity and flexibility

The network architecture shall be modular and flexible enough to not preclude a broad range of implementation and deployment options. For example, a deployment could follow a centralized, fully distributed, or hybrid architecture. The access network may be decomposed in many ways, and multiple types of decomposition topologies may coexist within a single access network. The architecture shall scale from the trivial case of a single operator with a single base station to a large-scale deployment by multiple operators with roaming agreements.

Support for variety of usage models

The architecture shall support the coexistence of fixed, nomadic, portable, and mobile usage models.[3] The architecture shall also allow an evolution path from fixed to nomadic to portability with simple mobility (i.e., no seamless handoff) and eventually to full mobility with end-to-end QoS and security support. Both Ethernet and IP services shall be supported by the architecture.

---

3. See Section 2.4.4 for usage models.

Decoupling of access and connectivity services

The architecture shall allow the decoupling of the access network and supported technologies from the IP connectivity network and services and consider network elements of the connectivity network agnostic to the IEEE 802.16e-2005 radio specifications. This allows for unbundling of access infrastructure from IP connectivity services.

Support for a variety of business models

The network architecture shall support network sharing and a variety of business models. The architecture shall allow for a logical separation between (1) the network access provider (NAP)—the entity that owns and/or operates the access network, (2) the network service provider (NSP)—the entity that owns the subscriber and provides the broadband access service, and (3) the application service providers (ASP). The architecture shall support the concept of virtual network operator and not preclude the access networks being shared by multiple NSPs or NSPs using access networks from multiple NAPs. The architecture shall support the discovery and selection of one or more accessible NSPs by a subscriber.

Extensive use of IETF protocols

The network-layer procedures and protocols used across the reference points shall be based on appropriate IETF RFCs. End-to-end security, QoS, mobility, management, provisioning, and other functions shall rely as much as possible on existing IETF protocols. Extensions may be made to existing RFCs, if necessary.

Support for access to incumbent operator services

The architecture should provide access to incumbent operator services through interworking functions as needed. It shall support loosely coupled interworking with all existing wireless networks (3GPP, 3GPP2) or wireline networks, using IETF protocols.

10.2 Network Reference Model

Figure 10.1 shows the WiMAX network reference model (NRM), which is a logical representation of the network architecture. The NRM identifies the functional entities in the architecture and the reference points (see Section 10.2.3) between the functional entities over which interoperability is achieved. The NRM divides the end-to-end system into three logical parts: (1) mobile stations used by the subscriber to access the work; (2) the access service network (ASN) which is owned by a NAP and comprises one or more base stations and one or more ASN gateways that form the radio access network; and (3) the connectivity service network (CSN), which is owned by an NSP, and provides IP connectivity and all the IP core network functions. The subscriber is served from the CSN belonging to the visited NSP; the home NSP is where the subscriber belongs. In the nonroaming case, the visited and home NSPs are one and the same.

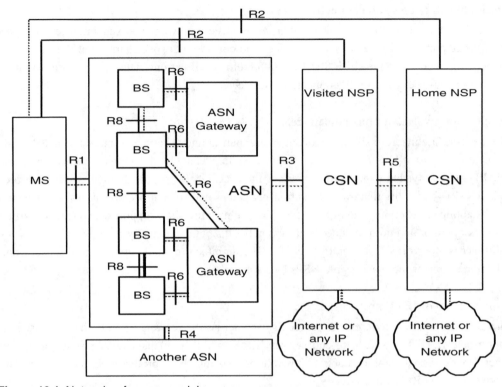

Figure 10.1 Network reference model

10.2.1 ASN Functions, Decompositions, and Profiles

The ASN performs the following functions:

- IEEE 802.16e–based layer 2 connectivity with the MS
- Network discovery and selection of the subscriber's preferred CSN/NSP
- AAA proxy: transfer of device, user, and service credentials to selected NSP AAA and temporary storage of user's profiles
- Relay functionality for establishing IP connectivity between the MS and the CSN
- Radio resource management (RRM) and allocation based on the QoS policy and/or request from the NSP or the ASP
- Mobility-related functions, such as handover, location management, and paging within the ASN, including support for mobile IP with foreign-agent functionality

The ASN may be decomposed into one or more *base stations* (BSs) and one or more *ASN Gateways* (ASN-GW) as shown in Figure 10.1. The WiMAX NRM defines multiple profiles for the ASN, each calling for a different decomposition of functions within the ASN. ASN profile B calls for a single entity that combines the BS and the ASN-GW. Profiles A and C split the functions between the BS and the ASN-GW slightly differently, specifically functions related to mobility management and radio resource management.

The BS is defined as representing one sector with one frequency assignment implementing the IEEE 802.16e interface to the MS. Additional functions handled by the BS in both profiles include scheduling for the uplink and the downlink, traffic classification, and service flow management (SFM) by acting as the QoS policy enforcement point (PEP) for traffic via the air interface, providing terminal activity (active, idle) status, supporting tunneling protocol toward the ASN-GW, providing DHCP proxy functionality, relaying authentication messages between the MS and the ASN-GW, reception and delivery of the traffic encryption key (TEK) and the key encryption key (KEK) to the MS, serving as RSVP proxy for session management, and managing multicast group association via Internet Group Management Protocol (IGMP) proxy. A BS may be connected to more than one ASN-GW for load balancing or redundancy purposes.

The ASN gateway provides ASN location management and paging; acts as a server for network session and mobility management; performs admission control and temporary caching of subscriber profiles and encryption keys; acts as an authenticator and AAA; client/proxy, delivering RADIUS/DIAMETER messages to selected CSN AAA; provides mobility tunnel establishment and management with BSs; acts as a client for session/mobility management; performs service flow authorization (SFA), based on the user profile and QoS policy; provides foreign-agent functionality; and performs routing (IPv4 and IPv6) to selected CSNs.

Table 10.1 lists the split of the various functional entities within an ASN between the BS and the ASN-GW, as per the ASN profiles defined by the WiMAX Forum.[4] Profile B has both BS and ASN-GW as an integrated unit. Profiles A and C are quite similar with the following exceptions. In profile A, the handover function is in the ASN-GW; in profile C, it is in the BS, with the ASN-GW performing only the handover relay function. Also, in profile A, the radio resource controller (RRC) is located in the ASN-GW, allowing for RRM across multiple BSs. This is similar to the BSC functionality in GSM and allows for better load balancing and spectrum management across base stations. In profile C, the RRC function is fully contained and distributed within the BS.

It should be noted that the ASN gateway may optionally be decomposed into two groups of functions: decision point (DP) functions and enforcement point (EP) functions. The EP functions include the bearer plane functions, and the DP functions may include non–bearer plane control functions. When decomposed in such a way, the DP functions may be shared across multiple ASN Gateways. Examples of DP functions include intra-ASN location management and paging, regional radio resource control and admission control, network session/mobility management (server), radio load balancing for handover decisions, temporary caching of subscriber profile and encryption keys, and AAA client/proxy. Examples of EP functions include mobility tunneling establishment and management with BSs, session/mobility management (client), QoS and policy enforcement, foreign agent, and routing to selected CSN.

---

4. The definition of each of these entities is provided in the appropriate subsequent sections.

Table 10.1 Functional Decomposition of the ASN in Various Release 1 Profiles

Functional Category	Function	ASN Entity Name		
		Profile A	Profile B	Profile C
Security	Authenticator	ASN-GW	ASN	ASN-GW
	Authentication relay	BS	ASN	BS
	Key distributor	ASN-GW	ASN	ASN-GW
	Key receiver	BS	ASN	BS
Mobility	Data path function	ASN-GW and BS	ASN	ASN-GW and BS
	Handover control	ASN-GW	ASN	BS
	Context server and client	ASN-GW and BS	ASN	ASN-GW and BS
	MIP foreign agent	ASN-GW	ASN	ASN-GW
Radio resource management	Radio resource controller	ASN-GW	ASN	BS
	Radio resource agent	BS	ASN	BS
Paging	Paging agent	BS	ASN	BS
	Paging controller	ASN-GW	ASN	ASN-GW
QoS	Service flow authorization	ASN-GW	ASN	ASN-GW
	Service flow manager	BS	ASN	BS

10.2.2 CSN Functions

The CSN provides the following functions:

- IP address allocation to the MS for user sessions.
- AAA proxy or server for user, device and services authentication, authorization, and accounting (AAA).
- Policy and QoS management based on the SLA/contract with the user. The CSN of the home NSP distributes the subscriber profile to the NAP directly or via the visited NSP.
- Subscriber billing and interoperator settlement.
- Inter-CSN tunneling to support roaming between NSPs.
- Inter-ASN mobility management and mobile IP home agent functionality.
- Connectivity infrastructure and policy control for such services as Internet access, access to other IP networks, ASPs, location-based services, peer-to-peer, VPN, IP multimedia services, law enforcement, and messaging.

10.2.3 Reference Points

The WiMAX NWG defines a reference point (RP) as a conceptual link that connects two groups of functions that reside in different functional entities of the ASN, CSN, or MS. Reference points are not necessarily a physical interface, except when the functional entities on either side of it are implemented on different physical devices. The WiMAX Forum will verify interoperability of all exposed RPs based only on specified normative protocols and procedures for a supported capability across an exposed RP.

Figure 10.1 shows a number of reference points defined by the WiMAX NWG. These reference points are listed in Table 10.2.

Release 1 will enforce interoperability across R1, R2, R3, R4, and R5 for all ASN implementation profiles. Other reference points are optional and may not be specified and/or verified for Release 1.

10.3 Protocol Layering Across a WiMAX Network

It is instructive to view the end-to-end WiMAX architecture using the logical representation shown in Figure 10.2. The architecture is quite similar to most other wide area IP access networks, where a link-layer infrastructure is used for concentrating traffic of individual users, with a separate entity providing an IP address to the end-user device for access to IP-based applications and services. Here, ASN is the link-layer infrastructure providing link concentration, and CSN is the infrastructure providing IP address and access to IP applications. The concentrated links are forwarded from the ASN-GW to the CSN via a managed IP network.

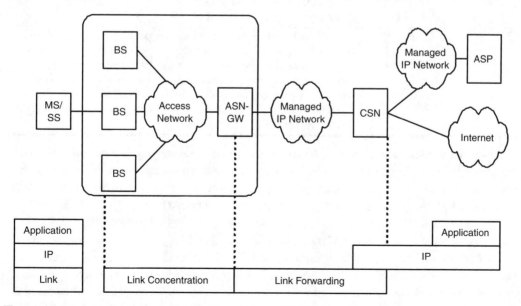

Figure 10.2 Logical representation of the end-to-end WiMAX architecture

Table 10.2 WiMAX Reference Points

Reference Point	End Points	Description
R1	MS and ASN	Implements the air-interface (IEEE 802.16e) specifications. R1 may additionally include protocols related to the management plane.
R2	MS and CSN	For authentication, authorization, IP host configuration management, and mobility management. Only a logical interface and not a direct protocol interface between MS and CSN.
R3	ASN and CSN	Supports AAA, policy enforcement, and mobility-management capabilities. R3 also encompasses the bearer plane methods (e.g., tunneling) to transfer IP data between the ASN and the CSN.
R4	ASN and ASN	A set of control and bearer plane protocols originating/terminating in various entities within the ASN that coordinate MS mobility between ASNs. In Release 1, R4 is the only interoperable interface between heterogeneous or dissimilar ASNs.
R5	CSN and CSN	A set of control and bearer plane protocols for interworking between the home and visited network.
R6	BS and ASN-GW	A set of control and bearer plane protocols for communication between the BS and the ASN-GW. The bearer plane consists of intra-ASN data path or inter-ASN tunnels between the BS and the ASN-GW. The control plane includes protocols for mobility tunnel management (establish, modify, and release) based on MS mobility events. R6 may also serve as a conduit for exchange of MAC states information between neighboring BSs.
R7	ASN-GW-DP and ASN-GW-EP	An optional set of control plane protocols for coordination between the two groups of functions identified in R6.
R8	BS and BS	A set of control plane message flows and, possibly, bearer plane data flows between BSs to ensure fast and seamless handover. The bearer plane consists of protocols that allow the data transfer between BSs involved in handover of a certain MS. The control plane consists of the inter-BS communication protocol defined in IEEE 802.16e and additional protocols that allow controlling the data transfer between the BS involved in handover of a certain MS.

Now, let us look more closely at the end-to-end protocol layering as data and control packets are forwarded from the MS to the CSN. The WiMAX architecture can be used to support both IP and Ethernet packets. IP packets may be transported using the IP convergence sublayer (IP-CS) over IEEE 802.16e or using the Ethernet convergence sublayer (ETH-CS) over IEEE 802.16e. Within the ASN the Ethernet or IP packets may be either routed or bridged. Routing over the ASN, may be done using IP-in-IP encapsulation protocols, such as GRE (generic routing encapsulation).

Figure 10.4 shows the protocol stack when using the IP convergence layer to transport IP packets over a routed ASN. If the ASN were a bridged network, the shaded layers (GRE, IP, link) in Figure 10.4 would be replaced with an Ethernet layer. The protocol stack when using the Ethernet convergence layer to transport IP packets over a routed ASN is shown in Figure 10.3.

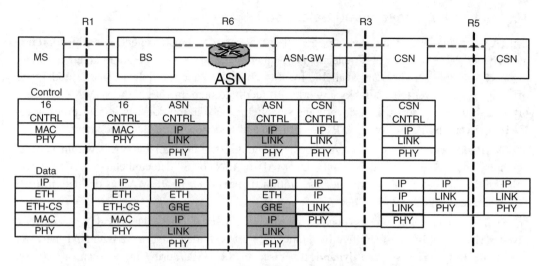

Figure 10.3 Protocol stack for Ethernet convergence sublayer with routed ASN

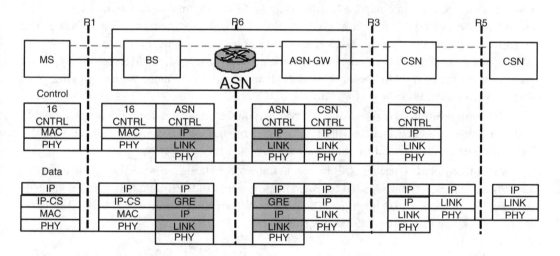

Figure 10.4 Protocol stack for IP convergence sublayer with routed ASN

In this case, if the ASN were a bridged network, the shaded layers would not be needed. It is also possible to transport Ethernet packets all the way up to the CSN by using the ETH-CS, in which case, an encapsulation protocol, such as GRE, may be used for forwarding from the ASN to the CSN. This type of Ethernet service may be used to provide end-to-end VLAN (virtual local area networking) services.

10.4 Network Discovery and Selection

WiMAX networks are required to support either manual or automatic selection of the appropriate network, based on user preference. It is assumed that an MS will operate in an environment in which multiple networks are available for it to connect to and multiple service providers are offering services over the available networks. To facilitate such operation, the WiMAX standard offers a solution for network discovery and selection. The solution consists of four procedures: NAP discovery, NSP discovery, NSP enumeration and selection, and ASN attachment.

NAP discovery: This process enables the MS to discover all available NAPs within a coverage area. The MS scans and decodes the DL MAP of ASNs on all detected channels. The 24-bit value of the "operator ID" within the base station ID parameter in DL MAP as defined in IEEE 802.16 serves as the NAP identifier.

NSP discovery: This process enables the MS to discover all NSPs that provide service over a given ASN. The NSPs are identified by a unique 24-bit NSP identifier, or 32-byte NAI (network access identifier). The MS can dynamically discover the NSPs during initial scan or network entry by listening to the NSP IDs broadcast by the ASN as part of the system identity information advertisement (SII-ADV) MAC management message. NSP-IDs may also be transmitted by the BS in response to a specific request by MS, using an SBC-REQ message. Alternatively, the MS could have a list of NSPs listed in its configuration. The NSP-IDs are mapped to an NSP realm either by using configuration information in the MS or by making a query to retrieve it from the ASN. If the preconfigured list does not match the network broadcast, the MS should rely on the information obtained from the network. If an NAP and an NSP have a one-to-one mapping, NSP discovery does not need to be performed.

NSP enumeration and selection: The MS may make a selection from the list of available NSPs by using an appropriate algorithm. NSP selection may be automatic or manual, with the latter being particularly useful for initial provisioning and for "pay-per-use" service.

ASN attachment: Once an NSP is selected, the MS indicates its selection by attaching to an ASN associated with the selected NSP and by providing its identity and home NSP domain in the form of a network access identifier. The ASN uses the realm portion of the NAI to determine the next AAA hop to send the MS's AAA packets.

10.5 IP Address Assignment

The Dynamic Host Control Protocol (DHCP) is used as the primary mechanism to allocate a dynamic *point-of-attachment* (PoA) IP address to the MS. Alternatively, the home CSN may allocate IP addresses to an ASN via AAA, which in turn is delivered to the MS via DHCP. In this case, the ASN will have a DHCP proxy function as opposed to a DHCP relay function. When an MS is an IP gateway or host, the standard requires that a PoA IP address be allocated to the gateway or the host, respectively. If the MS acts as a layer 2 bridge (ETH-CS), IP addresses may be allocated to the hosts behind the MS. For fixed access, the IP address must be allocated from the CSN address space of the home NSP and may be either static or dynamic. For nomadic, portable, and mobile access dynamic allocation from either the home or the visited CSN is allowed, depending on roaming agreements and the user subscription profile and policy.

To support IPv6, the ASN includes an IPv6 access router (AR) functionality, and the MS obtains a globally routable IP address from the AR. When using mobile IPv6, the MS obtains the care-of address (CoA) from the ASN, and a home address (HoA) from the home CSN. The MS may use either the CoA or the HoA as its PoA address, depending on whether it routes packets directly to correspondent nodes (CNs) or via the home agent (HA) in the CSN. When using IPv6, static IP address, stateful autoconfiguration based on DHCPv6 (RFC 3315), or stateless address autoconfiguration (RFC 2462) is allowed. When Mobile IPv6 is used, the HoA is assigned via stateless DHCP. For stateful configuration, the DHCP server is in the serving CSN, and a DHCP relay may exist in the network path to the CSN. For stateless configuration, the MS will use neighbor discovery or DHCP to receive network configuration information.

One known issue with the use of IPv6 in WiMAX stems from the lack of link-local multicast support in IEEE 802.16e air-interface. IPv6 has several multicast packets, such as neighbor solicitation, neight advertisement, router solicitation and router advertisement, that have a link-local scope. Since, packet transmission in IEEE 802.16e is based on a connection identifier (CID) as opposed to the 48-bit hardware MAC address as is assumed by conventional IPv6 and RFC 2464, there is a need to define new mechanisms to share multicast CIDs among multicast group members in a WiMAX network.

10.6 Authentication and Security Architecture

The WiMAX authentication and security architecture is designed to support all the IEEE 802.16e security services, using an IETF EAP-based AAA framework. In addition to authentication, the AAA framework is used for service flow authorization, QoS policy control, and secure mobility management. Some of the WiMAX Forum specified requirements that the AAA framework should meet are as follows:

- Support for device, user, and mutual authentication between MS/SS and the NSP, based on Privacy Key Management Version 2 (PKMv2) as defined in IEEE 820.16e-2005.
- Support for authentication mechanisms, using a variety of credentials, including shared secrets, subscriber identity module (SIM) cards, universal SIM (USIM), universal integrated circuit card (UICC), removable user identity module (RUIM), and X.509 certificates, as long as they are suitable for EAP methods satisfying RFC 4017.
- Support for global roaming between home and visited NSPs in a mobile scenario, including support for credential reuse and consistent use of authorization and accounting through the use of RADIUS in the ASN and the CSN. The AAA framework shall also allow the home CSN to obtain information, such as visited network identity, from the ASN or the visited CSN that may be needed during AAA.
- Accommodation of mobile IPv4 and IPv6 security associations (SA) management.

• Support for policy provisioning at the ASN or the CSN by allowing for transfer of policy-related information from the AAA server to the ASN or the CSN.

10.6.1 AAA Architecture Framework

The WiMAX Forum recommends using the AAA framework based on the *pull model*, interaction between the AAA elements as defined in RFC 2904. The pull sequence consists of four basic steps.

1. The supplicant MS sends a request to the network access server (NAS) function in the ASN.
2. The NAS in the ASN forwards the request to the service provider's AAA server. The NAS acts as a AAA client on behalf of the user.
3. The AAA server evaluates the request and returns an appropriate response to the NAS.
4. The NAS sets up the service and tells the MS that it is ready.

Figure 10.5 shows the pull model as applied to the generic WiMAX roaming case. Here, steps 2 and 3 are split into two substeps, since the user is connecting to a visited NSP that is different from the home NSP. For the nonroaming case, the home CSN and the visited CSN are one and the same. It should be noted that the NAP may deploy a AAA proxy between the NAS(s) in the ASN and the AAA in the CSN, especially when the ASN has many NASs, and the CSN is in another administrative domain. Using an AAA proxy enhances security and makes configuration easier, since it reduces the number of shared secrets to configure between the NAP and the foreign CSN.

In the WiMAX architecture, the AAA framework is used for authentication, mobility management, and QoS control. The NAS is a collective term used to describe the entity that performs the roles of authenticator, proxy-MIP client, foreign agent, service flow manager, and so on. The NAS resides in the ASN, though the implementation of the various functions may reside in different physical elements within the ASN.

10.6.2 Authentication Protocols and Procedure

The WiMAX network supports both user and device authentication. An operator may decide to implement either one or both of these authentications. For user and device authentication, the IEEE 802.16e-2005 standard specifies PKMv2 with EAP. PKMv2 transfers EAP over the air interface between the MS and the BS in ASN. If the authenticator does not reside in the BS, the BS acts as an authentication relay agent and forwards EAP messages to the authenticator over an authentication relay protocol that may run over an R6 interface. The AAA client at the authenticator encapsulates the EAP in AAA protocol packets and forwards them via one or more AAA proxies[5] to the AAA server in the CSN of the home NSP. EAP runs over RADIUS between the

---

5. One or more AAA brokers may exist between the authenticator and the AAA server in roaming scenarios.

AAA Server and the authenticator in ASN. Depending on the type of credential, a variety of EAP

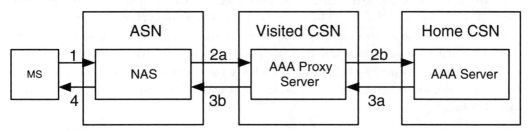

Figure 10.5 Generic AAA roaming model

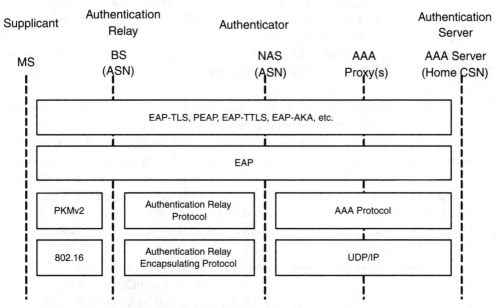

Figure 10.6 Protocol stack for user authentication in WiMAX

schemes, including EAP-AKA (authentication and key agreement), EAP-TLS, EAP-SIM, and EAP-PSK (preshared key), may be supported. It is also possible to optionally secure the transport of end-to-end user authentication within a tunnel by using protocols such as tunneled transport-layer security (TTLS). Figure 10.6 depicts the protocol stack for PKMv2 user authentication.

When both user and device authentications need to be performed and both authentications terminate in different AAA servers, PKMv2 double-EAP mode is used. Here, user EAP authentication follows device authentication before the MS is allowed access to IP services. If the same AAA server is used for both, the process could be shortened by doing joint device and user authentication.

Device credentials typically take the form of a digital certificate or a preprovisioned preshared secret key. It is also possible to dynamically generate a secret key from a built-in X.509 certificate. The EAP device identifier may be a MAC address or a NAI in the form of MAC_address @ NSP-domain. A master session key (MSK) appropriate for device authentication is generated once the appropriate credential is determined. Both device and user authentication must generate an MSK.

Figure 10.7 shows the PKMv2 procedures followed after initial network entry before a service flow can be set up between the MS and the WiMAX network. The various steps involved are as follows:

1. **Initial 802.16e network entry and negotiation:** After successful ranging, the MS and the ASN negotiate the security capabilities, such as PKM version, PKMv2 security capabilities, and authorization policy describing PKMv2 EAP only or PKMv2 double-EAP. In order to initiate an EAP conversation, the MS may also send a PKMv2-EAP-Start message to initiate EAP conversations with the ASN. Once an active air link is set up between the BS and the MS, a link activation is sent over R6 to the authenticator to begin the EAP sequence.

2. **Exchange of EAP messages:** EAP exchange begins with an EAP-Identity-Request message from the EAP authenticator to the EAP supplicant, which is the MS. If the EAP authenticator is not in the BS, an authentication relay protocol over R6 may be used for communication between the Authenticator and the BS. The MS responds with an EAP-Response message to the authenticator, which forwards all the responses from the MS to the AAA proxy, which then routes the packets based on the associated NAI realm to a remote AAA authentication server, using RADIUS. After one or more EAP request/response exchanges, the authentication server determines whether the authentication is successful and notifies the MS accordingly.

3. **Establishment of the shared master session key and enhanced master session key:** An MSK and an enhanced MSK (EMSK) are established at the MS and the AAA Server as part of a successful EAP exchange. The MSK is then also transferred from the AAA server to the authenticator (NAS) in the ASN. Using the MSK, the authenticator and the MS both generate a *pairwise master key* (PMK) according to the IEEE 802.16e-2005 specifications. The MS and the AAA server use to the EMSK to generate mobile keys.

4. **Generation of authentication key:** Based on the algorithm specified in IEEE 802.16e-2005, the AS and the MS generate the *authentication key* (AK).

5. **Transfer of authentication key:** The AK and its context are delivered from the *key distributor* entity in the authenticator to the *key-receiver* entity in the serving BS. The key receiver caches this information and generates the rest of the IEEE 802.16e-2005–specified keys from it.

6. **Transfer of security associations:** SAs are the set of security information that the BS and one or more of its MS share in order to support secure communications. The shared information includes TEK and initialization vectors for cipher-block chaining (CBC). SA trans-

fer between the BS and the MS is done via a three-way handshake. First, the BS transmits the SA-TEK Challenge message, which identifies an AK to be used for the SA, and includes a unique challenge. In the second step, the MS transmits an SA-TEK Request message after receipt and successful HMAC/CMAC verification of an SA challenge from the BS. The SA-TEK-Request message is a request for SA descriptors identifying the SAs the requesting MS is authorized to access and their particular properties. In the third step, the BS transmits the SA-TEK Response message identifying and describing the primary and static SAs the requesting MS is authorized to access.

7. **Generation and transfer of traffic encryption keys:** Following the three-way hand-shake, the MS requests the BS for two TEKs each for every SA. The BS randomly generates a TEK, encrypts it using the secret symmetric *key encryption key* (KEK), and transfers it to the MS.

8. *Service flow creation:* Once the TEKs are established between the MS and the BS, service flows are created, using another three-way handshake. Each service flow is then mapped onto an SA, thereby associating a TEK with it.

10.6.3 ASN Security Architecture

Within the ASN, the security architecture consists of four functional entities: (1) authenticator, which is the authenticator defined in the EAP specifications (RFC 4017); (2) authentication relay, which is the functional entity in a BS that relays EAP packets to the authenticator via an authentication relay protocol; (3) key distributor, which is the functional entity that holds the keys (both MSK and PSK[6]) generated during the EAP exchange; and (4) key receiver, which holds the authentication key and generates the rest of the IEEE 802.16e keys. Figure 10.8 shows the two deployment models for these security-related functional blocks within the ASN. One is the integrated model, whereby all the blocks are within the BS, as in ASN profile B. The alternative model has the authenticator and the key distributor in a separate stand-alone entity or in the ASN-GW (as in ASN Profile A and C). For the integrated model, the Authentication Relay Protocol and the AK Transfer Protocol (see Section 7.4.5 of [3]) are internal to the BS. For the stand-alone model, they are exposed and must comply with the standards.

10.7 Quality-of-Service Architecture

The QoS architecture framework developed by the WiMAX Forum extends the IEEE 802.16e QoS model by defining the various QoS-related functional entities in the WiMAX network and the mechanisms for provisioning and managing the various service flows and their associated policies. The WiMAX QoS framework supports simultaneous use of a diverse set of IP services, such as differentiated levels of QoS on a per user and per service flow basis, admission control, and bandwidth management. The QoS framework calls for the use of standard IETF mechanisms for managing policy decisions and policy enforcement between operators.

---

6. MSK is delivered by the AAA Server, and the PSK is generated locally.

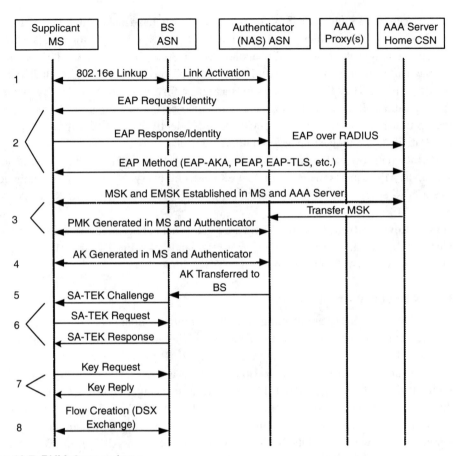

Figure 10.7 PKMv2 procedures

The WiMAX QoS framework supports both static and dynamic provisioning of service flows, though Release 1 specifications support only static QoS. In the static case, the MS is not allowed to change the parameters of the provisioned service flows or create new service flows dynamically. Dynamic service flow creation in WiMAX follows three generic steps.

1. For each subscriber, the allowed service flows and associated QoS parameters are provisioned via the management plane.
2. A service flow request initiated by the MS or the BS is evaluated against the provisioned information, and the service flow is created, if permissible.
3. Once a service flow is created, it is admitted by the BS, based on resource availability; if admitted, the service flow becomes active when resources are assigned.

Dynamic service flow creation, however, is not part of Release 1 but may be included in Release 1.5.

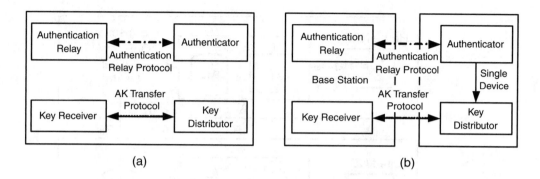

Figure 10.8 ASN security architecture and deployment models: (a) integrated deployment model and (b) stand-alone deployment model

Figure 10.9 shows the proposed QoS functional architecture as proposed by the WiMAX NWG. This architecture supports the dynamic creation, admission, activation, modification, and deletion of service flows. The important functional entities in the architecture are as follows:

Policy function. The policy function (PF) and a database reside in the home NSP. The PF contains the general and application-dependent policy rules of the NSP. The PF database may optionally be provisioned by an AAA server with user-related QoS profiles and policies. The PF is responsible for evaluating a service request it receives against the provisioned. Service requests to the PF may come from the service flow authorization (SFA) function or from an application function (AF), depending on how the service flows are triggered.

AAA server. The user QoS profiles and associated policy rules are stored in the AAA server. User QoS information is downloaded to an SFA at network entry as part of the authentication and authorization procedure. The SFA then evaluates incoming service requests against this downloaded user profile to determine handling. Alternatively, the AAA server can provision the PF with subscriber-related QoS information. In this case, the home PF determines how incoming service flows are handled.

Service flow management. The service flow management (SFM) is a logical entity in the BS that is responsible for the creation, admission, activation, modification, and deletion of 802.16e service flows. The SFM manages local resource information and performs the admission control (AC) function, which decides whether a new service flow can be admitted into the network.

Service flow authorization. The SFA is a logical entity in the ASN. A user QoS profile may be downloaded into the SFA during network entry. If this happens, the SFA evaluates the incoming service request against the user QoS profile and decides whether to allow the flow. If the user QoS profile is not with the SFA, it simply forwards the service flow request to the PF for decision making. For each MS, one SFA is assigned as the *anchor SFA* for a given session and is responsible for communication with PF. Additional SFAs may exist in an NAP that relays QoS-related primitives and applies QoS policy for that MS. The relay SFA that directly communicates with

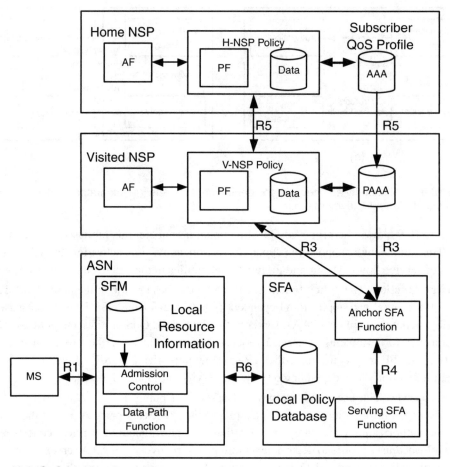

Figure 10.9 QoS functional architecture

the SFM is called the *serving SFA*. The SFAs may also perform ASN-level policy enforcement, using a local policy function (LPF) and database. The LPF may also be used for local admission control enforcement.

    Application function. The AF is an entity that can initiate service flow creation on behalf of a user. An example of an AF is a SIP proxy client.

10.8 Mobility Management

The WiMAX mobility-management architecture was designed to

- Minimize packet loss and handoff latency and maintain packet ordering to support seamless handover even at vehicular speeds

- Comply with the security and trust architecture of IEEE 802.16 and IETF EAP RFCs during mobility events
- Supporting macro diversity handover (MOHO) as well as fast base station switching (FBSS)[7]
- Minimize the number of round-trips of signaling to execute handover
- Keep handover control and data path control separate
- Support multiple deployment scenarios and be agnostic to ASN decomposition
- Support both IPv4- and IPv6-based mobility management and accommodate mobiles with multiple IP addresses and simultaneous IPv4 and IPv6 connections
- Maintain the possibility of vertical or intertechnology handovers and roaming between NSPs
- Allow a single NAP to serve multiple MSs using different private and public IP domains owned by different NSPs
- Support both static and dynamic home address configuration
- Allow for policy-based and dynamic assignment of home agents to facilitate such features as route optimization and load balancing

The WiMAX network supports two types of mobility: (1) *ASN-anchored mobility* and (2) *CSN-anchored mobility.* ASN-anchored mobility is also referred to as intra-ASN mobility, or micromobility. In this case, the MS moves between two data paths while maintaining the same anchor foreign agent at the northbound edge of the ASN network. The handover in this case happens across the R8 and/or R6 reference points. ASN-anchored handover typically involves migration of R6, with R8 used for transferring undelivered packets after handover. It is also possible to keep the layer 3 connection to the same BS (anchor BS) through the handover and have data traverse from the anchor BS to the serving BS throughout the session. CSN-anchored mobility is also referred to as inter-ASN mobility, or macromobility. In this case, the MS changes to a new anchor FA—this is called FA migration—and the new FA and CSN exchange signaling messages to establish data-forwarding paths. The handover in this case happens across the R3 reference point, with tunneling over R4 to transfer undelivered packets. Figure 10.10 illustrates the various possible handover scenarios supported in WiMAX.

Mobility management is typically triggered when the MS moves across base stations based on radio conditions. It may also be triggered when an MS wakes up from idle mode at a different ASN or when the network decides to transfer R3 end points for an MS from the serving FA to a new FA for resource optimization.

In many cases, both ASN- and CSN-anchored mobility may be triggered. When this happens, it is simpler and preferable to initiate the CSN-anchored mobility after successfully completing the ASN-anchored mobility.

---

7. Note that the initial WiMAX system profiles do not support MDH and FBSS.

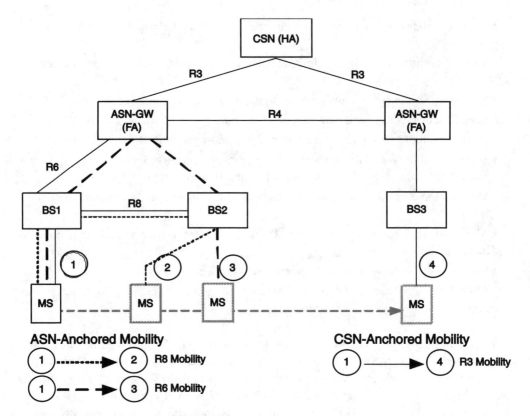

Figure 10.10 Handover scenarios supported in WiMAX

10.8.1 ASN-Anchored Mobility

ASN-anchored mobility supports handoff scenarios in which the mobile moves its point of attachment from one BS to another within the same ASN. This type of movement is invisible to the CSN and does not have any impact at the network- or IP-layer level. Implementing ASN-anchored mobility does not require any additional network-layer software on the MS.

The WiMAX standard defines three functions that together provide ASN-anchored mobility managment: data path function, handoff function, and context function.

1. The *data path function (DPF)* is responsible for setting up and managing the bearer paths needed for data packet transmission between the functional entities, such as BSs and ASN-GWs, involved in a handover. This includes setting up appropriate tunnels between the entities for packet forwarding, ensuring low latency, and handling special needs, such as multicast and broadcast. Conceptually, four DPF entities are defined in WiMAX: (1) *anchor data path function,* which is the DPF at one end of the data path that anchors the data path associated with the MS across handovers; (2) *serving data path function,* which

is the DPF associated with the BS that currently has the IEEE 802.16e link to the MS; (3) *target data path function*, which is the DPF that has been selected as the target for the handover; and (4) *relaying data path function*, which mediates between serving, target, and anchor DPFs to deliver the information.

2. The *handoff (HO) function* is responsible for making the HO decisions and performing the signaling procedures related to HO. The HO function supports both mobile and network-initiated handovers, FBSS, and MDHO. Note, however, that FBSS and MDHO are not supported in Release 1. Like the DPF, the HO function is also distributed among many entities, namely, *serving HO function*, *target HO function,* and *relaying HO function.* The serving HO function controls the overall HO decision operation and signaling procedures, signaling the target HO function to prepare for handover and informing the MS. Signaling between the serving and target HO functions may be via zero or more relaying HO functions. A relaying HO may modify the content of HO messages, and impact HO decisions. The WiMAX standard defines a number of primitives and messages such as HO Request, HO Response, and HO Confirm, for communication among the HO functions.

3. The *context function* is responsible for the exchange of state information among the network elements impacted by a handover. During a handover execution period, there may be MS-related state information in the network and network-related state information in the MS that need to be updated and/or transferred. For example, the target BS needs to be updated with the security context of the MS that is being handed in. The context function is implemented using a client/server model. The *context server* keeps the updated session context information, and the *context client,* which is implemented in the functional entity that has the IEEE 802.16e air link, retrieves it during handover. There could be a *relaying context function* between the context server and context client. The session-context information exchanged between the context client and server may include MS NAI, MS MAC address, anchor ASN-GW associated with the MS, SFID and associated parameters, CID, home agent IP address, CoA, DHCP Server, AAA server, security information related to PKMv2 and proxy MIP, and so on.

Two different types of bearers called Type 1 and Type 2 may be used for packet transfer between DPFs. Type 1 forwards IP or Ethernet packets, using layer 2 bridging (e.g., Ethernet or MPLS) or layer 3 routing (e.g. GRE tunnel) between two DPFs. Type 2 forwards IEEE 802.16e MSDUs, appended with such additional information as connection ID of the target BS, and ARQ parameters using layer 2 bridging or layer 3 routing. It is likely that most ASN deployments will prefer type 1 over type 2 for simplicity. Data path forwarding may be done at different levels of granularity, such as per service flow per subscriber, per subscriber, or per functional entity. Forwarding of individual streams can be done easily, using tagging that is supported by the forwarding technologies, such as GRE, MPLS, or 802.1Q used within the ASN.

The DPF includes mechanisms to guarantee data integrity and synchronization during handoff. Data that requires integrity may be buffered in either the originator or the terminator DPF, either only during HO periods or always. Another method to achieve data integrity, applicable only for downstream traffic, is multicasting at the anchor DPF to the serving DPF and one or more target DPFs. Synchronization of data packets during HO may be achieved by using sequence numbers attached to the MSDU in the ASN data path. Alternatively, the anchor DPF can buffer the data during HO; when a final target BS is identified, the serving BS may be asked to return to the anchor DPF all unsent packets and to the target BS all unacknowledged packets.

10.8.2 CSN-Anchored Mobility for IPv4

CSN-anchored mobility refers to mobility across different ASNs, in particular across multiple foreign agents. For Release 1, the WiMAX specifications limit CSN anchored mobility to between FAs belonging to the same NAP. CSN-anchored mobility involves mobility across different IP subnets and therefore requires IP-layer mobility management. As discussed in Chapter 7, mobile IP (MIP) is the IETF protocol for managing mobility across IP subnets. Mobile IP is used in WiMAX networks to enable CSN-anchored mobility.

The WiMAX network defines two types of MIP implementations for supporting CSN anchored mobility. The first one is based on having a MIP client (MN) at the MS, and the other is based on having a proxy MIP in the network that implements the MN in the ASN on behalf of the MS. With proxy MIP, IP mobility is transparent to the MS, which then needs only a simple IP stack. Coexistence of proxy MIP and client MIP in a network is also supported. When both are supported in the network, the MS should support either mobile IP with client MIP or regular IP with proxy MIP.

10.8.2.1 Client MIP-Based R3 Mobility Management

Here, the MS has a MIP-enabled client that is compliant with the IETF mobile IP standard (RFC 3334). The client in this case gets a CoA from an FA within the ASN. As the client moves across FA boundaries, the client becomes aware of the movement via agent advertisements, does a MIP registration with the new FA, and gets a new CoA from it.

Figure 10.11 shows the network elements and protocol stacks involved when using a client MIP implementation. The MS gets its HoA from the HA in the CSN of the home NSP or the visited NSP. In case the HA is assigned in the visited CSN, MIP authentication occurs between the visited HA and the home AAA, with the security exchanges being transparent to the visited AAA server. On the other hand, when the HA is assigned by the home CSN, both the HA address and, optionally, the DHCP server address or the MS home address are appended to the AAA reply of the home AAA server.

In addition to RFC 3344, client MIP is also expected to support several MIP extensions in WiMAX networks. These include reverse tunneling, based on RFC 3024; FA challenge/response, based on RFC 3012; NAI extensions based on RFC 2794; and mobile IP vendor-specific extensions, based on RFC 3115.

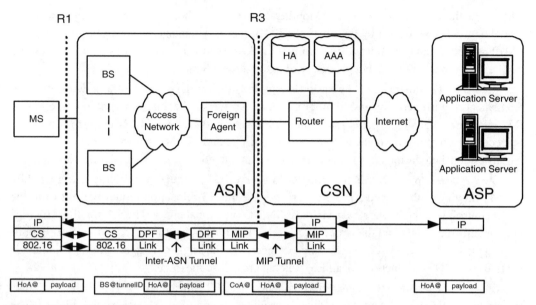

Figure 10.11 Client MIP data plane protocol stack

10.8.2.2 Proxy MIP R3 Mobility Management

Proxy MIP (PMIP) is an embodiment of the standard MIP framework wherein an instance of the MIP stack is run in the ASN on behalf of an MS that is not MIP capable or MIP aware. Using proxy MIP does not involve a change in the IP address of the MS when the user moves and obviates the need for the MS to implement a MIP client stack.

The functional entity in the WiMAX network that runs proxy MIP instances on behalf of the various clients is called the *proxy MIP mobility manager*. Within a proxy MIP mobility manager, a unique PMIP client corresponding to an MS is identified by the user NAI, which is typically the same as that used for access authentication. The proxy MIP mobility manager is also typically colocated with an authenticator function. The MIP registration to set up or update the forwarding path of the MS on the HA is performed by the proxy MIP client on behalf of the MS. The MIP-related information required to perform MIP registrations to the HA are retrieved via the AAA messages exchanged during the authentication phase. This information consists of the HA address, the security information to generate the MN AAA and MN HA authentication extension, and either the DHCP server address or the HoA address.

The foreign-agent behavior in proxy MIP is somewhat different from that of the standard RFC 3344 mobile IP. Specifically, the destination IP addresses for the control and data planes are different. With the PMIP approach, the MIP signaling needs to be directed to a PMIP client within the PMIP mobility manager, but the user data still needs to be sent to the MS over the corresponding R6 or R4 data path. To accomplish this, an odd-numbered MN-HA SPI (security parameter index) is used to flag PMIP usage in all MIP signaling messages between the PMIP manager and the FA. Messages originated by the PMIP manager will set the IP packet source address to the address of the PMIP Mobility Manager. Replies to messages tagged with the PMIP flag will be returned by the FA to the PMIP mobility manager instead of to the MS. The PMIP mobility manager address is not directly linked to an R3 mobility session of the MS and can be changed at any time independently of an ongoing R3 mobility session.

Figure 10.12 illustrates of the various steps involved in the process of a network-initiated R3 reanchoring triggered by an MS mobility event. Arrows on the top from right to left represent R3, inter/intra-ASN data paths, and intra-ASN data paths after a CSN-anchored mobility, respectively. To minimize delay and packet loss during handoff, a temporary R4 data-forwarding path is established during an R3 handover and is maintained until it is successfully completed.

Step 1 shows the ASN mobility trigger. In Figure 10.12, where the mobility is triggered by the MS moving out of a cell's radio coverage area, the R3 relocation request is sent from the RRM controller to the PMIP server via the ASN HO functional entity. The R3 mobility HO trigger contains the MSID and the target FA address. The PMIP client instance in the PMIP mobility manager will initiate FA reanchoring on receiving the trigger. Trigger receipt is indicated back to the ASN HO function via an R3_Relocate.response message.

Steps 2–5 show the MIP registration sequence, which begins when the PMIP client initiates registration to the target FA whose address is provided as part of the R3_Relocation.request. The target FA then forwards the registration request to the HA, which updates the MS binding, relocates the R3, and signals the PMIP client of the successful update. The ASN functional entity is then notified of the successful R3 relocation, using an R3_Relocation.confirm message.

Figure 10.12 shows only the case of MS movement-triggered mobility management, using PMIP architecture. For similar process flows for other mobility cases, refer to the WiMAX Forum NWG Stage 2 document [3].

10.8.3 CSN Anchored Mobility for IPv6

As mentioned earlier, WiMAX network architecture supports both IPv4 and IPv6. The CSN anchored mobility management for IPv6 differs from the IPv4 case, owing to differences between mobile IPv4 and mobile IPv6. The key differences are discussed in Section 7.4.3 of Chapter 7. WiMAX supports CSN anchored mobility for IPv6 based mobile stations using CMIPv6. The absence of an FA and support for route optimization in IPv6 implies that the CMIPv6 operates using a *colocated CoA* (CCoA), which is typically obtained by the MS using stateless auto-configuration (RFC2462) or DHCPv6 (RFC 3315). The CCoA is communicated to the HA via a binding update, and also to the CN. Binding updates to the CN allows it to directly communicate with the MN without having to traverse via the HA. If the CN does not keep a binding update cache, it reverts to sending packets to the MN via the HA using normal mobile IP tunneling.

Like in the case of IPv4, CSN anchored R3 mobility is initiated by the network, and may be triggered by an MS mobility event, by an MS waking up from idle mode, or a by a network resource optimization need.

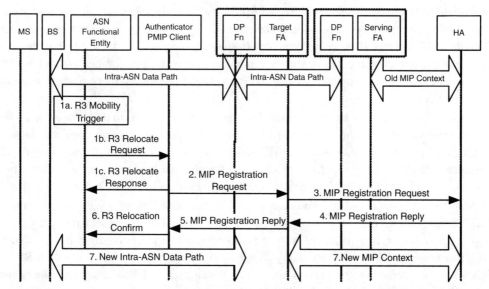

Figure 10.12 Mobility event triggering a network-initiated R3 reanchoring (PMIP)

WiMAX requires the use of protocols specified in RFC 3775, RFC 4285, and RFC 4283 for IPv6 mobility. Support for IPSec and IKEv2 per internet draft [5] and for DHCP option for home information discovery [6] are also required. The architecture supports HA assignment by either the home or the visited NSP based on the roaming agreements between the NSPs. If the visited NSP's HA is assigned, MIPv6 authentication takes place between the visited HA and the Home AAA server, without involving the visited AAA proxy. Alternatively, the Home NSP could assign its HA to the user in a AAA reply message during authentication.

10.9 Radio Resource Management

The RRM function is aimed at maximizing the efficiency of radio resource utilization and is performed within the ASN in the WiMAX network. Tasks performed by the WiMAX RRM include (1) triggering radio-resource-related measurements by BSs and MSs, (2) reporting these measurements to required databases within the network, (3) maintaining one or more databases related to RRM, (4) exchanging information between these databases within or across ASNs, and (5) making radio resource information available to other functional entities, such as HO control and QoS management.

The WiMAX architecture decomposes the RRM function into two functional entities: the *radio resource agent* and the *radio resource controller*. The radio resource agent (RRA) resides in each BS and collects and maintains radio resource indicators, such as received signal strength, from the BS and all MSs attached to the BS. The RRA also communicates RRM control information, such as neighbor BS set and their parameters, to the MSs attached to it. The radio resource controller (RRC) is a logical entity that may reside in each BS, in ASN-GW, or as a stand-alone server in the ASN. The RRC is responsible for collecting the radio resource indicators from the various RRAs attached to it and maintaining a "regional" radio resource database. When the RRC and the RRA are implemented in separate functional entities, they communicate over the R6 reference point. Multiple RRCs may also communicate with one another over the R4 reference point if implemented outside the BS and over the R8 reference point if integrated within the BS.

Each RRA in the BS is also responsible for controlling its radio resources, based on its own measurement reports and those obtained from the RRC. Control functions performed by the RRA include power control, MAC and PHY supervision, modification of the neighbor BS list, assistance with the local service flow management function and policy management for service flow admission control, and assistance with the local HO functions for initiating HO.

Standard procedures and primitives are defined for communication between the RRA and the RRC. The procedures may be classified as one of two types. The first type, called *information reporting procedures*, is used for delivery of BS radio resource indicators from the RRA to the RRC, and between RRCs. The second type, called *decision-support procedures* from RRC to RRA is used for communicating useful hints about the aggregated RRM status that may be used by the BS for various purposes. Defined RRM primitives include those for requesting and reporting link-level quality per MS, spare capacity available per BS, and neighbor BS radio resource status. Future enhancements to RRM may include additional primitives, such as for reconfiguring subchannel spacing, burst-selection rules, maximum transmit power, and UL/DL ratio.

Figure 10.13 shows two generic reference models for RRM as defined in WiMAX. The first one shows the split-RRM model, where the RRC is located outside the BS; the second one shows the RRC colocated with RRA within the BS. In the split-RRM case, RRAs and RRC interact across the R6 reference point. In the integrated RRM model, the interface between RRA and RRC is outside the scope of this specification, and only the information reporting procedures, represented with dashed lines, are standardized. The decision-support procedures, shown as solid lines between RRA and RRC in each BS, remain proprietary. Here, the RRM in different BSs may communicate with one another using an RRM relay in the ASN-GW. The split model and the integrated model are included as part of ASN profiles A and C, respectively (see Table 10.1.

10.10 Paging and Idle-Mode Operation

In order to save battery power on the handset, the WiMAX MS goes into idle mode[8] when it is not involved in an active session. Paging is the method used for alerting an idle MS about an incoming message. Support for paging and idle-mode operation are optional for nomadic and portability usage models but mandatory for the full-mobility usage model.[9] The WiMAX architecture mandates that paging and idle-mode features be compliant with IEEE 802.16e.

---

8. For definition of idle mode, see Section 9.6.2.
9. For usage models, see Section 2.4.4.

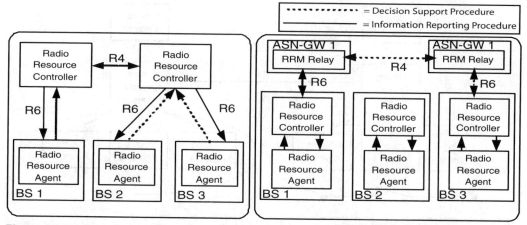

Figure 10.13 Generic reference models for RRM: (a) split RRM and (b) integrated RRM

The WiMAX paging reference model, as shown in Figure 10.14, decomposes the paging function into three separate functional entities: the paging agent, the paging controller, and the location register. The *paging controller* (PC), is a functional entity that administers the activity of idle-mode MS in the network. It is identified by PC ID (6 bytes) in IEEE 802.16e and may be either colocated with the BS or separated from it across an R6 reference point. For each idle-mode MS, WiMAX requires a single PC containing the location information of the MS. This PC is referred to as the *anchor PC*. Additional PCs in the network may, however, participate in relaying paging and location management messages between the PA and the anchor PC. These additional PCs are called relay PCs. The *paging agent* (PA) is a BS functional entity that handles the interaction between the PC and the IEEE 802.16e–specified paging-related functions.

One or more PAs can form a *paging group* (PG) as defined in IEEE 802.16e. A PG resides entirely within a NAP boundary and is provisioned and managed by the network operator. A PA may belong to more than one PG, and multiple PGs may be in an NAP. The provisioning and management of PG are outside the scope of the WiMAX standard.[10]

The *location register* (LR) is a distributed database that maintains and tracks information about the idle mobiles. For each idle-mode MS, the information contained in the LR includes its current paging group ID, paging cycle, paging offset, and service flow information. An instance of the LR is associated with every anchor PC, but the interface between them is outside the scope of the current WiMAX specification. When an MS moves across paging groups, location update occurs across PCs via R6 and/or R4 reference points, and the information is updated in the LR associated with the anchor PC assigned to the MS.

---

10. The tradeoffs involved in paging-group design are discussed in Section 7.4.1.

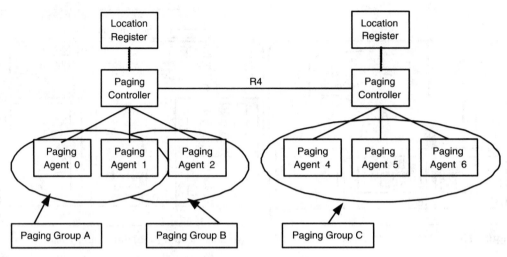

Figure 10.14 WiMAX paging network reference model

10.11 Summary and Conclusions

This chapter presented an overview of the WiMAX network architecture as defined by the WiMAX Forum Network Working Group.

- The WiMAX Forum NWG has developed a network reference model that provides flexibility for implementation while at the same time providing a mechanism for interoperability.
- The network architecture provides a unified model for fixed, nomadic, and mobile usage scenarios.
- The security architecture of WiMAX supports the IEEE 802.16e MAC privacy services, using an-EAP based AAA framework that supports global roaming.
- The WiMAX architecture defines various QoS-related functional entities and mechanisms to implement the QoS features supported by IEEE 802.16e.
- The WiMAX architecture supports both layer 2 and layer 3 mobility. Layer 3 mobility is based on mobile IP and can be implemented without the need for a mobile IP client.
- The WiMAX architecture defines two generic reference models for radio resource management: one with and the other without an external controller for managing the BS resources.
- The network architecture supports paging and idle-mode operation of mobile stations.

10.12 Bibliography

[1] WiMAX Forum. Recommendations and requirements for networks based on WiMAX Forum certified™ products. Release 1.0, February 23, 2006.

[2]   WiMAX Forum. Recommendations and requirements for networks based on WiMAX Forum certi-
        fied™ products. Release 1.5, April 27, 2006.

[3]   WiMAX Forum. WiMAX end-to-end network systems architecture. Stage 2: Architecture tenets, ref-
        erence model and reference points. Release 1.0, V&V Draft, August 8, 2006. www.wimaxforum.org/
        technology/documents.

[4]   WiMAX Forum. WiMAX end-to-end network systems architecture. Stage 3: Detailed protocols and
        procedures. Release 1.0, V&V Draft, August 8, 2006. www.wimaxforum.org/technology/documents.

[5]   V. Devarapalli and F. Dupont. Mobile IPv6 Operation with IKEv2 and the revised IPsec Architecture,
        draft-ietf-mip6-ikev2-ipsec-07.txt. MIP6 Working Group Internet-Draft, October, 2006.

[6]   H. J. Jang et al. DHCP Option for Home Information Discovery in MIPv6, draft-jang-mip6-hiopt-
        00.txt. MIP6 Working Group Internet-Draft, June, 2006.

CHAPTER 11

Link-Level Performance
of WiMAX

T he goal of any communication system is to reliably deliver information bits from the transmitter to the receiver, using a given amount of spectrum and power. Since both spectrum and power are precious resources in a wireless network, it should come as no surprise that efficiency is determined by the maximum rate at which information can be delivered using the least amount of spectrum and power. Since each information bit must reach the intended receiver with a certain amount of energy—over the noise level—a network's power efficiency and bandwidth efficiency cannot be maximized at the same time; there must be a trade-off between them. Thus, based on the nature of the intended application, each wireless network chooses an appropriate trade-off between bandwidth efficiency and power efficiency. Wireless networks intended for low-data-rate applications are usually designed to be more power efficient, whereas wireless networks intended for high-data-rate applications are usually designed to be more bandwidth efficient. Most current wireless standards, including WiMAX (IEEE 802.16e-2005), provide a wide range of modulation and coding techniques that allow the system to continuously adapt from being power efficient to bandwidth efficient, depending on the nature of the application. The amount of available spectrum for licensed operation is usually constrained by the allocations provided by the regulatory authority. Thus, in the given spectrum allocation, most cellular communication systems strive to maximize capacity while using the minimum amount of power.

Due to the complex and nonlinear nature of most wireless systems and channels, it is virtually impossible to determine the exact performance and capacity of a wireless network, based on analytical methods. Analytical methods can often be used to derive bounds on the system capacity in channels with well-defined statistical properties, such as flat-fading Rayleigh channels or AWGN channels. Computer simulations, on the other hand, not only provide more accurate results but can also model more complex channels and incorporate the effects of implementation imperfections, such as performance degradation owing to channel estimation and tracking errors [19].

A complete PHY and MAC simulation of an entire wireless network consisting of multiple base stations (BSs) and multiple mobile stations (MSs) is prohibitive in terms of computational complexity. Thus, it is common practice to separate the simulation into two levels: link- level simulations and system-level simulations. Link-level simulations model the behavior of a single link over short time scales and usually involve modeling all aspects of the PHY layer and some relevant aspects of the MAC layer. These simulations are then used to arrive at abstraction models that capture the behavior of a single link under given radio conditions. Often, these abstraction models are represented in terms of bit error rate (BER) and block error rate (BLER) as a function of the signal-to-noise ratio (SNR). The abstracted model of a single link can then be used in a system-level simulator that models an entire network consisting of multiple BSs and MSs. Since in a system-level simulation, each link is statistically abstracted, it is sufficient to model only the higher protocol-layer entities, such as the MAC, radio resource management (RRM), and mobility management.

In the first section of this chapter, we a describe a link-level simulation methodology that can be used for broadband wireless systems, such as WiMAX. In the next section, we examine the link-level performance of WiMAX in a static non-fading AWGN channel for both convolutional codes and turbo codes and compare it to the Shannon capacity of a single input/single output (SISO) WiMAX system. Next, we provide link-level performance results of WiMAX in fading channels for a SISO configuration. These results show the benefits of various PHY and MAC features, such as hybrid-ARQ, and subcarrier permutation schemes. Next, we consider various multiple input/multiple output (MIMO) configurations. We first provide link-level results for various open- and closed-loop diversity techniques for WiMAX, highlighting the various benefits and trade-offs associated with such multiantenna techniques. Next, we provide link-level results for the various open-loop and closed-loop spatial-multiplexing techniques. Finally, we examine link level results for some commonly used nonlinear receivers structures, such as *ordered successive interference cancellation* (O-SIC) [8] and *maximum-likelihood detection* (MLD) [11, 21] in order to highlight their performance benefits.

11.1 Methodology for Link-Level Simulation

As discussed previously, link-level simulations are used to study the behavior of a single communication link under varying channel conditions. These results can be used to judge the potential benefit of various PHY features, such as subcarrier permutation schemes, receiver structure, and multiantenna techniques in various radio frequency conditions. The link-level results are expressed in terms of BER and BLER. Sometimes, we also use the average number of transmissions required per FEC block: for example, to understand the benefits of hybrid-ARQ (HARQ) techniques. For the results presented here, we consider only the case of a single user over a single subchannel. Systemwide behavior of a WiMAX network with multiple users across multiple cells is presented in Chapter 12. The link-level simulator, as shown in Figure 11.1, consists of a *transmitter* and a *receiver*.

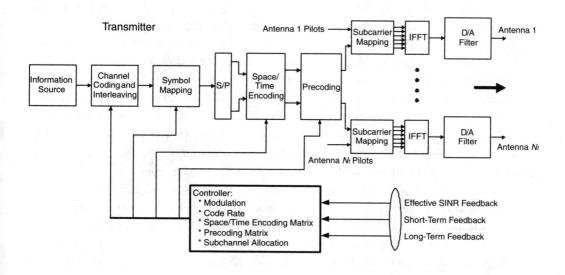

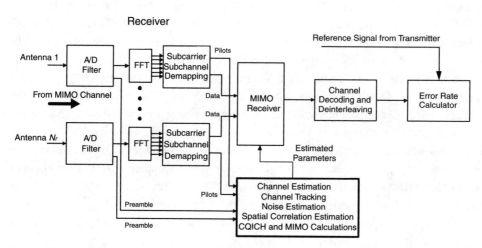

Figure 11.1 Link-level simulator for WiMAX

The transmitter is responsible for all digital and analog domain processing of the signal before it is sent over the wireless channel: channel encoding, interleaving, symbol mapping, and space/time encoding. When closed-loop MIMO is used, the transmitter also applies a linear precoding matrix and/or an antenna-selection matrix, if applicable. To create the signal in the frequency domain, the transmitter maps the data and pilot signals of each subchannel onto the

OFDM subcarriers based on the subcarrier permutation[1] scheme and the subchannel index. Then the time-domain signal is created by taking an inverse discrete fourier transform of the frequency-domain signal, which is then passed through the pulse-shaping filter to create an analog-domain representation of the signal. The pulse-shaping filter typically oversamples the signal by a factor of 4–16 to model the signal in the analog domain. The transmitter also selects various transmission parameters, such as modulation constellation, code rate, number of parallel streams, rank of the precoding matrix, and the subchannel index, based on feedback provided by the receiver. Feedback errors are not modeled in the link-level simulation results presented here.

In the context of this chapter, the receiver has two main functions: to estimate the transmitted signal and to provide feedback that allows the transmitter to adapt the transmission format according to channel conditions. At the receiver, the analog-domain signal from the channel is first converted to its digital-domain representation, using a pulse-shaping filter. The receiver uses a filter that is matched to the transmitter's pulse-shaping filter to perform this conversion. Then the time domain-signal is converted to a frequency-domain signal, using a discrete Fourier transform, which is then mapped onto the various subchannels, based on the subcarrier permutation scheme. In order to invert the effect of the channel, the receiver first forms an estimate of the MIMO channel matrix. The downlink frame preamble or the MIMO midambles are used for frequency synchronization[2] and to form an initial channel estimate. The dedicated pilots are then used for tracking/updating the MIMO channel over the subsequent OFDM symbols. Next, the estimated MIMO channel and the received signal are provided to the MIMO receiver,[3] which then develops soft-likelihood estimates of the signal. The soft-likelihood estimates are used by the channel decoder to ultimately compute hard-decision estimates of the transmitted signal. In the case of convolutional codes, a soft-output Viterbi algorithm (SOVA) is used to generate the hard decision. In the case of turbo codes, a MAX LogMAP algorithm is used to generate the hard-decisions. The hard decision bits at the receiver are then compared with the transmitted bits to develop BER and BLER statistics. The receiver also calculates the effective SNR[4] per subchannel and provides that information to the transmitter, using the 6-bit CQICH channel. The periodicity of the SNR feedback can be varied according to the Doppler spread of the channel. When closed-loop MIMO is used, the receiver also uses the CQICH channel or the fast feedback channel to provide feedback needed for closed-loop MIMO or beamforming.

A very important component of link-level simulations for WiMAX is the MIMO multipath fading channel. For the purposes of the link-level simulation results presented here the channel

---

1. The concept of subcarrier permutation is discussed in Section 8.6.
2. A frequency offset of +/– 500Hz is modeled in the link-level simulations.
3. Linear MIMO receivers, such as MMSE, and nonlinear receivers, such as V-BLAST and MLD, are modeled explicitly in link-simulation results presented in this chapter.
4. The average SNR per subchannel is usually not a good metric for link adaptation. An effective SNR based on an exponentially effective SNR map (EESM) is usually considered a better metric than average SNR.

models the effect of multipath fading and adds other-cell/sector interference. The other-cell interference in link-level simulations is modeled as filtered[5] AWGN.

The multipath fading between each pair of transmit and receive antennae is modeled as a tap-delay line (see Section 3.2). The received signal at the ith receive antenna can thus be written as

$$r_i(t) = \sum_{k=1}^{N_t} \sum_{l=1}^{L_k} h_{i,\,k}(t) x_k(t - \tau_{l,\,k}) + z_i(t), \tag{11.1}$$

where k is the transmit antenna index, Nt is the total number of transmit antennas, $x_k(t)$ is the signal transmitted from the kth antenna at time t, and $z_i(t)$ is the other-cell interference. Additionally, l is the multipath index, $\tau_{l,k}$ is the delay of the lth path—relative to the first arriving path—from the kth antenna, and L_k is the total number of multipath components as seen from kth antenna.

For the simulation results presented in this chapter, we modeled the channel between each pair of transmit and receive antennas as a SISO multipath channel. The ITU pedestrian and vehicular multipath channel profiles, as described in Section 12.1, are used because they are considered good representations of the urban and suburban macro-cellular environments. The spatial channel model (SCM) [1] developed by 3GPP is also a good representative of the MIMO multipath channel in macrocellular environments. In this chapter, however, link-level results for the SCM are not presented, since most of the literature for link- and system-level performance of competitive wireless technologies, such as 1xEV-DO and HSDPA, is available for ITU channel models. Thus, the ITU channel models provide a better data point for comparative analysis with such wireless technologies.

The ITU channel models, unlike the SCM, do not model any correlation between the fading waveforms across the transmit and receive antennas; those models were developed primarily to model SISO channels. Because the spacing between the various antenna elements at the transmitter and the receiver are of an order of a few wavelengths, in a wireless channel with a finite number of scatterers the fading waveforms across the antenna elements are expected to be correlated. In order to incorporate the effect of such correlation, we first generate the MIMO multipath channel between the various pairs of transmit and receive antennas independently, without correlation. Then correlation is added, using coloring matrices $\mathbf{Q}_t$ and $\mathbf{Q}_r$ for transmit and receive ends, respectively:

$$\mathbf{H}_l t = \mathbf{Q}_r \mathbf{H'}_l t \mathbf{Q}_t^H \tag{11.2}$$
$$\mathbf{Q}_r \mathbf{Q}_r^H = \mathbf{R}_r$$
$$\mathbf{Q}_t \mathbf{Q}_t^H = \mathbf{R}_t$$

where $\mathbf{H}_l(t)$ and $\mathbf{H'}(t)$ are the correlated and uncorrelated MIMO channel matrix for the lth path, respectively, at time t. The spatial-correlation matrices $\mathbf{R}_t$ and $\mathbf{R}_r$ capture the correlation between

---

5. The transmitter pulse-shaping filter is used to generate the filtered AWGN for modeling other-cell interference. In system-level simulations, as discussed in Chapter 12, other-cell interference is modeled as an OFDM signal.

the channel across the various transmit and receive antennas. Since the angular-scattering model of the wireless channel is not explicitly captured by the ITU channels, we assume that the spatial correlation is an exponential function of distance. Thus, for a linear array of antenna elements with equal spacing, $\mathbf{R}_t$ and $\mathbf{R}_r$ can be expressed as a function of ρ_t and ρ_r, the correlation between two adjacent antennas at the transmit and receive ends, respectively:

$$\mathbf{R}_t = \begin{bmatrix} 1 & \rho_t & \rho_t^2 & \rho_t^3 \\ \rho_t & 1 & \rho_t & \rho_t^2 \\ \rho_t^2 & \rho_t & 1 & \rho_t \\ \rho_t^3 & \rho_t^2 & \rho_t & 1 \end{bmatrix} \qquad \mathbf{R}_r = \begin{bmatrix} 1 & \rho_r & \rho_r^2 & \rho_r^3 \\ \rho_r & 1 & \rho_r & \rho_r^2 \\ \rho_r^2 & \rho_r & 1 & \rho_r \\ \rho_r^3 & \rho_r^2 & \rho_r & 1 \end{bmatrix}. \qquad (11.3)$$

The coloring matrices $\mathbf{Q}_r$ and $\mathbf{Q}_t$ can be obtained by Choleski factorization of the correlation matrices $\mathbf{R}_t$ and $\mathbf{R}_r$, respectively. Table 11.1 shows the various parameters and assumptions used for the link-level simulation results.

11.2 AWGN Channel Performance of WiMAX

The Shannon capacity [18] of a communication system is a theoretical bound that no real communication system can exceed given the SNR and bandwidth constraints. Thus, how close a real-world communication system comes to this bound is often used as a measure of its efficiency.

Since in an AWGN channel, the receiver does not need to mitigate the effects of the channel, performance is limited only by the modulation and channel coding used. Thus, the performance in an AWGN channel relative to Shannon capacity can be used as a benchmark to understand the inherent limitations of a communication system, such as WiMAX. AWGN channel performance can also be used to determine the SNR threshold for adaptive modulation and coding. The system can use these thresholds to determine the appropriate choice of modulation and coding formats for a given SNR in a fading channel.

A fundamental assumption behind Shannon's channel capacity is that the transmitter has an arbitrarily large set of continuously varying modulation alphabets and FEC codewords that can be used to transmit the information. However, a real communication system must operate within limited combinations of available modulation alphabets and suboptimal codes. For example, in the case of WiMAX, the only available modulation alphabets are QPSK, 16 QAM, and 64 QAM, as described in Section 8.4. Although it can adapt the modulation according to the current SNR, the transmitter must choose from one of these three modulation alphabets. In this section, we provide a derivation of a modified capacity of a system constrained to the finite choice of modulation alphabets, which we believe is a more appropriate theoretical bound to be compared against the capacity of WiMAX.

Figure 11.2 shows a communication system consisting of an information source, a channel, and a detector. The information signal x entering the channel is a sequence of amplified symbols belonging to a given modulation alphabet. These symbols are amplified such that the total energy per symbol is E_s. An AWGN noise z is added to the signal by the channel before presenting it to

Table 11.1 Downlink and Uplink Link-Level Simulation Parameters

Parameter	Value
Channel bandwidth	10MHz
Number of subcarriers	1,024
Subcarrier permutation	PUSC and band AMC
Cyclic prefix	1/8
Frame length	5msec
Channel coding	Convolutional and turbo
Channel decoder	SOVA for convolutional codes and Max LogMAP for turbo codes
Hybrid-ARQ	Type I (Chase combining) and type II (incremental redundancy)
Maximum number of H-ARQ transmissions	4
Subpacket ID for type II H-ARQ	0, 1, 2, 3
Carrier frequency	2,300MHz
Multipath channel	Ped B, Ped A, and Veh A
MS speed	3km/hr for Ped B and Ped A, 30kmph and 120kmph for Veh A
Transmit-antenna correlation (ρ_t)	0 and 0.5
Receive-antenna correlation (ρ_r)	0 and 0.5
SNR feedback interval	1 frame (5msec)
SNR feedback delay	2 frames (10msec)
MIMO feedback duration	1 frame (5msec)
MIMO feedback delay	2 frames (10msec)
Overampling factor (digital to analog)	8
Total number of FEC code blocks simulated	15,000

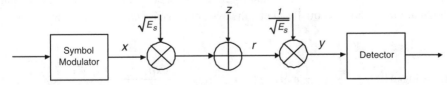

Figure 11.2 A communication system

the receiver. Before performing a symbol detection, the receiver scales back the received signal such the total energy per symbol is 1. The received signal can be written as

$$r = x\sqrt{E_s} + z \tag{11.4}$$
$$y = \text{De}[r],$$

where De[r] represents the decision-making criteria on: In this case, a symbol slicer. The capacity [18] of such a system normalized to the channel bandwidth, or the spectral efficiency, is the difference between the self-entropy of the transmitted sequence and the conditional entropy of the received sequence to the transmitted sequence:

$$= H(x) - H_y(x) = \int \int p(x, y)\log_2\left(\frac{p(x, y)}{p(y)}\right)dxdy - \int p(x)\log_2(p(x)), \tag{11.5}$$

where $H(x)$ is the self-entropy of the transmitted sequence, and $H_y(x)$ is the conditional entropy of the received sequence to the transmitted sequence. Also, $p(x)$ and $p(y)$ are the probability density functions of x and y, respectively, and $p(x,y)$ is the joint probability density function of x and y. In this case since the transmitted and the detected information symbols belong to a discrete modulation constellation, such as QPSK, 16 QAM, or 64 QAM, the integrations in Equation (11.5) can be replaced by a summation over all possible transmitted and detected symbols. The system capacity is thus given by

$$C = \sum_{m=1}^{M} p_m\log_2 p_m + \frac{1}{2\pi}\sum_{m=1}^{M}\sum_{n=1}^{M}\int_{r \in S_n} p_m\exp(-\gamma(r-x_m)^2)\log_2\left|\frac{\exp(-\gamma(r-x_m)^2)}{\sum_{k=1}^{M}\exp(-\gamma(r-x_k)^2)}\right|dr \tag{11.6}$$

where M is the total number of modulation symbols, p_m is the probability of the mth symbol, σ^2 is the noise variance, γ is the SNR (E_s/σ^2), and S_n is the decision region of symbol y_n. Since the decision regions of the modulation symbols are disjoint and together span the entire 2D complex space, the summation of the individual integrals over the decision region of each symbol y_n can be replaced by a single integral over the entire 2D complex space. Also, if we assume that each of the modulation symbols is equally likely, the system capacity can be written as

$$C = \log_2 M + \frac{1}{2\pi}\sum_{m=1}^{M}\int \frac{1}{M}\exp(-\gamma(r-x_m)^2)\log_2\left|\frac{\exp(-\gamma(r-x_m)^2)}{\sum_{k=1}^{M}\exp(-\gamma(r-x_k)^2)}\right|dr . \tag{11.7}$$

Since WiMAX can use only the limited set of modulation alphabets as defined in the standard, the modulation-constraint capacity is a more appropriate benchmark for the capacity of a WiMAX system to be compared against. Figure 11.3 shows the comparison of the Shannon capacity with the capacity of a system constrained to QPSK, 16 QAM, and 64 QAM modulations.

We can use link-level simulation results to compare the performance of a WiMAX system in a static channel with the modulation-constrained capacity. We present link-level simulation results for the spectral efficiency of a WiMAX link in an AWGN channel. For the purposes of results presented here, spectral efficiency is defined as the net throughput divided by the total spectrum and is expressed in units of bps/Hz. The spectral efficiency for each modulation and code rate is calculated using the following expression:

$$C = C_{max}(1 - BLER),$$
(11.8)

where C_{max} is the maximum normalized capacity of the modulation and code rate if the received symbol have no errors, and $BLER$ is the block error rate of the FEC code blocks. Although most of the results have been presented for turbo codes (CTCs), some results for convolutional codes (CCs) have also been presented to highlight their relative performance. *From here on, results only for turbo codes are presented.* Table 11.2 shows the sizes of the FEC code blocks in units of bits and slots and the maximum spectral efficiency for the simulated modulation formats and code rates.

Figure 11.4 and Figure 11.5 show the block error rates and the spectral efficiencies, respectively, for each modulation and code rate. The spectral-efficiency curves can be used to judge the performance of a WiMAX link relative to the Shannon capacity and to determine the link-adaptation thresholds. Figure 11.5 also shows the modulation-constrained capacity with a 3dB shift, which is the same capacity expression as shown in Equation (11.5) but with a 3dB shift, to the right.

This 3dB shifted capacity appears to be a good fit to the WiMAX capacity curve and can often be used for *back-of-the-envelope* approximation for overall capacity of a WiMAX network. Figure 11.6 shows compares the BER performance of turbo codes and convolutional codes. The performance gap between convolutional codes and the turbo codes is about 1dB to 1.5dB, depending on the chosen BER value. This performance gap is further increased if larger FEC block sizes are used. It has been shown that the coding gain of turbo codes increases linearly with the size of the interleaver [15], that is, the block size; for convolutional codes, the coding gain is independent of the block size. At large block sizes and high code rates, turbo codes provide a gain of 2dB to 3dB over convolutional codes of the same code rate.

11.3 Fading Channel Performance of WiMAX

In the previous section, we compared the capacity of a WiMAX link and the Shannon capacity. As mentioned earlier, the AWGN channel results also provide SNR thresholds for adaptive modulation and coding that can be used to optimize the performance of a WiMAX system. However,

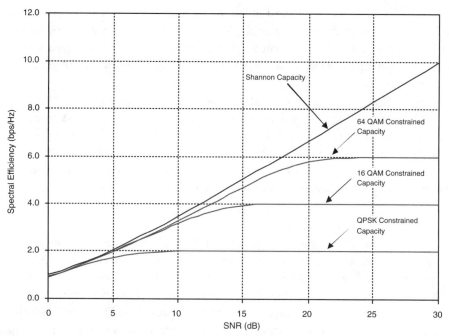

Figure 11.3 Shannon capacity and modulation constrained Shannon capacity

Table 11.2 FEC Block Sizes

Modulation and Code Rate	FEC Block Size		Maximum Spectral Efficiency (C_{max}) (bps/Hz)
	Bits	**Slots**	
QPSK R1/2	48	1	1.0
QPSK R3/4	72	1	1.5
16 QAM R1/2	96	1	2.0
16 QAM R3/4	144	1	3.0
64 QAM R1/2	144	1	3.0
64 QAM R2/3	192	1	4.0
64 QAM R3/4	216	1	4.5

a real-world wireless channel is rarely AWGN in nature, especially in non-line-of-sight environments. Since a WiMAX network is expected to operate primarily in NLOS conditions, it is important to understand and characterize WiMAX performance in the fading channels that are typical representatives of expected deployment scenarios.

In order to *undo* the effect of the channel in multipath fading channels, the receiver must first estimate the channel response. In the context of a MIMO system, this implies estimating the

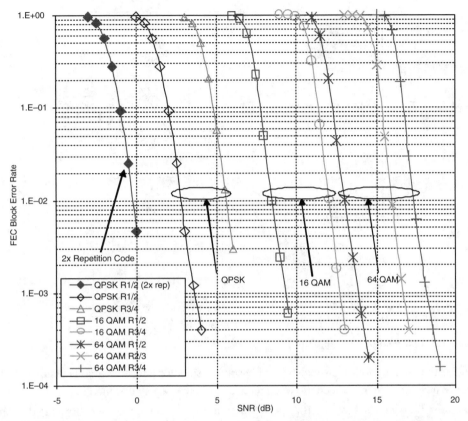

Figure 11.4 FEC block error rate of turbo codes in static channel (AWGN channel)

time-varying amplitude and phase of each multipath tap between each transmit- and receive-antenna pair. This is denoted as $h_{i,k}(t)$ in Equation (11.1). In most wireless systems that use coherent modulation schemes, this is achieved by using an embedded pilot signal that is known a priori at the receiver. This embedded pilot can be time multiplexed, such as TDMA systems, code multiplexed, such as CDMA systems, or frequency multiplexed such as OFDM systems. In all these cases, a certain amount of system time, power, and frequency must be allocated to the pilot signal to allow for proper and reliable channel estimation at the receiver. Clearly, having a large portion of system resources dedicated to the pilot signal will allow for more reliable channel estimation at the receiver but will also reduce the amount of resources available for carrying data payload. On the other hand, reducing the amount of resources for the pilot signal increases the resources available to carry data payload at the cost of less reliable channel estimation at the receiver. A system/protocol designer must strike a balance between the amount of resources dedicated to the pilot signal and for carrying data.

Channel estimation is a significant part of any *real* receiver's operation and has a significant impact on performance. Although most of the results presented in this chapter are based on realistic

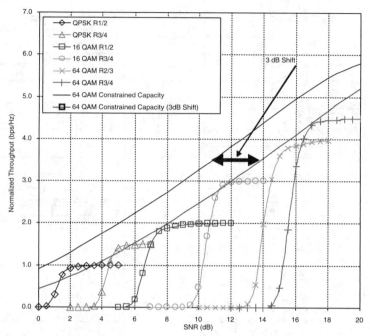

Figure 11.5 WiMAX spectral efficiency

channel-estimation algorithms, we also provide some results for a receiver that is assumed to have perfect channel knowledge, a scenario often referred to as the *perfect CSI* model.

Figures 11.7–11.10 show the BER as a function of SNR at the receiver for the Ped A and B channels. As discussed in Section 12.1 in greater detail, the Ped B channel has a larger delay spread and hence a smaller coherence bandwidth than the Ped A channel. Since in a fading channel, the BER is dominated by the occurrence of the *deep fades,* a small coherence bandwidth implies that when the signal fades in one part of the spectrum, it very likely can be retrieved from another part of spectrum. This *frequency-selective* fading of the Ped B channel allows for a form of signal diversity commonly referred to as frequency diversity. A comparison of the BER for any given modulation and coding scheme (MCS) shows that the frequency diversity of the Ped B channel results in a considerably lower error rate, especially for the PUSC subcarrier permutation[6] mode, since it is designed to take advantage of frequency diversity.

Figures 11.7–11.10 also highlight the benefit of the band AMC subcarrier permutation at pedestrian speeds. In the case of band AMC operations, we assume that the receiver provides the channel-quality feedback, using the CQICH channel once every 5msec frame. Since the best subchannel can be allocated to the MS based on the feedback, the performance of band AMC is significantly better than PUSC, particularly in Ped A channel where the PUSC subcarrier per-

---

6.   PUSC and other subcarrier permutation schemes are discussed in Section 8.6.

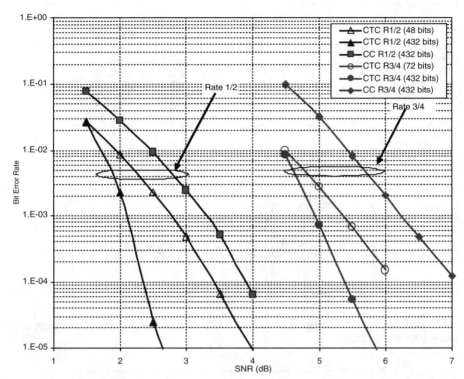

Figure 11.6 Comparison of turbo codes of short and long block lengths with convolutional codes for QPSK modulation

mutation is unable to extract the benefit of frequency diversity, owing to the large coherence bandwidth of the channel. As the results show, the gain from band AMC subcarrier permutation is also dependent on the code rate because higher code rates are more sensitive to the occurrence of deep fades, which can be mitigated to a certain extent by the use of band AMC by allocating the subchannels to the MS in the part of the spectrum that is not experiencing fade.

In the Ped B channel, band AMC provides a 2dB to 2.5dB link gain for the R1/2 code rates and a link gain of 4dB to 5dB for the R3/4 code rates. On the other hand, in the Ped A channel, band AMC provides a link gain of 5dB to 7dB for both R1/2 and R3/4 code rates. Note that the link gain is determined by the BER point at which it is evaluated; that is, the link gain at 1 percent BER point is different from that at 0.1 percent BER point.

Figure 11.11 and Figure 11.12 show the performance of PUSC and the band AMC subcarrier modes in the Veh A channel for MS speeds of 30kmph and 120kmph, respectively. Unlike in the case of pedestrian channels, PUSC provides some benefit over the band AMC in vehicular channels. Depending on the speed, modulation, and code rate, PUSC provides a link gain of 1dB to 2.5dB relative to band AMC. In the case of band AMC subcarrier permutation, the subchannel allocation is done based on the CQICH feedback, which is an indicator of the state of the

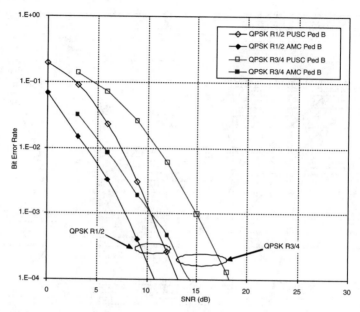

Figure 11.7 BER versus SNR for band AMC and PUSC modes in Ped A and B channels for QPSK modulations with turbo codes

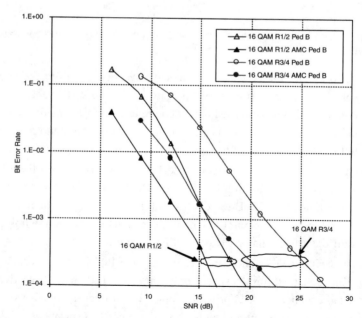

Figure 11.8 BER versus SNR for band AMC and PUSC modes in Ped B channel for 16 QAM modulations with turbo codes

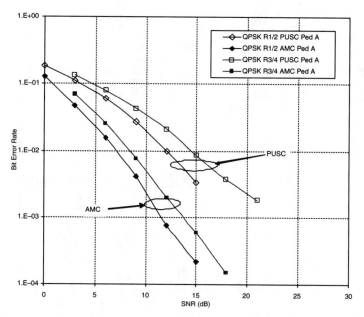

Figure 11.9 BER versus SNR for band AMC and PUSC modes in Ped A channel for QPSK modulations with turbo codes

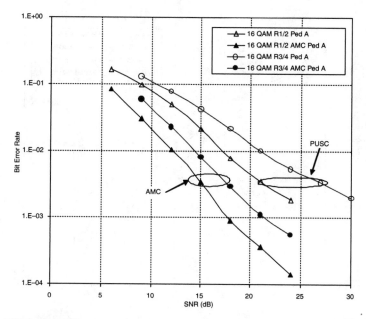

Figure 11.10 BER versus SNR for band AMC and PUSC modes in Ped A channel for 16 QAM modulations with turbo codes

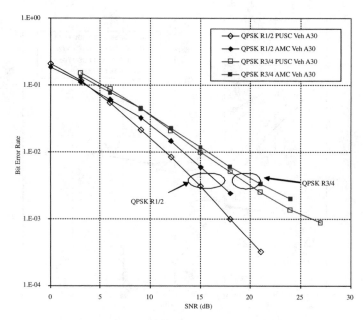

Figure 11.11 BER versus SNR for band AMC and PUSC modes in Veh A channel with 30kmph speeds for QPSK modulation

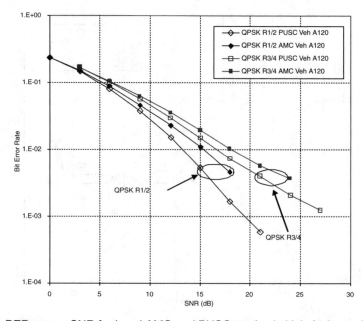

Figure 11.12 BER versus SNR for band AMC and PUSC modes in Veh A channel with 120kmph speeds for QPSK modulation

channel in the previous frame. At vehicular speeds, however, the feedback based on the previous frame is not an accurate measure of the channel state in the current frame; thus, CQICH feedback–based subchannel allocation performs poorly.

In the case of a pedestrian channel, the feedback duration of 5 msec is significantly smaller than the coherence time of the channel (~150msec), but in the case of vehicular channel at 120kmph this feedback duration is larger than the coherence time of the channel (~3msec) and hence the feedback is unreliable. For QPSK modulation, there appears to be a link loss of 1dB to 1.5dB in going from a pedestrian channel to a vehicular channel at 120kmph. Similarly this loss for 16 QAM modulation is around 2dB to 2.5dB. The link loss results from the channel Doppler spread, which is proportional to the speed of the MS. This implies that at higher speeds, as the channel changes at a faster rate, it is more difficult for the receiver to track it.[7] Since higher-order modulation is more sensitive to channel-estimation errors than lower-order modulations are, the link loss at higher speeds is larger in the case of 64 QAM and 16 QAM than in QPSK modulation. In the next section, we discuss channel estimation and channel tracking for WiMAX and its impact on receiver performance.

Table 11.3 shows the gains of band AMC subcarrier permutation over PUSC subcarrier permutation in various fading channels. For the vehicular channels, the band AMC mode appears to have a poorer performance than the PUSC mode, due to the unreliable nature of the CQICH in a fast-changing channel. Also, 16 QAM experiences a larger performance degradation at vehicular speeds than QPSK does. Further, the performance gap between AMC and PUSC increases with the Doppler spread of the channel.

Table 11.3 Band AMC Gain[a] over PUSC Subcarrier Permutation

	QPSK				16 QAM			
	10^{-2} BER (dB)		10^{-4} BER (dB)		10^{-2} BER (dB)		10^{-4} BER (dB)	
	R 1/2	R 3/4	R 1/2	R 3/4	R 1/2	R 3/4	R 1/2	R 3/4
Ped A	5.0	6.5	6.5	11.0	5.0	6.5	6.0	12.0
Ped B	2.5	5.0	2.5	5.0	2.5	5.0	2.5	5.0
Veh A 30	−1.0	< −.5	−2.0	−1.0	−3.0	−2.0	−4.5	−4.0
Veh A 120	−2.0	−1.0	−2.5	−2.0	−3.5	−2.5	−5.5	−4.5

a. A negative gain implies a link loss in going from band AMC to PUSC subcarrier permutation.

11.3.1 Channel Estimation and Channel Tracking

Figure 11.13 and Figure 11.14 show the BER as a function of SNR for the PUSC subcarrier mode in vehicular channel with speeds of 30kmph and 120kmph, respectively, for a realistic

---

7. Note that the accuracy of the channel-tracking algorithm can be improved by increasing the number of pilot subcarriers in the standard, but that would reduce data rate.

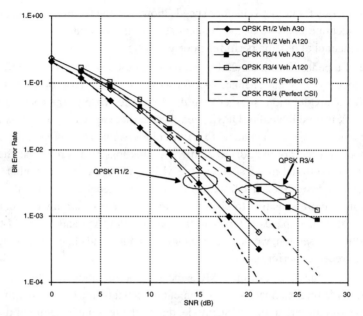

Figure 11.13 Performance of PUSC in Veh A channel with 30kmph speed for real receivers and receivers with perfect CSI

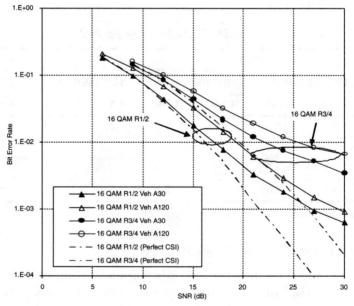

Figure 11.14 Performance of PUSC in Veh A channel with 120kmph speed for real receivers and receivers with perfect CSI

channel-estimation algorithm that can be used in WiMAX. These figures also shown the BER for a receiver with perfect CSI; that is, the channel response over all the subcarriers of all the OFDM symbols is known a priori at the receiver.

To arrive at an initial estimate of the channel response, we use the frequency-domain *linear minimum mean square error* (LMMSE) channel estimator with partial information about the channel covariance.[8] The DL frame preamble or the MIMO midambles are used for this purpose. Over the subsequent OFDM symbols, the channel is tracked, using a Kalman filter–based estimator. Thus, at vehicular speeds, if reliable channel tracking is not performed, considerable degradation in the performance of the system occurs.

The channel response of a multipath tap when sampled at a given instance in time can be related to its previous samples as

$$\tilde{h}_l(n+1) = \alpha_l \tilde{h}_l(n) + \sqrt{(1-\alpha_l^2)p_l}\, \tilde{z}_l(n), \tag{11.9}$$

where l is the tap index, α_l is the temporal correlation between two consecutive samples of the tap, p_l is the average power of the tap, and $\tilde{z}_l$ is a sequence of independent and identically distributed complex Gaussian random variables with zero mean and unit power. If we assume each multipath tap to be an independent Rayleigh channel, the correlation α_l between two consecutive samples of a tap is $J_0(2\pi f_l \Delta t)$, where f_l is the Doppler frequency of the tap, Δt is the elapsed time between two samples, and J_0 is the zero-order Bessel function of the first kind. In the frequency domain, the channel response over the kth subcarrier can thus be written as

$$h_k(n+1) = \alpha h_n(n) + \sqrt{1-\alpha^2} \sum_{l=1}^{L} \sqrt{p_l}\, \tilde{z}_l(n) \exp\left(-\frac{2\pi j f_k \tau_l}{N}\right). \tag{11.10}$$

We assume that the Doppler frequency of each tap is the same. Although this assumption is not necessary, it is used here without any loss of generality. In Equation (11.10), if we substitute the sequence $\tilde{z}_l$ by its time-reversed version $z_l = \tilde{z}_{L-l}$, the last term can be written as the lth term of the convolution between the power delay profile ($\sqrt{p_l}$) and the time-reversed random sequence z_l. Since $\tilde{z}_l$ is a sequence of independent and identically distributed random variables, the time-reversed version has exactly the same statistical properties as the original sequence. In the frequency domain after the DFT, the convolution between the power-delay profile and the time-reversed sequence appears as a product. This allows Equation (11.11) to be written in a more compact matrix notation as the following:

---

8. An LMMSE receiver with full channel covariance in not considered here, since a real receiver does not possess a priori information about the channel covariance. However, an LMMSE receiver with partial channel covariance based on the knowledge of the RMS delay spread is expected to be realistic. A knowledge of the RMS delay spread can bound the channel covariance, which can improve the fidelity of the channel estimates.

$$h(n+1) \ = \ \alpha h(n) + \sqrt{1-\alpha^2} \mathbf{R} z(n),\tag{11.11}$$

where $\mathbf{R}$ is the covariance matrix of the frequency-domain channel, and z is a vector of independent and identically distributed random variables, with the same statistics as $\tilde{z}_l$. This recursive relationship between the channel samples can be tracked using a Kalman filter.

Figure 11.15 compares the actual channel and the channel estimated by this tracking algorithm in a Veh A 120 kmph channel for various SNR values. Since we use the DL preamble for an initial channel estimate that is then tracked in subsequent OFDM symbols, the results shown here are for the last symbol of the DL subframe, where the estimates are likely to be most erroneous. At 0dB SNR, the reliability of the estimates is quite poor, as expected, particularly in subchannels that are experiencing a fade. However, at 20dB SNR, the reliability of the channel estimate is quite good.

Although the reliability of the channel-estimation algorithm improves as SNR increases, the difference between ideal and real channel-estimation schemes seem to be more prominent in higher SNR values, (Figure 11.10). The reason is that at low SNR, the error occurrences are dominated by noise and interference, not by channel estimation error, which starts to play an important role in the occurrence of detection errors only at high SNR.

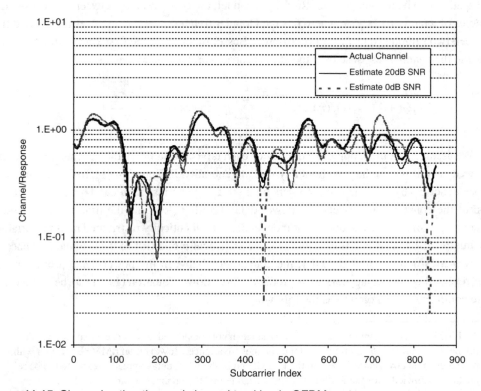

Figure 11.15 Channel estimation and channel tracking in OFDM systems

From an information theoretic perspective, supported by the results in Figure 11.13 and Figure 11.14, it is clearly not possible to design a real receiver that will perform at par with a receiver that has exact channel state information. In a real system with limited time, frequency, and power dedicated to the pilot signal, it is not possible to eliminate channel-estimation error. Such errors can be reduced by increasing the resources given to the pilot signal, which, however, comes at the cost of reducing the resources for data and thus system capacity. In order to maximize the capacity of an OFDM system, such as WiMAX, one needs to carefully balance the division of resources in terms of time, frequency, and power between the pilot signals and the data signals [10].

11.3.2 Type I and Type II Hybrid-ARQ

HARQ is an error-correction technique that has become an integral part of most current broadband wireless standards, such as 1xEV-DO, WCDMA/HSDPA, and IEEE 802.16e-2005. Unlike in conventional ARQ techniques at layer 2 or above, where all transmissions are decoded independently, subsequent retransmissions in the case of HARQ are jointly decoded with all the previous transmissions to reduce the probability of decoding error.

In type I HARQ, also known as *Chase combining,* all HARQ retransmissions are identical to the first transmission. The soft-reliability values of the current retransmission are combined with all previous transmissions before decoding the data. In a noise-limited scenario, optimum combining of the soft-reliability values from multiple retransmissions is equivalent to performing maximum ratio combining (MRC), which reduces the decoding-error probability by increasing the SNR.

In the case of type II HARQ, also known as *incremental redundancy,* the puncturing patterns of each subsequent transmissions are different from those of the earlier transmissions. Thus, when each retransmission is combined with all the previous transmissions, the code rate is reduced (implementation of type II HARQ is discussed in Section 8.2.). As shown in Section 11.2, the performance of turbo codes and convolutional codes is sensitive to the degree of puncturing; thus, the decoding-error probability is reduced as the code rate decreases with each subsequent retransmission.

In order to quantify the benefit of HARQ techniques, we use as the metric of performance[9] the average number of retransmissions required per FEC block to decode it without errors. As shown in Figure 11.16 and Figure 11.17, both type I and type II HARQ techniques provide a significant benefit at low SNR. Type II HARQ provides the highest gain, particularly for higher code rates because the type II HARQ code rate is reduced with each new retransmission, thus providing a significant benefit over type I HARQ. At high SNR, there is no apparent benefit from HARQ, since most of the FEC blocks are decoded without error in the first transmission. Table 11.4 shows the gains from type I and type II HARQ compared to conventional ARQ techniques.

---

9. When HARQ is used in an actual WiMAX system, all the FEC blocks within the data region of the MS are retransmitted because CRC is applied only to the entire data region, not per individual FEC block. For the results presented in this chapter, we assume that each FEC block is a data region and can be retransmitted independently of other FEC blocks. This assumption, however, does not change the results presented in this section, and the effect of type I and type II HARQ schemes is expected to be same when multiple FEC blocks constitute the data region of the MS.

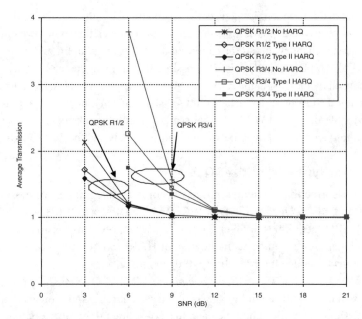

Figure 11.16 Average number of transmissions for QPSK modulation in PUSC mode in Ped B channel with type I and type II hybrid-ARQ

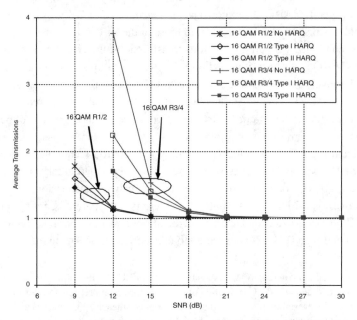

Figure 11.17 Average number of transmissions for 16 QAM modulation in PUSC mode in Ped B channel with type I and type II hybrid-ARQ

Table 11.4 HARQ Gain for QPSK and 16 QAM Modulation

	Type I HARQ		Type II HARQ	
	10^{-1} BLER (dB)	10^{-2} BLER (dB)	10^{-1} BLER (dB)	10^{-2} BLER (dB)
QPSK R1/2	0.5	0	0.75	0
QPSK R3/4	0.75	0	1.0	0
16 QAM R1/2	0.75	0	1.5	1.0
16 QAM R3/4	1.0	0	1.5	1.5

11.4 Benefits of Multiple-Antenna Techniques in WiMAX

Support for multiantenna technology is a key feature distinguishing WiMAX from other broadband wireless technologies, such as 1xEV-DO and HSDPA. Although both 1xEV-DO and HSDPA, along with their respective evolutions, support multiantenna techniques, such as spatial multiplexing and transmit diversity, WiMAX has a much more advanced framework for supporting various forms of open- and closed-loop techniques. In order to fully appreciate the potential of WiMAX as a broadband wireless system, it is imperative to understand the link-level performance gains that result from such multiantenna techniques.

In this section, we provide link-level results of WiMAX in MIMO fading channels. *Unless otherwise stated, the results shown here are for the band AMC subcarrier permutation in the Ped B channel.* First, we provide results for transmissions with a single stream (matrix A); that is, the multiple antennas at the transmitter and the receiver are used for diversity only. Next, we provide results for transmissions with two streams (matrix B); that is, the multiple antennas are used for pure spatial multiplexing or for a combination of spatial multiplexing and diversity.

11.4.1 Transmit and Receive Diversity

Diversity is a technique in which multiple copies of the signal are created at the receiver. In the context of MIMO, multiple antennas are used at the receiver and/or the transmitter. With single-stream transmission, the antennas at the transmitter can be used in an open-loop fashion—without CSI feedback from the receiver—using space/time block codes or in a closed-loop fashion (with CSI feedback from the receiver) using beamforming. See Chapter 8 the allowed open-loop and closed-loop diversity modes in WiMAX.

Figure 11.18 and Figure 11.19 show the link-level results of WiMAX in a *single-input multiple-output* (SIMO) channel. The BER as a function of the SNR for a SISO mode indicated as 1×1 and two SIMO modes indicated as 1×2 and 1×4 with two and four receive antennas, respectively, are shown. In the case of receive diversity a MMSE detection is used. At low SNR (high BER), two-antenna receive diversity provides a 3dB gain, and four-antenna receive diversity provides a 6dB gain, which corresponds to the array gain for the two and four antenna cases, respectively. At higher SNR, both two and four receive antennas provide additional diversity gain.

Table 11.5 shows the diversity gain due to multiple antennas at the receiver for band AMC.

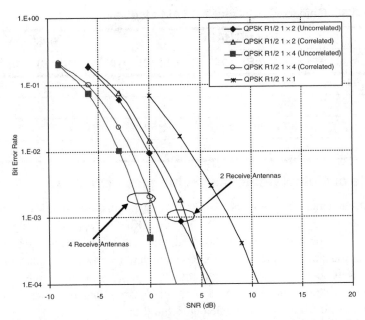

Figure 11.18 Average BER QPSK R1/2 with band AMC in a Ped B multipath channel with correlated and uncorrelated fading

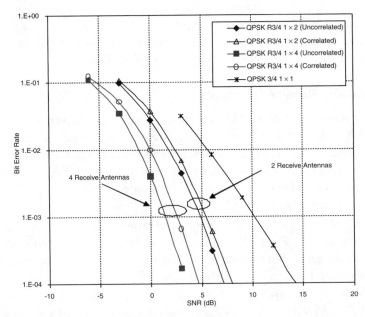

Figure 11.19 Average BER QPSK R1/2 with band AMC in a Ped B multipath channel with correlated and uncorrelated fading

Table 11.5 Receive Diversity Gain for Band AMC in Ped B Channel

	Uncorrelated Fading		Correlated Fading	
	10^{-2} BER (dB)	10^{-4} BER (dB)	10^{-2} BER (dB)	10^{-4} BER (dB)
QPSK R1/2 1×2	4	6.0	3.0	5.0
QPSK R3/4 1×2	4	7.0	3.5	6.0
QPSK R1/2 1×4	7	9.0	5.5	7.5
QPSK R3/4 1×4	7	10.5	6.0	9.0

Figure 11.18 and Figure 11.19 also provide the BER when the multipath fadings at the various receive antennas are correlated. A complex correlation of 0.5 is modeled as described in Section 11.1. The gain due to multiple antennas at low SNR does not seem to depend explicitly on this correlation because at low SNR, the gain is predominantly owing to array gain, which does not depend on the fading correlation. At higher SNR, however, the multiantenna gain is reduced by 1dB to 0.5dB, owing to the correlation in the fading waveform. Lower code rates are more sensitive to this correlation than are higher code rates.

Figure 11.20 and Figure 11.21 provide link-level results for various possible open-loop and closed-loop transmit diversity schemes in WiMAX. The open-loop diversity considered here is the 2×2 Alamouti pace/time block cde (STBC). In the case of band AMC subcarrier permutation, the benefit of STBC seems to be marginal, especially with correlated fading because STBC hardens the channel variation that band AMC is designed to exploit. On the other hand PUSC subcarrier permutation, as shown in Figure 11.22 and Figure 11.23 benefits significantly from 2×2 STBC.

In Figure 11.19 and Figure 11.20, the results for closed transmit diversity are shown for the cases with two and four antennas at the transmitter, indicated as 2×1 and 4×1, respectively. The CSI feedback used here is based on quantized MIMO channel feedback from the receiver, as explained in Chapter 8. A single feedback comprising of quantized MIMO channel coefficients is provided for each band AMC subchannel once every frame. The transmitter uses the quantized channel feedback to calculate a beamforming (precoding) vector for the subchannel.

Figure 11.20 and Figure 11.21 also provide the results for a four antenna closed loop transmit diversity scheme, shown as perfect CSI, with uncorrelated fading, whereby the transmitter is assumed to have perfect knowledge of the CSI for each subcarrier and calculates a beamforming (precoding) vector for *each subcarrier*. The performance of this scheme represents a limiting case that other open-loop and closed-loop transmit diversity schemes for WiMAX can be compared against. Table 11.6 shows the link gains of various closed-loop and open-loop transmit diversity schemes for WiMAX.

11.4.2 Open-Loop and Closed-Loop MIMO

A key attribute that allows WiMAX to provide high data rates is the ability to spatially multiplex more than one stream, or layer, of data over the same time and frequency resources simultaneously.

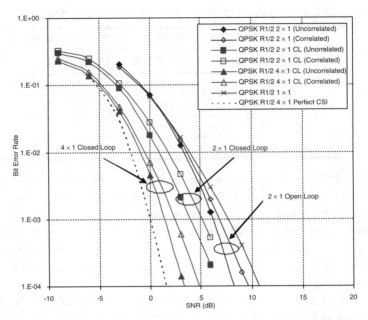

Figure 11.20 Average BER for open-loop and closed-loop transmit diversity for QPSK R1/2 with band AMC in a Ped B multipath channel correlated and uncorrelated fading

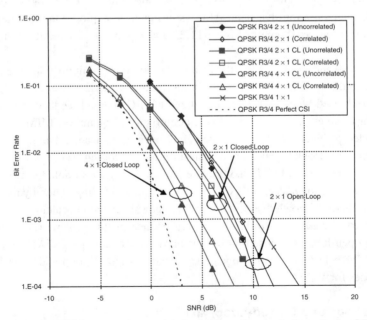

Figure 11.21 Average BER for open-loop and closed-loop transmit diversity for QPSK R3/4 with band AMC in a Ped B multipath channel with correlated and uncorrelated fading

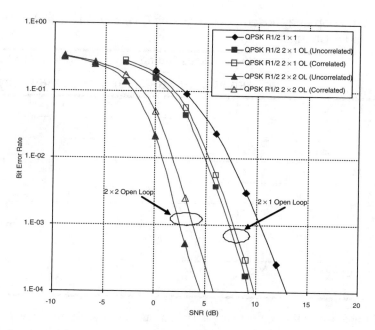

Figure 11.22 Average BER for transmit and receive diversity for QPSK R1/2 PUSC in a Ped B multipath channel with uncorrelated fading

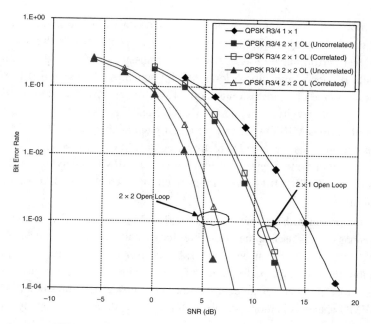

Figure 11.23 Average BER for transmit and receive diversity for QPSK R3/4 PUSC in a Ped B multipath channel with uncorrelated fading

Table 11.6 Open-Loop and Closed-Loop Transmit Diversity Gain Relative to SISO for Band AMC in Ped B Channel

	Uncorrelated Fading		Correlated Fading	
	10^{-2} BER (dB)	10^{-4} BER (dB)	10^{-2} BER (dB)	10^{-4} BER (dB)
QPSK R1/2 2 × 1 (STBC)	0.5	2.5	0.0	1.0
QPSK R3/4 2 × 1 (STBC)	0.5	3.5	0.25	2.0
QPSK R1/2 2 × 1 Closed loop	3.0	4.5	2.0	3.0
QPSK R3/4 2 × 1 Closed loop	2.5	4.5	2.0	3.5
QPSK R1/2 4 × 1 Closed loop	5.0	7.5	4.5	6.0
QPSK R3/4 4 × 1 Closed loop	5.0	7.5	4.5	6.0
QPSK R1/2 4 × 1 Closed loop (perfect CSI)	6.0	9.25	N/A	N/A
QPSK R3/4 4 × 1 Closed loop (perfect CSI)	6.5	11.0	N/A	N/A

In the case of single-user MIMO, the multiple streams are intended for the same receiver; in the case of multiuser MIMO, the multiple streams are intended for different receivers. The high rank of the MIMO channel created by the multiple antennas allows the receiver to spatially separate the multiple layers.

So far, we have considered transmission formats with only a single stream for a single user; in other words, the multiple antennas have been used for diversity only. In this section, we investigate the link-level performance of WiMAX for transmissions with two data streams to highlight the benefits of open- and closed-loop MIMO schemes. Although IEEE 802.16e-2005 allows for transmission of up to four streams, only two streams have been considered in this section.[10] In the interest of brevity, only the QPSK R1/2 and R3/4 modes are considered, but the overall benefits of various MIMO schemes are equally applicable to the 16 QAM and 64 QAM modes. The link-level results presented here are based on a MMSE MIMO receiver with realistic channel-estimation algorithms. Benefits of more advanced MIMO receivers, such as successive interference cancellation (SIC) [8] or maximum-likelihood detection (MLD) [21] are presented in the next section.

For the results presented here, the baseline is a 2 × 2 open-loop MIMO scheme, which consists of two antennas that are used at the transmitter for spatially multiplexing the two data streams. The receiver in the baseline case is an MMSE MIMO receiver with two antennas. Figure 11.24 and Figure 11.25 show the link-level performance of the baseline case and various other open-loop schemes with various numbers of antennas at the receiver and the transmitter. The benefit of higher-order MIMO channels is more prominent for higher code rates, since they are more sensi-

---

10. The various transmission formats for WiMAX are discussed in Sections 8.8 and 8.9.

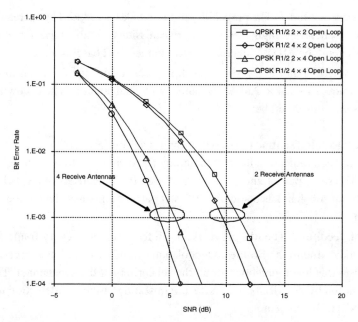

Figure 11.24 Bit error rate for band AMC QPSK R1/2 in Ped B channel with dual streams (matrix B) for open-loop MIMO schemes

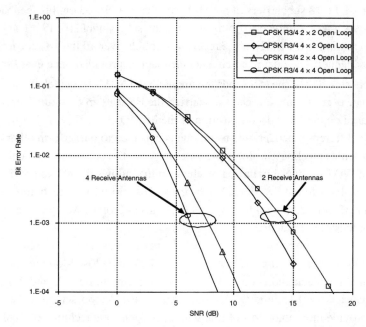

Figure 11.25 Bit error rate for band AMC QPSK R3/4 in Ped B channel with dual streams (matrix B) for open-loop MIMO schemes

tive to the occurrence of fades. The probability of these fades is reduced by increasing the number of antennas, thus benefitting higher code rate transmissions significantly. Table 11.7 shows the gains of various open-loop MIMO schemes relative to the 2×2 baseline case.

Figure 11.26 and Figure 11.27 show the link-level results for the open-loop and various closed-loop techniques for a 4×2 MIMO channel with dual streams. The following four closed-loop techniques are considered here:

1. **Antenna selection feedback.** The MS provides a 3-bit feedback once every frame for each subchannel, indicating the combination of two antennas to be used for the DL transmission. The same pair of antennas is used for all the subcarriers of a subchannel; however, different subchannels could use different pairs of antennas, depending on the channel condition.

2. **Codebook feedback.** The MS provides a 6-bit feedback once every frame for each subchannel, indicating to the BS the codebook entry to be used for linear precoding [12, 13]. The BS uses this linear precoder for all the subcarriers of the subchannel. The codebook entry is chosen by the MS, based on the minimization of the postdetection *mean square error* (MSE) of both streams.

3. **Quantized channel feedback.** the MS quantizes the complex coefficients of the MIMO channel and sends them to the BS. A single quantized feedback is provided once every frame for all the 18 subcarriers of an AMC subchannel. Based on this feedback, the BS then chooses a unitary precoder to be used for the subchannel [16, 17]. The precoder is chosen to minimize the MSE of the received symbols over all the 18 subcarriers of the AMC subchannel. The quantized channel feedback is provided once every frame.

4. **Per subcarrier SVD.** The MS sends the unquantized MIMO channel of each subcarrier to the BS once every frame. For each subcarrier, the BS uses an optimum linear precoding matrix based on the SVD decomposition of the MIMO channel [9, 14]. Since each subcarrier uses a different precoder, this technique is expected to outperform other closed-loop techniques that can choose a single precoder for an entire subchannel or a bin. It should be noted that WiMAX does not have a mechanism that allows the MS to provide a MIMO channel feedback to the BS for each subcarrier. This closed-loop technique is presented only as a performance bound for any practical closed-loop MIMO technique in WiMAX and is not feasible in practice.

As the results show in Figure 11.26 and Figure 11.27, the closed-loop techniques based on quantized channel feedback and codebook feedback perform within 1dB and 2dB, respectively, of the per subcarrier SVD technique. Although these closed-loop schemes are suboptimal at best, they can provide more than 5dB of link gain over open-loop techniques. Table 11.8, shows the link gains for various closed-loop MIMO techniques in WiMAX for a 4×2 MIMO configuration with dual streams.

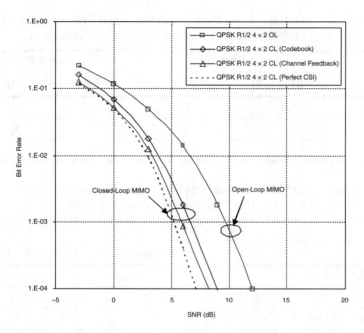

Figure 11.26 Bit error rate for band AMC QPSK R1/2 in Ped B channel with dual streams (matrix B) for closed-loop MIMO schemes

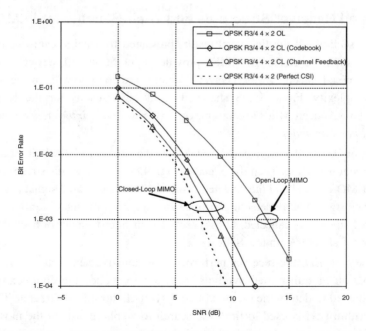

Figure 11.27 Bit error rate for band AMC QPSK R3/4 in Ped B channel with dual streams (matrix B) for closed-loop MIMO schemes

Table 11.7 Open-Loop MIMO Gains Relative to the Open-Loop Baseline Case for Band AMC in a Ped B Multipath Channel with Dual Streams (Matrix B)

	Code Rate 1/2		Code Rate 3/4	
	10^{-2} BER (dB)	10^{-4} BER (dB)	10^{-2} BER (dB)	10^{-4} BER (dB)
4×2 MIMO	0.75	2.0	0.75	2.5
2×4 MIMO	5.0	6.5	5.0	8.0
4×4 MIMO	6.0	8.0	6.5	10.0

Table 11.8 Closed-Loop MIMO Gains Relative to the Open-Loop Baseline Case for Band AMC in a Ped B 4×2 MIMO Channel with Dual Streams

	Code Rate 1/2		Code Rate 3/4	
	10^{-2} BER (dB)	10^{-4} BER (dB)	10^{-2} BER (dB)	10^{-4} BER (dB)
Antenna selection feedback				
Codebook feedback	2.5	3.5	3.0	4.4
Quantized channel feedback	3.25	4.5	3.75	5.5
Optimal per subcarrier SVD	4.0	5.5	4.5	6.5

11.5 Advanced Receiver Structures and Their Benefits for WiMAX

In the previous section, all the link-level results presented for dual streams were based on an MMSE receiver structure. Although MMSE provides a good trade-off between complexity and performance, more advanced MIMO receiver structures are possible with an acceptable level of increase in complexity. Figure 11.28 shows the link-level results for the baseline MMSE receiver and two advanced MIMO receivers: *ordered successive interference cancellation* and *maximum-likelihood detection*.

In the case of O-SIC, the receiver [8] first detects the stream with the highest SNR, based on the MMSE detection scheme. Then the expected signal belonging to this stream is regenerated, based on its MIMO channel and the detected symbols. The regenerated signal is then subtracted from the received signal before detecting the next stream. Since the interference from all the previously detected streams is canceled, O-SIC provides an improvement in overall performance, particularly for the streams with low SNR.

In the case of MLD, the receiver performs an exhaustive search to determine the most likely combination of transmitted symbols. In order to reduce the complexity, an MMSE receiver is first used to determine the most likely symbols for all the streams. Then a *sphere-decoding* algorithm [21] is used to limit the search to a sphere around the most likely symbols. The radius of the sphere can be adjusted to achieve a tradeoff between complexity and performance. Although the MLD is the optimum *noniterative* algorithm for MIMO receivers,

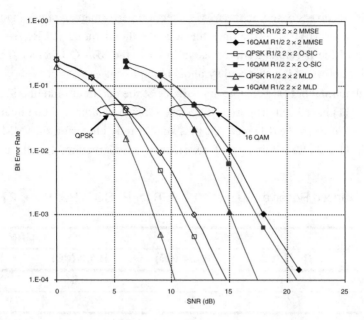

Figure 11.28 Bit error rate for QPSK R1/2 with PUSC in a Ped B 2 × 2 MIMO channel for various MIMO receiver structures

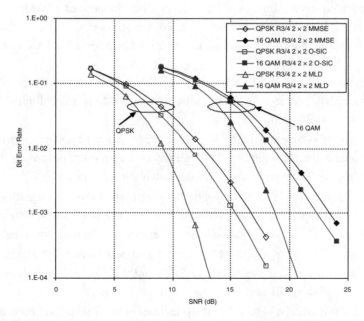

Figure 11.29 Bit error rate for QPSK R3/4 with PUSC in a Ped B 2 × 2 MIMO channel for various MIMO receiver structures

iterative MIMO receivers based on the MAP detection perform even better than MLD receivers do. An iterative MAP receiver uses the log likelihood ratios (LLR) from the channel decoder output of the previous iteration as an input to the MIMO receiver. Thus, with each iteration, the reliability of the received symbols is improved.

Several suboptimal but low-complexity variants of the maximum-likelihood receivers, such as QRM-MLD [11] have been proposed. It has been shown that these suboptimal receivers perform within a dB of the full MLD receiver, thus significantly outperforming MMSE and O-SIC receivers. Table 11.9 shows the link gain for various receiver structures over the baseline MMSE receiver.

Table 11.9 Advanced Receiver Gain at 10^{-4} BER for PUSC in Ped B 2 × 2 MIMO Channel

	QPSK		16QAM	
	R 1/2 (dB)	R 3/4 (dB)	R 1/2 (dB)	R 3/4 (dB)
O-SIC Receiver	1.0	2.0	0.8	1.5
MLD Receiver	4.5	6.0	3.5	5.5

11.6 Summary and Conclusions

This chapter provided some estimates of the link-level performance of a WiMAX and its dependence on various physical-layer parameters and receiver structures. Based on these results, we can derive the following high-level conclusions on the behavior of a WiMAX system.

- The capacity curve of a WiMAX link is within 3dB of the Shannon capacity curve at low to moderate SNRs. At high SNRs, the capacity of a WiMAX link is limited by allowed modulation constellations.
- The optional turbo codes provide a significant performance advantage over the mandatory convolutional codes. The additional complexity of the decoder for turbo codes is well justified by their performance benefit over the convolutional codes.
- In fading channels, the band AMC subcarrier permutation provides significant performance benefit over the PUSC subcarrier permutation at low speeds (< 10kmph). However, at moderate to high speeds, PUSC subcarrier permutation outperforms band AMC.
- Multiantenna techniques give WiMAX a significant performance advantage over other broadband wireless techniques, such as HSDPA and 1xEV-DO. Closed-loop multiantenna techniques provide > 5dB of gain at low speeds (< 10kmph).
- Advanced MIMO receivers based on the principles of maximum-likelihood detections and their derivatives provide additional link gains in excess of 5dB compared to linear receivers, such as MMSE.

11.7 Bibliography

[1] 3GPP. Spatial channel models for MIMO simulations, v6.1.0. TR 25.996. September 2003.

[2] S. Alamouti. A simple transmit diversity technique for wireless communications. *IEEE Journal on Selected Areas of Communication,* 16(8), October 1998.

[3] L. Bahl, J. Jelinek, J. Cocke, and F. Raviv. Optimal decoding of linear codes for minimising symbol error rate. *IEEE Transactions on Information Theory,* 20, March 1974.

[4] C. Berrou and A. Glavieux. Near optimum error correcting coding and decoding: Turbo codes. *IEEE Transactions Communication,* 44(10), October 1996.

[5] C. Berrou and M. Jezequel. Nonbinary convolutional codes and turbo coding. *Electronics Letters,* 35(1), January 1999.

[6] C. Berrou, A. Glavieux, and P. Thitimajshima. Near Shannon limit error-correcting codes: Turbo codes. *Proceedings of the IEEE International Communication Conference,* 1993.

[7] O. Edfors, M. Sandell, J.-J. van de Beek, S. Wilson, and P. Borjesson. OFDM channel estimation by singular value decomposition. *Proceedings of the IEEE Vehicular Technical Conference,* April 1996.

[8] G. Foschini. Layered space-time architecture for wireless communication in a fading environment when using multi-element antennas. *Bell Systems Technical Journal,* 41, 1996.

[9] G. Foschini and M. Gans. On limits of wireless communication in a fading environment when using multiple antennas. *Wireless Personal Communication,* 6(3), March 1998.

[10] T. Kim and J. Andrews. Optimal pilot-to-data power ration for MIMO-OFDM. *Proceedings of IEEE Globecom,* December 2005.

[11] K. Kim and J. Yue. Joint channel estimation and data detection algorithms for MIMO OFDM. *Proceedings of the Asilomar Conference of Signals, Systems, and Computers,* November 2002.

[12] D. Love and R. Heath. Limited feedback unitary precoding for orthogonal space time block codes. *IEEE Transitions on Signal Processing,* 53(1), January 2005.

[13] D. Love, R. Heath, and T. Strohmer. Grassmannian beamforming for multiple input multiple output wireless systems. *IEEE Transactions on Information Theory,* 49, October 2003.

[14] A. Paulraj, R. Nabar, and D. Gore. *Introduction to Space-Time Wireless Communications,* Cambridge University Press, 2003.

[15] M. Rodrigues, I. Chatzgeorgiou, I. Wassell, and R Carrasco. On the performance of turbo codes in quasi-static fading channels. *Procceedings of the International Symposium on Information Theory,* September 2005.

[16] H. Sampath and A. Paulraj. Linear precoding for space-time coded systems. *Proceedings of the Asilomar Conference on Signals, Systems, and Computers,* November 2001.

[17] H. Sampath, P. Stoica, and A. Paulraj. Generalized linear precoder and decoder design for MIMO channels using weighted MMSE criterion. *IEEE Transactions on Communications,* 49(12), December 2001.

[18] C. E. Shannon. A mathematical theory of communication. *Bell Systems Technical Journal,* 27, July and October 1948.

[19] W. Tranter, K. Shanmugam, T. Rappaport, and K. Kosbar. *Principles of Communication System Simulation with Wireless Applications.* Prentice Hall, 2003.

[20] M. Valenti and B. Woerner. Performance of turbo codes in interleaved flat fading channels with estimated channel state information. *Proceedings, IEEE Vehicular Technical Conference,* May 1998.

[21] E. Viterbo and J Boutros. A universal lattice code decoder for fading channel. *IEEE Transactions on Information Theory,* 45, July 1997.

[22] B. Vucetic and J. Yuan. *Space-Time Coding.* Wiley, 2003.

CHAPTER 12

System-Level Performance
of WiMAX

T he link-level simulation and analysis results presented in Chapter 11 describe the performance of a single WiMAX link, depending on the choice of various physical-layer features and parameters. The results also provide insight into the benefits and the associated trade-offs of various signal-processing techniques that can be used in a WiMAX system. These results, however, do not offer much insight into the overall system-level performance of a WiMAX network as a whole. The overall system performance and its dependence on various network parameters, such as frequency-reuse pattern, cell radius, and antenna patterns, are critical to the design of a network and the viability of a business case. In this chapter, we provide some estimates of the system-level performance of a WiMAX network, based on simulations.

In the first section of this chapter, we describe the broadband wireless channel and its impact on the design of a wireless network. Next, we describe the system-simulation methodology used to generate the system-level performance results of a WiMAX network. Finally, we discuss the system-level performance of a WIMAX network under various network configurations. These results illustrate the dependency of system-level performance on network parameters, such as frequency reuse; type of antenna used in the mobile station (MS);[1] environmental parameters, such as the multipath power-delay profile; and the traffic model, such as VoIP, FTP, and HTTP. We also offer some results pertaining to system-level benefits of open-loop and closed-loop MIMO features that are part of the IEEE 802.16e-2005 standards.

---

1. Since WiMAX can also be used for fixed networks, we consider two MS form factors. The first is a handheld form factor with omnidirectional antennas, which is representative of a mobile use case. The second is a desktop form factor with directional antennas, which is representative of a fixed use case with an indoor desktop modem.

12.1 Wireless Channel Modeling

The validity of simulation-based performance analysis of wireless systems depends crucially on having accurate and useful models of the wireless broadband channel. We therefore begin with a brief overview of how wireless broadband channels are modeled and used for the performance analysis presented in this chapter.

For the purposes of modeling, it is instructive to characterize the radio channel at three levels of spatial scale. As discussed in Section 3.2, the first level of characterization is at the largest spatial scale, with a mathematical model used to describe the distance-dependent decay in power that the signal undergoes as it traverses the channel. These *median pathloss models* are useful for getting a rough estimate of the area that can be covered by a given radio transmitter. Since radio signal power tends to decay exponentially with distance, these models are typically linear on a logarithmic decibel scale with a slope and intercept that depend on the overall terrain and clutter environment, carrier frequency, and antenna heights. Median pathloss models are quite useful in doing preliminary system designs to determine the number of base stations (BSs) required to cover a given area. Widely used median pathloss models derived from empirical measurements are the Okumura-Hata model, the COST-231-Hata model, the Erceg model, and the Walfisch-Ikegami model, which are discussed in the chapter appendix.

The second level of characterization is modeling the local variation in received signal power from the median-distance-dependent value. Section 3.2.2 introduced shadow fading and highlighted the various aspects of the dynamic wireless channel, such as terrain, foliage, and large obstructions, that cause it. In this section, we describe the effect of shadow fading on the coverage and capacity of a wireless network and how it impacts the network design process. Measurements have shown that these large-scale variations from the median-distance-dependent value can be modeled as a random variable having a lognormal distribution with a standard deviation σ_S around the median value. Clearly, the system design and BS deployment should account for this lognormal shadowing, and this is usually done by adding a shadow-fading margin, S, to the link budget and accepting the fact that some users will experience outage at a certain percentage of locations, owing to shadowing. Having a large shadow-fading margin will lower the outage probability at the cost of cell radius. This implies that more BSs are needed to cover a given geographical area if the shadow fading margin is increased. For a given shadow margin in the link budget, the outage probability at the edge of the cell is related to the standard deviation of the lognormal fading statistics via the Q-function as

$$\text{Outage}_{\text{celledge}} = Pr\{\chi \geq S\} = Q\left(\frac{S}{\sigma_S}\right), \qquad (12.1)$$

where S is the shadow-fading margin, χ is the instantaneous shadow fade, and σ_S is the standard deviation of the shadow-fading process. How this translates to an outage probability averaged across the entire area of the cell is a more complex relationship that depends on the median pathloss model—more specifically, on the pathloss exponent, α, as well as σ_S. For the case of $S = 0\,\text{dB}$

or 50 percent cell-edge outage probability, it can be shown that coverage probability over the entire cell is given by

$$\text{Outage}_{\text{cellarea}} = \frac{1}{2}\left[1 + \exp\left(\frac{1}{b^2}\right)Q\left(\frac{1}{b}\right)\right], \qquad \text{where } b = \frac{10\alpha\log e}{\sigma_S\sqrt{2}}. \qquad (12.2)$$

The relationship is even more complex if S is not equal to $0\,\text{dB}$ and the cell-edge outage is less than 50 percent. For example, for $\alpha = 4$ and $\sigma_S = 8\,\text{dB}$, a 25 percent cell-edge outage probability will translate to a 6 percentage area outage, or 94 percentage area coverage. Similarly if $\alpha = 2$ and $\sigma_S = 8\,\text{dB}$, a 25 percentage cell-edge outage probability will translate to a 9 percentage area outage, or 91 percentage area coverage. Typical cellular designs aim for a 90 percent to 99 percent coverage probability, which often requires a shadow margin of $6\,\text{dB}$–$12\,\text{dB}$. Determining the median pathloss and the shadow fading using these models is critical for network design and planning, as it often directly dictates the BS density required to provide reliable signal quality to the desired area of coverage.

The third level of spatial scale at which a radio channel can be characterized is the variation in signal strength observed over a small scale. As discussed in Section 3.4, the phenomenon of multipath propagation means that the amplitude of the received radio signal can vary significantly (several tens of dBs) over very small distances on the order of wavelengths or inches. A good understanding of multipath fading and its impact on system performance is required to design a wireless network.

As explained in Section 3.2, multipath channels are often modeled using tap-delay lines with noninfinitesimal amplitude response over a span of v taps:

$$h(t,\tau) = h_0(t)\delta(\tau - \tau_0) + h_1(t)\delta(\tau - \tau_1) + \ldots + h_{v-1}(t)\delta(\tau - \tau_{v-1}). \qquad (12.3)$$

Here, t indicates the time variable and captures the time variability of the impulse response of each multipath component modeled typically as Rayleigh or Rician fading, and τ indicates the delay associated with each multipath. Empirical multipath channels are often specified using the number of taps v and the relative average power and delay associated with each tap. For purposes of modeling in a simulation environment, the most frequently used power-delay profiles are those specified by ITU. ITU has specified two multipath profiles, A and B, for vehicular, pedestrian, and indoor channels. Channel B has a much longer delay spread than channel A and is generally accepted as a good representative of urban macro-cellular environment. Channel A, on the other hand, is accepted as a good representative of rural macrocellular environment. Channel A is also recommended for microcellular scenarios in which the cell radius is less than 500m. The specified values of delay and the relative power associated with each of these profiles are listed in Table 12.1. Most simulation results presented in this chapter are based on the pedestrian channel B (referred to as Ped B) model, since it is commonly accepted as representative of an environment suitable for broadband wireless communications. Some results, however, have been provided for other multipath channels such as pedestrian channel A (referred to as Ped A) to illustrate the impact of multipath propagation on the system-level behavior of a WIMAX network.

Table 12.1 ITU Multipath Channel Models

Tap Number	Delay (nsec)	Relative Power (dB)	Delay (nsec)	Relative Power (dB)
		Vehicular (60kmph—120 kmph)		
		Channel A	**Channel B**	
1	0	0	0	−2.5
2	310	−1	300	0
3	710	−9	8,900	−12.8
4	1,090	−10	12,900	−10.0
5	1,730	−15	17,100	−25.2
6	2,510	−20	20,000	−16.0
		Pedestrian (<= 3 kmph)		
		Channel A	**Channel B**	
1	0	0	0	0
2	110	−9.7	200	−0.9
3	190	−19.2	800	−4.9
4	410	−22.8	1,200	−8.0
5			2,300	−7.8
6			3,700	−23.9
		Indoor		
		Channel A	**Channel B**	
1	0	0	0	0
2	50	−3	100	−3.6
3	110	−10	200	−7.2
4	170	−18	300	−10.8
5	290	−26	500	−18.0
6	310	−32	700	-25.2

12.2 Methodology for System-Level Simulation

Link-level simulations usually model a single link and study the small-scale behavior of the system that is affected by instantaneous variations in the channel. Also, link-level simulations usually model the wireless channel only over a small area and/or over a small time duration. In order to determine the overall performance and capacity of a wireless network, such as WiMAX, *system-level simulations* that model the network with multiple BSs and MSs are required.

System-level simulations usually model the wireless channel based on median propagation loss and shadow fading—channel variation over large scales—to the fullest extent. However, in order to increase the accuracy of the results by capturing small-scale variations in the channel, system-level simulations can also model the multipath fading—channel variation over small scales. In the case of a WiMAX, it is imperative to model the behavior of the channel in the frequency domain over both short and long time scales to the multicarrier nature and its MIMO features.

12.2.1 Simulator for WiMAX Networks

Simulating a wireless network that consists of a very large number of cell sites is often computationaly prohibitive and inefficient. Therefore, the system simulator used to generate the results presented in this chapter consists of only two tiers of cell sites that are present in a hexagonal grid, as shown in Figure 12.1.

Owing to the finite size of the simulated network, cell sites that lie toward the edge of the simulated network (outer tier) have missing neighbor cell sites, which causes the other-cell cochannel interference to be modeled inaccurately in these cells. This *edge effect* mandates that statistics related to network performance indicators, such as data rate and throughput, should be sampled only in the center cell, where the modeling of the other-cell interference is accurate enough. One solution to edge effect is the *wraparound* approach, which allows the simulator to model the interference and collect statistics even in the cells at the edge of the network. The system simulator used to generate the results presented in this chapter implements such a wraparound to mitigate the edge effect.

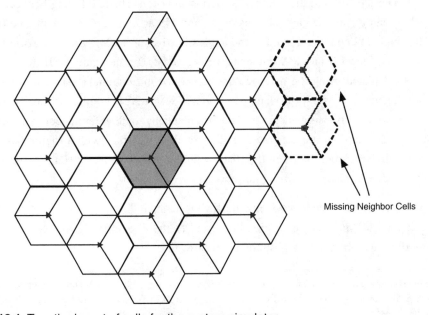

Missing Neighbor Cells

Figure 12.1 Two-tier layout of cells for the system simulator

The system simulator takes multiple Monte Carlo snapshots of the network to observe ergodic samples of the network and to determine how it behaves over long time scales [2]. Each Monte Carlo snapshot samples the behavior of the network over a 5msec frame. During each snapshot, the simulator randomly distributes various MSs in each sector and analyzes the expected instantaneous behavior of the network in terms of data rate, cell throughput, and outage probability. Information related to the location of each MS and the state of its traffic buffer are purged at the beginning of the next Monte Carlo snapshot of the network. This way, each Monte Carlo sample is random and completely independent of the previous sample. In each Monte Carlo snapshot, the simulator takes several steps to calculate the cell throughput and user data rates.

12.2.1.1 Computation of Time-Domain MIMO Channel

The instantaneous channel as observed by each MS from all the BSs in the network is calculated as

$$H'_{n,\ k}(\tau) \ = \ \sqrt{\overline{PL}(d_{n,\ k})s_{n,\ k}g_b(\theta_{n,\ k})g_m(\tilde{\theta}_{n,\ k})_n} \ \sum_{i=1}^{D} H_{n,\ k}(\tau - \tau_i) \ , \tag{12.4}$$

where $\overline{PL}(d_{n,\ k})$ is the median pathloss between the nth MS and the kth BS; $H_{n,k}(\tau)$ is the instantaneous fast-fading component of the MIMO channel; D is the total number of paths—depends on the path-delay profile—τ_i is the delay of the ith multipath relative to the first path; $d_{n,k}$ is the distance between the MS and the BS, $s_{n,k}$ is the instantaneous shadow fading between the MS and the BS, g_b and g_m are the gain patterns of BS and MS antennas, respectively; and $\theta_{n,\ k}$ and $\tilde{\theta}_{n,\ k}$ are the angle of departure at the BS and angle of arrival at the MS with respect to the boresight directions of the BS and MS antennas, respectively. The fast-fading MIMO component $H_{n,k}(\tau-\tau_i)$ is calculated using the methodology explained in Section 11.1.1, and is a complex matrix with dimension $N_r \times N_t$ for each path index i, where N_r is number of receive antennas, and N_t is the number of transmit antennas. The instantaneous shadow-fading components $s_{n,k}$ for each Monte Carlo snapshot are generated as i.i.d. random variables with a lognormal distribution. Appropriate correlation between the shadow fading observed by a given MS from various BSs, is also modeled,[2] using a *coloring matrix*.

12.2.1.2 Computation of Frequency-Domain MIMO Channel

The time-domain MIMO channel is then converted to the frequency-domain MIMO channel, using a Fourier transformation, as given by the following:

---

2. Since the shadow fading between a BS and an MS is partially dependent on the local neighborhood of the MS, it is expected that the shadow fading at a given MS from a different BS is correlated to a certain degree.

$$\tilde{H}'_{n,\,k}(l) = \frac{1}{N_{sub}} \sum_{i=1}^{D} H'_{n,\,k}(\tau_i) \exp(-2\pi\Delta f(l - N_{sub}/\,2)\tau_i, \tag{12.5}$$

where l is the subcarrier index, Δf is the frequency separation between two adjacent subcarriers, and N_{sub} is the total number of subcarriers. If transmit precoding is used at the transmitter, as is done in the case of closed-loop MIMO or beamforming, the net channel in the frequency domain can be written as

$$\tilde{H}'_{n,\,k}(l) = \frac{1}{N_{sub}} P_n(l) \sum_{i=1}^{D} H'_{n,\,k}(\tau_i) \exp(-2\pi\Delta f(l - N_{sub}/\,2)\tau_i, \tag{12.6}$$

where $P_n(l)$ is the precoding matrix (or vector) of the nth BS for the lth subcarrier.[3]

12.2.1.3 Computation of per Subcarrier SINR

Once the instantaneous channel response from each BS for the given MS is known in the frequency domain, the system simulator calculates the SINR per subcarrier. This calculation is performed for both matrix A and matrix B usage, since the system simulator does not possess a prior knowledge of the spacetime coding matrix used (see Chapter 8). In the case of matrix A, the simulator assumes a linear ML receiver for the 2×2 Alamouti block code; and in the case of matrix B, the simulator assumes a linear MMSE receiver. The post-detection SINR per subcarrier for the nth MS is thus given by

$$\gamma_A(n,\,l) = \frac{\text{Norm}(\Theta_{A,\,n}(l)\tilde{H}'_{n,\,m}(l))}{\beta N_0 W + \displaystyle\sum_{k=1,\,k\neq m}^{K} \text{Norm}(\Theta_{A,\,n}(l)\tilde{H}'_{n,\,k}(l))}$$

$$\gamma_B(n,\,l,\,ss) = \frac{\text{Norm}(D(ss)\Theta_{B,\,n}(l)\tilde{H}'_{n,\,m}(l))}{\beta N_0 W + \displaystyle\sum_{k=1,\,k\neq m}^{K} \text{Norm}(D(ss)\Theta_{B,\,n}(l)\tilde{H}'_{n,\,k}(l))}. \tag{12.7}$$

In Equation (12.7), Norm indicates the Forbenius norm of the matrix, and $\gamma_A(n,l)$ is the post-detection SINR for the lth subcarrier of the nth MS if matrix A space/time encoding is used. Similarly, $\gamma_B(n,l,1)$ and $\gamma_B(n,l,2)$ are the post-detection SINR of the first and the second streams, respectively, for the lth subcarrier of the nth MS and second streams, respectively, if matrix B space/time encoding is used. In the equation, β is the noise figure of the receiver, N_0 is the noise

---

3. The precoder used for a given subcarrier and the instantaneous MIMO channel depends on the closed-loop MIMO mode used and the optimization criteria used by the precoder, such as maximum capacity, minimum MSE.

power spectral density, and W is the noise bandwidth, which is the same as the channel bandwidth, assuming that a matched Nyquist filter is used at the receiver. The variables $\Theta_{A, n}(l)$ and $\Theta_{B, n}(l)$ are the linear estimation matrices for the lth subcarrier of the nth user, assuming matrix A and matrix B usage respectively. The matrices $D(ss)$ in are given by

$$D(1) = \begin{bmatrix} 1 & 0 \\ 0 & 0 \end{bmatrix} \qquad D(2) = \begin{bmatrix} 0 & 0 \\ 0 & 1 \end{bmatrix}. \qquad (12.8)$$

Thus, when used inside a Norm operator, $D(1)$ provides the SINR for the first stream. When used inside the Norm operator, $D(2)$ provides the SINR for the second stream.

12.2.1.4 Computation of per Subchannel Effective SINR

Next, the simulator calculates the *effective* SINR for each subchannel, based on the postprocessing SINR per subcarrier. The effective SINR is an AWGN-equivalent SNR of the instantaneous channel realization and can be calculated in different ways. Owing to the frequency selectivity of broadband channels, the average SINR over all the subcarriers that constitute a given subchannel is not a good indicator of the effective SINR, since averaging fails to capture the variation of the SINR over all the subcarriers. Several metrics, such as EESM (exponentially effective SINR map), ECRM (effective code rate map), and MIC (mean instantaneous capacity), are widely accepted as better representatives of the effective SINR and capture the variation of SINR in the subcarrier domain. In this chapter, the MIC metric is used, since it is considerably simpler to implement than the EESM and ECRM methods. In the MIC method, the SINR of each subcarrier is first used to calculate the instantaneous Shannon capacity of the subcarrier. Then the instantaneous Shannon capacity of all the subcarriers are added to calculate an effective instantaneous Shannon capacity of the subchannel. The effective instantaneous Shannon capacities of the subchannel is then converted back to an effective SINR for the subchannel of interest. Thus, the effective SINR for subchannel s of the nth user for matrix A and matrix B is given by

$$\tilde{\gamma}_{A, n}(s) = (2)^{\left(\frac{1}{48} \sum\limits_{i=1}^{48} \log_2(1 + \gamma_A(n, l_{i, s})) \right)} - 1 \qquad (12.9)$$

$$\tilde{\gamma}_{B, n}(s, ss) = (2)^{\left(\frac{1}{48} \sum\limits_{i=1}^{48} \log_2(1 + \gamma_B(n, l_{i, s}, ss)) \right)} - 1$$

12.2.1.5 Link Adaptation and Scheduling

The effective per subchannel SINR is then used by the scheduler to allocated the slots[4] in the DL and UL subframes among all the MSs that have traffic in the buffer. The effective SINR is also

---

4. A slot in WiMAX is the smallest quanta of PHY resources that can be allocated to an MS. A slot usually consists of one subchannel by one, two, or three OFDM symbols, depending on the subcarrier permutation scheme. For a more detailed definition of a slot, refer to Section 8.7.

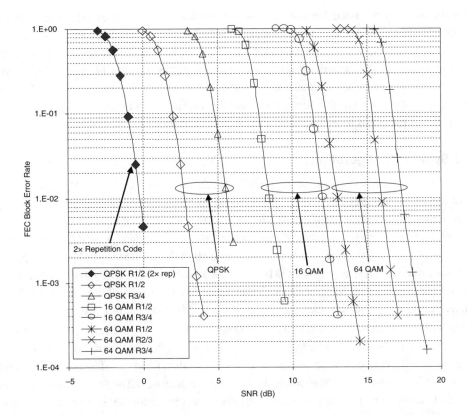

Figure 12.2 FEC block error rate for turbo codes in AWGN channel

used to determine the optimum modulation, code rate, FEC code block size, and space/time encoding matrix (matrix A or matrix B) for each scheduled MS.

As discussed in Chapter 11, the link-level simulation results are used to determine the SINR thresholds for the selection of optimum modulation and code rate (see Figure 12.2). The slots allocated to each MS are then divided into one or more FEC code blocks, based on the code block segmentation (see Section 8.1). The AWGN link-simulation results are used to calculate the block error probability of each FEC code block. Based on the block error probability, the system simulator performs a Bernoulli toss to determine which of the FEC blocks are received erroneously and need to be retransmitted. When H-ARQ, is used the retransmissions of an FEC code block are combined with the previous transmissions to determine the block error rate.

12.2.1.6 Computation of per Sector Throughput and User Data Rate

The per sector throughput and per user data rate are then calculated, based on the size and the average number of transmissions needed for each of the FEC code blocks over multiple Monte Carlo snapshots. Table 12.2 shows the various parameters and assumptions used for the system-level simulation results. In the DL subframe, the system simulator assumes that

11 OFDM symbols are used for the various control messages, such as DL frame preamble, FCH, DL-MAP, UL-MAP, DCD, and UCD, and the MSs are not scheduled over the first 11 DLs. Thus, the calculated average throughput and data rate do not include any MAP overheads and represent the effective layer 1 capacity available for data plane traffic. However, overheads due to the MAC header, convergence sublayer, mobility management, QoS management, and so on, are not explicitly modeled. As a result, the net layer 3 capacity available for IP traffic is expected to be less that what is indicated here.

12.2.2 System Configurations

The IEEE 802.16e-2005 standard offers a wide choice of optional PHY and MAC features, and it is not within the scope of this book to provide system-level performance results for all the possible combinations of such features. As shown in Table 12.3, this chapter considers only four configurations, based on the number of transmit and receive antennas. A SISO antenna configuration has not been considered in this chapter, since all WiMAX BSs and MSs are expected to have at least two antennas as per the WiMAX Forum's Mobile WiMAX System Profile. In order to preserve a competitive edge of 3G cellular networks, such as HSDPA and 1xEv-DO, the WiMAX Forum has decided to make the support for MIMO mandatory in all devices.

The basic configuration consists of a 2×2 open-loop MIMO in the DL and a 1×2 open-loop MIMO in the UL. In this basic configuration, collaborative MIMO between two MSs is also allowed in the UL, where two MS are allocated the same slots for their UL transmission, and the BS is able to separate the two MSs using the two receive antennas. The three enhanced configurations add various flavors of higher-order open-loop and closed-loop MIMO on top of the basic configuration. Enhanced configuration 1 increases the number of receive antennas in the DL from two to four thus providing higher order receive diversity in the DL. The UL in enhanced configuration 1 is essentially the same as that of the basic configuration. Enhanced configuration 2 increases the number of transmit antennas in DL from two to four over the basic configuration, thus providing higher-order transmit diversity. The UL in enhanced profile 2 also increases the number of receive antennas from two to four. Finally, enhanced configuration 3 uses 4×2 closed-loop MIMO in the DL, increases the number of transmit antennas in the UL from one to two, and increases the number of receive antennas in the UL from two to four.

Enhanced configuration 3 uses the antenna selection and the quantized channel-feedback-based closed loop MIMO in the DL, (see Section 8.9). Channel state information (CSI) is provided by the MS, using the fast-feedback channel. The MS provides a single feedback for the CSI over two bands (18 subcarriers) once every 10 msec. The BS uses the quantized feedback to calculate an optimum precoding or beamforming matrix that minimizes the mean square error (MSE) at the output of the receiver. (See Chapter 5 for the MMSE calculation of the precoding matrix.) In the case of the basic, enhanced 1, and enhanced 2 configurations, two MSs are simultaneously scheduled over the same subchannels. In this $2 \times N_r$ collaborative MIMO, N_r is the number of receive antennas at the BS. However, in the case of the enhanced 3 configuration,

Table 12.2 System-Level Simulation Parameters

Parameter	Value
Number of sites	19
Number of sectors per site	3
Site-to-site distance	2,000m
Frequency reuse	(1,1,3) and (1,3,3)
Channel bandwidth	10MHz
UL/DL duplexing scheme	TDD (28 symbols for DL, 9 symbols for UL, 11 symbols for frame overhead)
Number of subchannels DL	48 2 × 3 band AMC subchannels, 30 PUSC subchannels
Number of subchannels UL	48 2 × 3 band AMC subchannels, 35 PUSC subchannels
Number of subcarriers	1,024
Subcarrier permutation mode	Band AMC and PUSC
Carrier frequency	2,300MHz
BS antenna gain	18dBi
BS antenna pattern	$g(\theta) = -\min\left[12\left(\left(\frac{\theta}{\theta_{3dB}}\right)^2, A_m\right)\right]$, where $A_m = 20dB$, $\theta_{3dB} = 70^{o}$
BS antenna height	30m
BS noise figure	4dB
BS cable loss	3dB
BS transmit power	43dBm per antenna element (an 8dB backoff is assumed for 64 QAM modulation)
MS antenna gain	0dBi for handheld and 6dBi for desktop
MS antenna height	1m
MS noise figure	8dB
MS cable loss	0dB
MS transmit power	27dBm per antenna element
Building penetration loss	0dB for outdoor (handheld) and 10dB for indoor (desktop)
Standard deviation of shadow fade	8dB
Correlation of shadow fade	0.5 intersite envelope correlation and 0.9 intrasite (sector to sector) envelope correlation
Pathloss model	Suburban COST 231 Hata model
Multipath channel	Ped B, Ped A
MIMO fading correlation	0.5 for MS antenna elements, 0.5 for BS antenna element
Receiver structure	MLD for matrix A and MMSE for matrix B
Traffic model	Full buffer (QoS of all users is identical)
Number of users per sector	40 simultaneous users
Scheduling algorithm	Proportional fairness and round-robin

Table 12.3 System Configurations

Parameter	Basic	Enhanced 1	Enhanced 2	Enhanced 3
FEC type	Turbo code	Turbo code	Turbo code	Turbo code
H-ARQ type	Type I	Type I	Type I	Type I
Channel bandwidth	10MHz	10MHz	10MHz	10MHz
Number of subcarriers	1,024	1,024	1,024	1,024
MIMO mode (DL)	Open loop 2×2	Open loop 2×4	Open loop 4×2	Closed loop 4×2
MIMO mode (UL)	Open loop 1×2	Open loop 1×2	Open loop 1×4	Open loop 2×4
UL collaborative MIMO	Yes	Yes	Yes	Yes

since each MS has two transmit antennas, the UL collaborative MIMO is referred to as a $4\times N_r$ collaborative MIMO.

12.3 System-Level Simulation Results

First, results for the basic configuration under varying combinations of system parameters, such as frequency-reuse pattern, subcarrier permutation, multipath channel, and scheduling algorithms, are provided. These combinations all illustrate the trade-offs among performance metrics, such as average user data rate, percentile user data rate, and cell throughput due to variation in such parameters. Next, results for the three enhanced configurations are presented to highlight the benefits of various forms of open-loop and closed-loop higher-order MIMO techniques in WiMAX. As shown in Table 12.2, the results are for a *full-buffer* traffic model. It is expected that the overall performance of a WiMAX network will be different for other traffic models.

The key network-performance indicator presented in this section is the *average UL and DL throughput per sector.* Although average throughput per sector is a good indicator of system capacity, it fails to capture the variation in data rate experienced by various MSs that are distributed throughout the coverage of any given sector. To capture this variation, we also present the fifth and tenth percentile DL data rates as an indicator of cell-edge behavior of the network. We do not present UL percentile data rates, because the UL cell-edge behavior is more dependent on the link budget than on the traffic load and intercell interference.

12.3.1 System-Level Results of Basic Configuration

Two MS form factors have been considered in these simulations: a handheld device with omni-directional antennas and a desktop device with low-gain directional antennas. The handheld device is representative of a mobile or a portable network; the desktop device, of a fixed network. In the case of a desktop device, the MS is equipped with multiple low-gain antennas; at any given instant, the receiver chooses the antenna(s) with the strongest signal. Such a feature in the MS gives the benefit of having a directional antenna without the need for the antenna to be manually oriented in order to get a strong signal. Most WiMAX desktop devices are expected to be equipped with six to eight such antennas, each with a gain of 3dBi to 6dBi.

Figure 12.3 and Figure 12.4 show the average throughputs per sector for the basic configuration in Ped B and Ped A environments, respectively. The average throughput per sector is slightly better in the case of a Ped A channel than in a Ped B channel because the Ped A channel provides better multiuser diversity due to larger variations in channel amplitude, which is exploited by the proportional fairness scheduler.

The overall per sector throughput in the case of (1,1,3) reuse is better when a directional antenna is used at the MS, since the amount of cochannel interference is reduced by the directional nature of the channel. However, in the case of (1,3,3) reuse, the additional directionality of the antenna at the MS in an interference-limited environment does not provide any significant benefit, since (1,3,3) frequency reuse provides a sufficient geographical separation of cochannel BSs. It should be noted that in the case of noise-limited design—a design with larger cell radii— the gain of the directional antenna at the MS would provide an improvement in the sector throughput even with (1,3,3) frequency reuse.

Figure 12.5 and Figure 12.6 show the probability distributions of per subchannel user DL data rate for the Ped B and Ped A environments, respectively. One can conclude that in the case of (1,3,3) reuse, the fifth and tenth percentile data rates are much higher than the case of (1,1,3) reuse. This happens because in the case of (1,1,3) reuse, a large percentage of MSs that are present toward the cell edge experience a low SINR, due to cochannel interference and thus a low data rate. Based on the per user data rate distribution it should be noted that although (1,1,3) reuse is more spectrally efficient, it is achieved at the price of poor performance at the cell edge. In order to achieve an acceptable cell-edge performance, (1,3,3) reuse or (1,1,3) reuse with segmentation is required. When segmentation is used, all the subchannel are divided into three groups, and each of the three sectors is allocated one group of subchannels. Segmentation thus achieves an effective (1,3,3) reuse.

The fifth and tenth percentile data rates can also be improved in the case of (1,1,3) reuse by using directional antennas at the MS, as shown in Figure 12.6. This controlled trade-off between network reliability and spectral efficiency allows a system designer to choose the appropriate network parameters, such as cell radius, frequency reuse, and antenna pattern, that will meet the design goal. *From here on, we limit our discussion to the handheld-device scenario with (1,1,3) frequency reuse.*

Table 12.4 and Table 12.5 summarize the throughput per BS and the fifth and tenth percentile data rates for the various scenarios. The throughput of all the sectors is combined to get the throughput of the BS. Since a total of 30MHz of spectrum is assumed, as per Table 12.2, in the case of (1,1,3) frequency reuse, we assume that each sector is allocated three 10MHz TDD channels. Although the average throughput channel is less in the case of (1,1,3) frequency reuse than for (1,3,3) reuse, the overall capacity is higher with (1,1,3) reuse, since each sector is allocated three channels as opposed to one channel in the case of (1,3,3) reuse. On the other hand, network reliability is significantly improved by going from (1,1,3) reuse to (1,3,3) reuse.

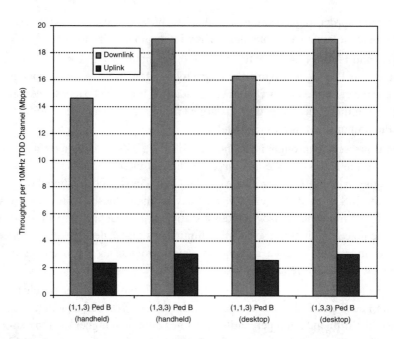

Figure 12.3 Downlink and uplink average throughput per sector for band AMC in Ped B

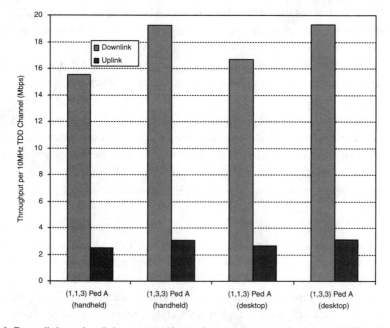

Figure 12.4 Downlink and uplink average throughput per sector for band AMC in Ped A

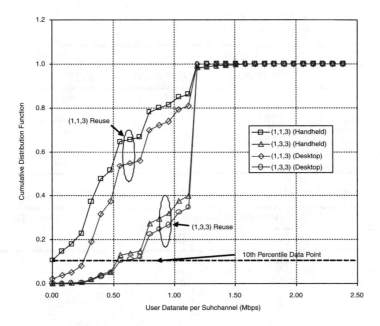

Figure 12.5 User DL data rate per band AMC subchannel for handheld and desktop devices in Ped B

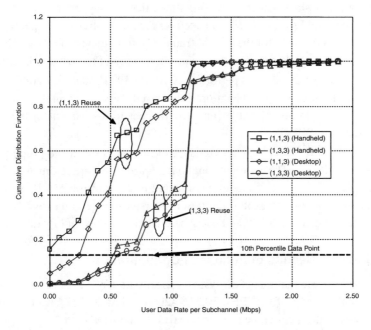

Figure 12.6 User DL data rate per band AMC subchannel for handheld and desktop devices in Ped A

Table 12.4 Average Throughput per BS Site for Handheld and Desktop Device with Band AMC Subcarrier Permutation in DL and UL

Scenario	Handheld Device		Desktop Device	
	DL (Mbps)	UL (Mbps)	DL (Mbps)	UL (Mbps)
(1,1,3) reuse in Ped B	131.47	21.13	146.77	23.59
(1,3,3) reuse in Ped B	57.09	9.18	57.08	9.17
(1,1,3) reuse in Ped A	139.88	22.48	150.26	24.15
(1,3,3) reuse in Ped A	57.62	9.26	57.86	9.30

Table 12.5 Fifth and Tenth Percentile Data Rate for Basic Configuration

Scenario	Handheld Device		Desktop Device	
	5% (Mbps)	10% (Mbps)	5% (Mbps)	10% (Mbps)
(1,1,3) reuse in Ped B	0.025	0.075	0.146	0.252
(1,3,3) reuse in Ped B	0.449	0.525	0.480	0.550
(1,1,3) reuse in Ped A	0.025	0.075	0.036	0.220
(1,3,3) reuse in Ped A	0.351	0.490	0.406	0.517

The DL throughput shown in this section is significantly higher than the UL throughout, since the number of OFDM symbols (28) allocated for the DL subframe is larger than the number of OFDM symbols (9) allocated for the UL subframe. Varying the number of symbols to be used for DL and UL subframes makes it possible to control the ratio of the DL and UL throughputs.

Figure 12.7 shows the throughput performance for the PUSC and the band AMC subcarrier permutations. Since no precoding or beamforming is used, depending on the multipath channel band AMC provides an improvement of only 14 percent to 18 percent in the overall sector throughput compared to PUSC.

Figure 12.8 shows the throughput performance of the round-robin and proportional-fairness (PF) scheduling algorithms. The sector throughout improves by approximately 25 percent by using a proportional fairness (PF) scheduler compared to a round-robin (RR) scheduler, due to the ability of the PF scheduler to exploit multiuser diversity to a certain extent. Table 12.6 summarizes the DL throughput for the PF and RR schedulers in various multipath environments.

12.3.2 System-Level Results of Enhanced Configurations

In this section, we estimate the impact on system capacity of a WIMAX network of some of the MIMO features that are part of the IEEE 802.16e-2005 standard. See Chapter 8 for a detailed description of the various open- and closed-loop MIMO schemes.

The average per sector throughput for the basic and various enhanced configurations is shown in Figure 12.9 and Figure 12.10. Based on these results, one can conclude that both

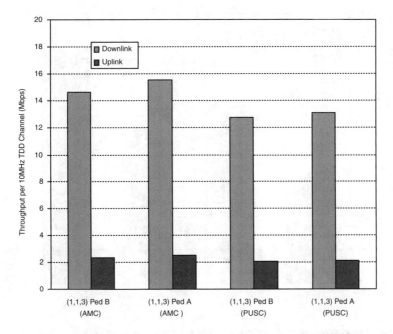

Figure 12.7 Comparison of PUSC and band AMC subcarrier permutation for handheld form factor

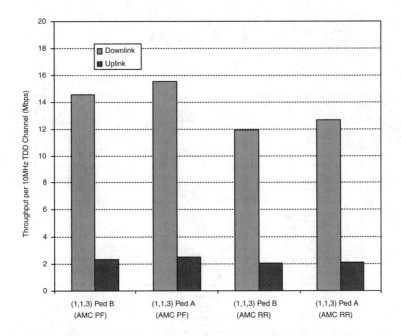

Figure 12.8 Comparison of proportional fairness and round-robin scheduling for handheld device

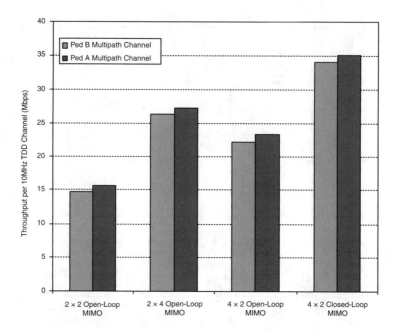

Figure 12.9 Downlink average throughput per sector for various MIMO configuration

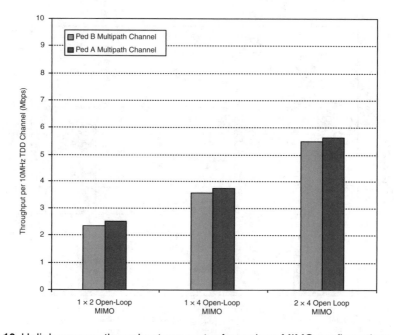

Figure 12.10 Uplink average throughput per sector for various MIMO configurations

Table 12.6 Average Throughput per Sector for Band AMC with PF and RR Schedulers

	Ped B		Ped A	
	DL (Mbps)	**UL (Mbps)**	**DL (Mbps)**	**UL (Mbps)**
Proportional-fairness scheduler	14.61	2.35	15.54	2.50
Round-robin scheduler	11.96	1.92	12.66	2.04

receive diversity and transmit diversity improve the average throughput of a WiMAX network. By increasing the number of transmit antennas from two to four, the per sector throughput improves by 50 percent. Similarly, by increasing the number of receive antennas from two to four, the per sector throughput is increased by 80 percent. However, based on Figure 12.11 and Figure 12.12, it is evident that for the basic 2×2 open-loop configuration, the fifth and tenth percentile DL data rates are not improved by increasing either transmit or receive diversity order. Thus, one can conclude that transmit diversity with antennas in DL is not sufficient to improve the cell-edge data rate in the case of (1,1,3) reuse. Although receive diversity with four antennas somewhat improves the tenth percentile data rate, it is still not sufficient to improve the cell-edge data rate when (1,1,3) frequency reuse is implemented.

However, when closed-loop MIMO with four antennas at the BS is used, the per sector throughput and the fifth and tenth percentile data rates are significantly improved. The average per sector throughput is improved by 130 percent, and the cell-edge data rate per subchannel is high enough to provide reliable broadband services.

Clearly, the 4×2 closed-loop MIMO feature provides significant improvement in the per sector throughput and percentile data rates, compared to both the open-loop 2×4 and open-loop 4×2 MIMO modes, because the transmitter is able to choose the optimum precoding matrix or beamforming vector, in order to increase the link throughput. In this case, we assume that a single precoding matrix or beamforming vector is chosen for each 2×3 band AMC subchannel.[5]

The DL simulation results shown in Figure 12.9–Figure 12.12 assume that feedback for the quantized MIMO channel is provided by the receiver once every frame (5msec). (See Section 8.9 for the quantized channel-feedback-based closed-loop MIMO solution). The UL enhanced configuration 3 uses a 2×4 open-loop MIMO. The increased performance over other enhanced profiles comes from increasing the number transmit antennas in the UL from one to two. The UL throughput results do not account for the fact that a part of the UL bandwidth is used by the closed-loop MIMO feedback.

Table 12.7 and Table 12.8 show the average throughputs per cell site and the percentile data rates for the various profiles. The biggest impact of the closed-loop MIMO appears to be on the percentile data rate (Table 12.8). Based on the system-level performance of a WiMAX network, one can conclude that a (1,1,3) frequency reuse will not be able to provide carrier-grade reliability and guaranteed data rate unless closed-loop MIMO features of IEEE 802.16e-2005 are used.

---

5. See Chapter 9 for a more detailed description of the band AMC subcarrier permutation mode.

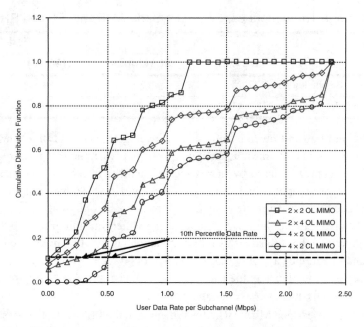

Figure 12.11 User DL data rate per subchannel for enhanced profile in Ped B

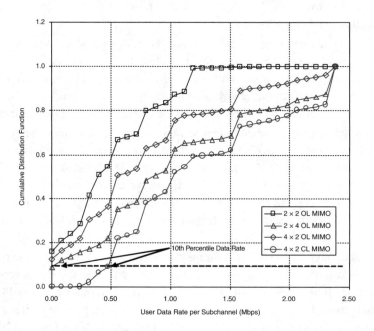

Figure 12.12 User DL data rate per subchannel for enhanced profile in Ped A

Table 12.7 Total Throughput per Cell Site for Band AMC in a Ped B Multipath Channel with 30MHz of Spectrum (1,1,3) Frequency Reuse

Profile	Ped B		Ped A	
	DL (Mbps)	UL (Mbps)	DL (Mbps)	UL (Mbps)
Basic configuration (2 × 2 open loop)	131.47	21.13	139.88	22.48
Enhanced configuration 1 (2 × 4 open loop)	236.79	21.13	245.07	22.48
Enhanced configuration 2 (4 × 2 open loop)	200.34	32.20	209.33	33.64
Enhanced configuration 3 (4 × 2 closed loop)	306.99	49.34	315.99	50.78

Table 12.8 Fifth and Tenth Percentile Data Rates per Subchannel for Band AMC in a Ped B Multipath Channel with (1,1,3) Frequency Reuse

Scenario	Ped B		Ped A	
	5% (Mbps)	10% (Mbps)	5% (Mbps)	10% (Mbps)
Basic configuration (2 × 2 open loop)	0.025	0.085	0.025	0.085
Enhanced configuration 1 (2 × 4 open loop)	0.075	0.206	0.065	0.198
Enhanced configuration 2 (4 × 2 open loop)	0.035	0.095	0.035	0.090
Enhanced configuration 3 (4 × 2 closed loop)	0.437	0.499	0.371	0.482

12.4 Summary and Conclusions

In this chapter, we provided some estimates of a WiMAX network system performance and its dependence on various parameters, such as frequency reuse, scheduling algorithm, subcarrier permutation, and MIMO. Based on these results, we can derive the following high-level conclusions on the behavior of a WiMAX network.

- Although a WiMAX network provides the highest per sector average throughput with a (1,1,3) frequency reuse, it does so at the cost of poor cell-edge performance. In order to achieve an acceptable level of user data rate at the cell edge, a (1,3,3) frequency reuse is required. If, however, sufficient spectrum is not available, a (1,1,3) frequency reuse with segmentation would also provide good cell-edge behavior.

- Scheduling algorithms and their ability to take advantage of multiuser diversity can lead to a significant improvement in the average throughput. We typically observe a 25 percent improvement in the cell throughput if multiuser diversity with a large number of users is used.
- Diversity, particularly at the receiver, provides significant gain in the average through-put—typically, 50 percent to 80 percent—but does not provide sufficient improvement in the cell-edge behavior to be able to use (1,1,3) frequency reuse without segmentation.
- Closed-loop MIMO with linear precoding and beamforming seems to provide a significant improvement in both the cell-edge experience and the average throughput. In going from an open-loop 2×2 MIMO configuration to a closed-loop 4×2 MIMO configuration, we observe an improvement of 125 percent to 135 percent in the per sector average throughput. The fifth and tenth percentile data rates are also significantly improved, indicating improvement in the cell-edge behavior. This would seem to indicate that with the closed-loop MIMO features of IEEE 802.16e-2005, a frequency reuse of (1,1,3) is usable.
- The overall spectral efficiency—defined as the average throughput per sector divided by the total frequency needed for the underlying frequency reuse—of a WiMAX network is quite high compared to the current generation of cellular networks. Even with a 2×2 open-loop MIMO configuration, a WiMAX network can achieve a spectral efficiency of 1.7bps/Hz with (1,1,3) reuse. The overall spectral efficiency in a pedestrian environment can increase to 3.9bps/Hz if closed-loop MIMO is implemented.

12.5 Appendix: Propagation Models

Median pathloss in a radio channel is generally estimated using analytical models based on either the fundamental physics behind radio propagation or statistical curve fitting of data collected via field measurements. For most of the practical deployment scenarios, particularly non-line-of-sight scenarios, statistical models based on empirical data are more useful. Although most of the statistical models for pathloss have been traditionally developed and tuned for a mobile environment, many of them can also be used for an NLOS fixed network with some modification of parameters. In the case of a line-of-sight-based fixed network, the *free-space* radio propagation model (see Chapter 3) can be used to predict the median pathloss. Since WiMAX as a technology has been developed to operate efficiently even in an NLOS environment, we focus extensively on this usage model for the remainder of the appendix. We describe a few of the pathloss models that are relevant to NLOS WiMAX deployments.

12.5.1 Hata Model

The Hata model is an analytical formulation based on the pathloss measurement data collected by Okumura in 1968 in Japan. The Hata model is one of the most widely used models for estimating median pathloss in macrocellular systems. The model provides an expression for median pathloss as a function of carrier frequency, BS and mobile station antenna heights, and the distance between the BS and the MS. The Hata model is valid only for the following range of parameters:

- 150MHz $\leq f \leq$ 500MHz
- 30m $\leq h_b \leq$ 200m
- 1m $\leq h_m \leq$ 10m
- 1km $\leq d \leq$ 20km

In these parameters, f is the carrier frequency in MHz, h_b is the BS antenna height in meters, h_m is the MS antenna height in meters, and d is the distance between the BS and the MS in km. According to the Hata model, the median pathloss in an urban environment is given by

$$\overline{PL}_{Urban} = 69.55 + 26.16\log_{10}f - 13.82\log_{10}h_b + (44.9 - 6.55h_b)\log_{10}d - a(h_m), \quad (12.10)$$

where $\overline{PL}_{Urban}$ is expressed in the dB scale, and $a(h_m)$ is the MS antenna-correction factor. For a large city with dense building clutter and narrow streets, the MS antenna-correction factor is given by

$$a(h_m) = 8.29[\log_{10}[1.54 \cdot f]]^2 - 0.8 \qquad f \leq 300\text{MHz}$$

$$a(h_m) = 3.20[\log_{10}[11.75 \cdot f]]^2 - 4.97 \qquad f \geq 300\text{MHz} \qquad (12.11)$$

For a small- or medium-size city, where the building-clutter density is smaller, the MS antenna-correction factor is given by

$$a(h_m) = (1.11\log_{10}f - 0.7)h_m - (1.56\log_{10}f - 0.8). \qquad (12.12)$$

For a suburban area, the same MS antenna-correction factor as used for small cities is applicable, but the median pathloss is modified to be

$$\overline{PL}_{Suburban} = \overline{PL}_{Urban} - 2\left[\log_{10}\left(\frac{f}{28}\right)\right]^2 - 5.4. \qquad (12.13)$$

For a rural area, the same MS antenna-correction factor as used for small cities is applicable, but the median pathloss is modified to be

$$\overline{PL}_{Rural} = \overline{PL}_{Urban} - 4.78[\log_{10}f]^2 - 18.33\log_{10}f - 40.98. \qquad (12.14)$$

The model may also be generalized to any clutter environment, such that the median pathloss is modified from that of a small urban city as

$$\overline{PL} = \overline{PL}_{Urban} + Clutter\ Offset. \qquad (12.15)$$

12.5.2 COST-231 Hata Model

The Hata model is widely used for cellular networks in the 800MHz/900MHz band. As PCS deployments begin in the 1,800MHz/1,900MHz band, the Hata model was modified by the European COST (Cooperation in the field of Scientific and Research) group, and the extended pathloss model is often referred to as the COST-231 Hata model. This model is valid for the following range of parameters:

- 150MHz $\leq f \leq$ 2000MHz
- 30m $\leq h_b \leq$ 200m
- 1m $\leq h_m \leq$ 10m
- 1km $\leq d \leq$ 20km

The median pathloss for the COST-231 Hata model is given by

$$\overline{PL} = 46.3 + 33.9\log_{10}f - 13.82\log_{10}h_b + (44.9 - 6.55\log_{10}h_b)\log_{10}d - a(h_m) + C_F. \quad (12.16)$$

The MS antenna-correction factor, $a(h_m)$, is given by

$$a(h_m) = (1.11\log_{10}f - 0.7)h_m - (1.56\log_{10}f - 0.8). \quad (12.17)$$

For urban and suburban areas, the correction factor C_F is 3dB and 0dB, respectively. The WiMAX Forum recommends using this COST-231 Hata model for system simulations and network planning of macrocellular systems in both urban and suburban areas for mobility applications. The WiMAX Forum also recommends adding a 10dB fade margin to the median pathloss to account for shadowing.

12.5.3 Erceg Model

The Erceg model is based on extensive experimental data collected at 1.9GHz in 95 macrocells across the United States [3]. The measurements were made mostly in suburban areas of New Jersey, Seattle, Chicago, Atlanta, and Dallas. The Erceg model is applicable mostly for fixed wireless deployment, with the MS installed under the eave/window or on the rooftop. The model, adopted by the IEEE 802.16 group as the recommended model for fixed broadband applications, has three variants, based on terrain type.

1. Erceg A is applicable to hilly terrain with moderate to heavy tree density.
2. Erceg B is applicable to hilly terrain with light tree density or flat terrain with moderate to heavy tree density.
3. Erceg C is applicable to flat terrain with light tree density.

The Erceg model is a slope-intercept model given by

$$PL = \overline{PL} + \chi = A + 10\alpha\log_{10}\left(\frac{d}{d_0}\right) + \chi, \tag{12.18}$$

where $\overline{PL}$ is the median pathloss, PL is the instantaneous attenuation, and χ is the shadow fade, A is the intercept and is given by free-space pathloss at the desired frequency over a distance of $d_0 = 100$ m:

$$A = 20\log_{10}\left(\frac{4\pi d_0 f}{C}\right), \tag{12.19}$$

and α is the pathloss exponent and is modeled as a random variable with a Gaussian distribution around a mean value of $A - Bh_b + Ch_b^{-1}$. The instantaneous value of the pathloss exponent is given by

$$\alpha = \left(A - Bh_b + \frac{C}{h_b}\right) + x\sigma_\alpha, \tag{12.20}$$

where x is a Gaussian random variable with zero mean and unit variance, and σ_α is the standard deviation of the pathloss exponent distribution. The parameters of the Erceg model, A, B, C, and σ_α for the various terrain categories, are given in Table 12.9.

Unlike the Hata model, which predicts only the median pathloss, the Erceg model has both a median pathloss and a shadow-fading component, χ, a zero-mean Gaussian random variable expressed as $y\sigma$ where y is a zero-mean Gaussian random variable with unit variance, and σ is the standard deviation of χ. The standard deviation σ is, in fact, another Gaussian variable with a mean of μ_S and a standard deviation of σ_S, such that $\sigma = \mu_S + z\sigma_S$, z being a zero-mean unit variance Gaussian random variable.

Strictly speaking, this base model is valid only for frequencies close to 1,900MHz, for an MS with omnidirectional antennas at a height of 2 meters and BS antenna heights between 10 meters and 80 meters. The base model has been expanded with correction factors to cover higher frequencies, variable MS antenna heights, and directivity. The extended versions of the Erceg models are valid for the following range of parameters:

- 1900MHz $\leq f \leq$ 3500MHz
- 10m $\leq h_b \leq$ 80m
- 2m $\leq h_m \leq$ 10m
- 0.1km $\leq d \leq$ 8km

The median pathloss formula for the extended version of the Erceg model is expressed as

$$\overline{PL} = A + 10\gamma\log\left(\frac{d}{d_0}\right) + \Delta PL_f + \Delta PL_{hMS} + \Delta PL_{\theta MS}. \tag{12.21}$$

Table 12.9 Parameters of Erceg Model

Parameters	Erceg Model A	Erceg Model B	Erceg Model C
A	4.6	4.0	3.6
B	0.0075	0.0065	0.005
C	12.6	17.1	20
s_a	0.57	0.75	0.59
μ_S	10.6	9.6	8.2
σ_S	2.3	3.0	1.6

The various correction factors in Equation (12.21) corresponding to frequency, MS height, and MS antenna directivity are given by

$$\Delta PL_f = 6\log\frac{f}{1900} \tag{12.22}$$

$$\Delta PL_{hMS} = -10.8\log\left(\frac{h_m}{2}\right) \quad \text{For Erceg A and B}$$

$$\Delta PL_{hMS} = -20\log\left(\frac{h_m}{2}\right) \quad \text{For Erceg C}$$

$$\Delta PL_{\theta MS} = 0.64\ln\left(\frac{\theta}{360}\right) + 0.54\left(\ln\left(\frac{\theta}{360}\right)\right)^2$$

where ΔPL_{MS} is often referred to as the antenna-gain reduction factor and accounts for the fact that the angular scattering is reduced owing to the directivity of the antenna. The antenna-gain reduction factor can be quite significant; for example, using a $20°$ antenna can contribute to a ΔPL_{MS} of 7 dB.

12.5.4 Walfish-Ikegami Model

The Hata model and its COST-231 extension are suitable for macrocellular environments, but not for smaller cells that have a radius less than 1 km. The Walfish-Ikegami model applies to these smaller cells and is recommended by the WiMAX Forum for modeling microcellular environments. The model assumes an urban environment with a series of buildings as depicted in Figure 12.13, with the building heights, interbuilding distance, street width, and so on, as parameters. In this model, diffraction is assumed to be the main mode of propagation, and the model is valid for the following ranges of parameters:

- $800\text{MHz} \leq f \leq 2000\text{MHz}$
- $4\text{m} \leq h_b \leq 50\text{m}$
- $1\text{m} \leq h_m \leq 3\text{m}$
- $0.2\text{km} \leq d \leq 5\text{km}$

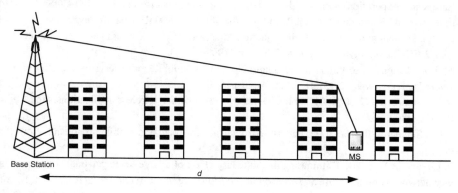

Figure 12.13 The Walfish-Ikegami model

The model is made up of three terms:

$$\overline{PL} = L_{fs} + L_{rts} + L_{msd},$$ (12.23)

where, L_{fs} is the free-space loss, L_{rts} is the rooftop-to-street diffraction loss, and L_{msd} is the multiscreen loss. The model provides analytical expressions for each of the terms for a variety of scenarios and parameter settings. For the standard NLOS case, with BS antenna height 12.5m, building height 12m, building-to-building distance 50m, width 25m, MS antenna height 1.5m, orientation 30° for all paths, and in a metropolitan center, the equation simplifies to

$$\overline{PL} = -65.9 + 38\log_{10}d + \left(24.5 + \frac{1.5f}{925}\right)\log_{10}f.$$ (12.24)

This equation is recommended by the WiMAX Forum to be used for system modeling. The use of an additional 10dB for fading margin is also recommended with this model.

The Walfish-Ikegami model also provides an expression for the urban canyon case, which has a LOS component between the BS and the MS. For the LOS case, the median pathloss expression is

$$\overline{PL} = -31.4 + 26\log_{10}d + 20\log_{10}f.$$ (12.25)

12.6 Bibliography

[1] 3GPP TSG-RAN-1. Effective SIR computation for OFDM system-level simulations. Document R1-03-1370, November 2003.

[2] 3GPP TSG-RAN1. System level simulation of OFDM—further considerations. Document R1-03-1303, November 2003.

[3] V. Erceg, et.al. An empirically based pathloss model for wireless channels in suburban environments. *IEEE Journal on Selected Areas of Communications,* 17(7), July 1999.

[4]   European Cooperation in the Field of Scientific and Technical Research EURO-COST 231. Urban transmission loss models for mobile radio in the 900 and 1800MHz bands, rev. 2. The Hague, 1991.

[5]   L. J. Greenstein and V. Erceg. Gain reductions due to scatter on wireless paths with directional antennas. *IEEE Communications Letters,* 3(6), June 1999.

[6]   M. Hata. Empirical formula for propagation loss in land mobile radio services. *IEEE Transactions on Vehicular Technology,* 29(3):317–325, August 1980.

[7]   IEEE. Standard 802.16.3c-01/29r4. Channel models for fixed wireless applications. tap:// www.ieee802.org/16.

[8]   IEEE. Standard 802.16e-2005, Part 16: Air interface for fixed and mobile broadband wireless access systems.

[9]   Y. Lin and V. W. Mark. Eliminating the boundary effect of a large-scale personal communication service network simulation. *ACM Transactions on Modeling and Computer Simulations,* 4(2), April 1994.

[10]   Y. Okumura, Field strength and its variability in UHF and VHF land-mobile radio service. *Review Electrical Communication Laboratory,* 16(9–10):825–873, September–October 1968.

[11]   A. Paulraj, R. Nabar, and D. Gore. *Introduction to Space-Time Wireless Communications,* Cambridge University Press, 2003.

[12]   J. W. Porter and J. A. Thewatt. Microwave propagation characteristics in the MMDS frequency band. *Proceedings of the ICC 2000 Conference,* June 2000.

[13]   T. S. Rappaport. *Wireless Communications: Principles and Practice,* 2nd ed. Prentice Hall, 2002.

[14]   W. H. Tranter, K. S. Shanmugam, T. S. Rappaport, and K. L. Kosbar. *Principles of Communication System Simulation with Wireless Applications.* Prentice Hall, 2002.

[15]   WiMAX Forum. Mobile WiMAX—Part 1: A technical overview and performance evaluation. June 2006.

[16]   WiMAX Forum Technical Working Group. WiMAX Forum mobile system profile, February 2006.

[17]   Y. R. Zheng and C. Xiao. Improved models for the generation of multiple uncorrelated Rayleigh fading waveforms, *IEEE Communications Letters,* 6(6), June 2002.

Acronyms

1xEV-DO	1x evolution—data optimized
1xEV-DV	1x evolution—data and voice
3DES	triple data encryption standard
3G	third generation
3GPP	third-generation partnership project
3GPP2	third generation partnership project-2
AAA	authentication, authorization, and accounting
AAS	advanced antenna systems
AC	admission control
ACE	active constellation extension
ADC	analog-to-digital converter
ADSL	asymmetric digital subscriber loop
AES	advanced encryption standard
AF	application function
AF	assured forwarding
AK	authentication key
AKA	authentication and key agreement
AM	amplitude modulation
AMC	adaptive modulation and coding
AoA	angle of arrival
AoD	angle of departure
API	application programing interface
AR	access router
ARQ	automatic repeat request
AS	angular spread
ASN	access services network
ASN-GW	ASN gateway
ASP	application service provider
ATM	asynchronous transfer mode
AWGN	additive white Gaussian noise
AWS	advanced wireless services
BE	best effort
BEP	bit error probability
BER	bit error rate
BGCF	breakout gateway control function
BLAST	Bell Labs layered spaced time
BLER	block error rate
BPSK	binary phase shift keying

BRS	broadband radio services
BS	base station
BSC	base station controller
BSN	block sequence number
BTS	base station transceivers
CBC	cipher-block chaining
CBR	constant bit rate
CC	convolutional coding
CCDF	complementary cumulative distribution function
CCI	cochannel interference
CDF	cumulative distribution function
CDMA	code division multiple access
CGI	common gateway interface
CHAP	challenge handshake authentication protocol
CID	connection identifier
CLT	central limit theorem
CM	cubic metric
CMAC	cipher-based message authentication code
CMAC	complex multiply and accumulate
CN	correspondent node
CoA	care-of address
COPS	common open policy service
CP	cyclic prefix
CPE	customer premise equipment
CPL	call-processing language
CQICH	channel-quality indicator channel
CRC	cyclic redundancy check
CQI	channel quality indicator
CS	convergence sublayer
CSCF	call session control function
CSI	channel state information
CSMA	carrier sense multiple access
CSN	connectivity services network
CTC	convolutional turbo code
DAC	digital-to-analog converter
DARS	digital audio radio services
DC	direct current
DCD	downlink channel descriptor
DCF	distributed coordination function
DECT	digital-enhanced cordless telephony
DDFSE	delayed-decision-feedback sequence estimation
DES	data encryption standard
DFE	decision-feedback equalizer
DFT	discrete Fourier transform

DHCP	dynamic host control protocol
DiffServ	differentiated services
DL	downlink
DNS	domain name system
DoA	direction of arrival
DOCSIS	data over cable service interface specification
DP	decision point
DPF	data path function
DRM	digital rights management
DS	delay spread
DSA	dynamic service allocation
DSC	dynamic service change
DSCP	DiffServ code point
DSD	dynamic service delete
DSL	digital subscriber line
DSP	digital-signal processing
DSTTD	double space/time transmit diversity
DVB-H	digital video broadcasting-handheld
EAP	extensible authentication protocol
ECRM	effective code rate map
EDGE	enhanced data rate for GSM evolution
EESM	exponentially effective SINR map
EF	expedited forwarding
EGC	equal gain combining
EIRP	effective isotopic radiated power
EMSK	enhanced master session key
EP	enforcement point
ErtPS	extended real-time packet service
ERT-VR	extended real-time variable-rate service
ESP	encapsulating security payload
ETH-CS	Ethernet convergence sublayer
ETRI	Electronics and Telecommunications Research Institute
ETSI	European Telecommunications Standards Institute
EVM	error vector magnitude
FA	foreign agent
FBSS	fast base station switching
FCC	Federal Communications Commission
FCH	frame control header
FDD	frequency division duplexing
FDMA	frequency division multiple access
FEC	forward error correction
FEC	forward equivalence class
FEQ	frequency-domain equalization
FER	frame error rate

FFT	fast Fourier transform
FHDC	frequency-hopping diversity code
FIB	forward information base
FIPS	Federal Information Processing Standard
FIR	finite impulse response
FM	frequency modulation
FSH	fragmentation subheader
FTP	file transfer protocol
FTTH	fiber-to-the-home
FUSC	full usage of subcarriers
FWA	fixed wireless access
GMH	generic MAC header
GPRS	GSM packet radio services
GRE	generic routing encapsulation
GSM	global system for mobile communications
GW	gateway
HA	home agent
HARQ	hybrid-ARQ
HDTV	high-definition television
HIPERMAN	high-performance metropolitan area network
HHO	hard handover
HMAC	hash-based message authentication code
HO	handover
HoA	home address
HPA	high-power amplifier
HSDPA	high-speed downlink packet access
HSPA	high-speed packet access
HSS	home subscriber server
HSUPA	high-speed uplink packet access
HTTP	hypertext transfer protocol
HUMAN	high-speed unlicensed metropolitan area network
IBO	input backoff
ICI	intercarrier interference
ICMP	Internet control message protocol
I-CSCF	interrogating call session control function
IDFT	inverse discrete Fourier transform
IEEE	Institute of Electrical and Electronics Engineers
IETF	Internet Engineering Task Force
IFFT	inverse fast Fourier transform
IGMP	Internet group management protocol
IM	instant messaging
IMS	IP multimedia subsystem
IN	intelligent network
IntServ	integrated services

IP	Internet protocol
IP-CS	IP convergence sublayer
IPsec	IP security
IP-TV	Internet protocol television
IS	integrated services
ISDN	integrated services digital network
ISI	inter-symbol interference
ITU	International Telecommunications Union
JAIN	Java for advanced intelligence network
KEK	key encryption key
LAN	local area network
LDAP	lightweight directory access protocol
LDPC	low-density parity codes
LDP-CR	label distribution protocol/constraint-based routing
LER	label-edge router
LLR	log liklihood ratio
LMOS	local multipoint distribution system
LMMSE	linear minimum mean square error
LOS	line of sight
LPF	local policy function
LR	location register
LS	least squares
LSB	least significant bit
LSP	label switched path
LSR	label switching router
LTE	long-term evolution
MAC	media access control
MAC	message-authentication code
MAN	metropolitan area network
MBS	multicast broadcast service
MC-CDMA	multicarrier CDMA
MCS	modulation and coding scheme
MD5	message-digest 5 algorithm
MDHO	macrodiversity handover
MIMO	multiple input multiple output
MIC	mean instantaneous capacity
MIP	mobile IP
MIP-HA	mobile IP home agent
MISO	multiple input/single output
ML	maximum likelihood
MLD	maximum likelihood detection
MLSD	maximum-likelihood sequence detection
MMDS	multichannel multipoint distribution services
MMS	multimedia messaging service

MMSE	minimum mean square error
MN	mobile node
MPDU	MAC protocol data unit
MPEG	Motion Picture Experts Group
MPLS	multiprotocol label switching
M-QAM	multilevel QAM
MRC	maximal ratio combining
MRT	maximum ratio transmission
MS	mobile station
MSB	most significant bit
MSDU	MAC service data unit
MSE	mean square error
MSK	master session key
MSL	minimum signal level
MSR	maximum sum rate
MUD	multiuser detection
NAI	network access identifier
NAP	network access provider
NAS	network access server
NAT	network address translation
NLOS	non–line-of-sight
NRM	network reference model
nrtPS	non–real-time polling service
NSP	network services provider
NTP	network timing protocol
NWG	Network Working Group
OBO	output backoff
OC	optimum combiner
OCI	other-cell interference
OFDM	orthogonal frequency division multiplexing
OFDMA	orthogonal frequency division multiple access
OSA	open systems architecture
OSI	open systems interconnect
O-SIC	ordered successive cancellation
OSS	operational support systems
OSTBC	orthogonal space/time block code
PA	paging agent
PAP	password authentication protocol
PAPR	peak-to-average-power ratio
PAR	peak-to-average ratio
PC	paging controller
PCS	personal communications services
P-CSCF	proxy call session control function
PDA	personal data assistant

PDF	probability density function
PDP	policy decision point
PDU	packet data unit
PEAP	protected extensible authentication protocol
PEP	policy enforcement point
PER	packet error rate
PF	proportional fairness; policy function
PG	paging group
PHB	per hop behavior
PHS	packet header suppression
PHSF	PHS field
PHSI	PHS index
PHSM	PHS mask
PHSV	PHS verify
PKI	public key infrastructure
PKM	privacy and key management
PM	phase modulation
PMIP	proxy mobile IP
PMK	pairwise master key
PN	pseudonoise
PoA	point of attachment
PPP	point-to-point protocol
PR	policy rule
PRC	proportional rate constraints
P/S	parallel to serial
PSH	packing subheader
PSK	preshared key
PSTN	public switched telephone network
PTS	partial transmit sequence
PUSC	partial usage of subcarriers
QoS	quality of service
QAM	quadrature amplitude modulation
QPSK	quadrature phase shift keying
RADIUS	remote access dial-in user service
RF	radio frequency
RFC	request for comments
RMS	root mean square
ROHC	robust header compression
RP	reference point
RR	radio resource
RR	round-robin
RRA	radio resource agent
RRC	radio resource controller
RRM	radio resource management

RS	Reed Solomon
RSA	Rivest-Shamir-Adleman
RSS	received signal strength
RSSE	reduced-state sequence estimation
RSSI	received signal strength indicator
RSVP	resource reservation protocol
RTCP	real-time control protocol
RTP	real-time transport protocol
rtPS	real-time polling service
RTT	roundtrip time
RUIM	removable user identity module
SA	security association
SC	selection combining
S-CSCF	serving call session control function
SCTP	stream control transport protocol
SDP	session description protocol
SDU	service data unit
SET	secure electronic transactions
SF	service flow; shadow fading
SFA	service flow authorization
SFBC	space/frequency block code
SFID	service flow identifier
SFM	service flow management
SGSN	serving GPRS support node
SH	subheader
SHA	secure hash algorithm
SIC	successive interference cancellation
SII	system identity information
SIM	subscriber identity module
SIMO	single input/multiple output
SINR	signal-to-interference-plus-noise ratio
SIP	session initiation protocol
SIR	signal-to-interference ratio
SISO	single input/single output
SLA	service-level agreement
SLM	selected mapping
SM	spatial multiplexing
SME	small and medium enterprise
SMS	short messaging service
SNDR	signal-to-noise and distortion ratio
SNR	signal-to-noise ratio
SOFDMA	sealable OFDMA
SOHO	small office/home office
SOVA	soft input/soft output

S/P	serial to parallel
SPI	security parameter index
SPID	subpacket identity
SPM	spatial-channel model
SPWG	Service Provider Working Group
SS	subscriber station
SSL	secure sockets layer
STBC	space/time block code
SUI	Standford University Interim
SVD	singular-value decomposition
TCP	transport control protocol
TD-SCDMA	time division/synchronous CDMA
TDD	time division duplexing
TDL	tap-delay line
TDM	time division multiplexing
TDMA	time division multiple access
TE	traffic engineering
TEK	traffic encryption key
TLS	transport-layer security
TOS	type of service
TR	tone reservation
TSD	transmit selection diversity
TTLS	tunneled transport layer security
TUSC	tile usage of subcarriers
UA	user agent
UCD	uplink channel descriptor
UDP	user datagram protocol
UGS	unsolicited grant services
UHF	ultrahigh frequency
UICC	universal integrated circuit card
UL	uplink
ULA	uniform linear array
UMTS	universal mobile telephone system
U-NII	unlicensed national information infrastructure
URL	universal resource locator
USIM	universal subscriber identity module
VDSL	very high data rate digital subscriber loop
VHF	very high frequency
VLAN	virtual local area networking
VoD	video on demand
VoIP	voice over Internet protocol
VCI	virtual circuit identifier
VPI	virtual path indicator
VPN	virtual private network

WAN	wide area network
WAP	wireless access protocol
WCDMA	wideband code division multiple access
WCS	wireless communications services
WiBro	wireless broadband
Wi-Fi	wireless fidelity
WiMAX	worldwide interoperability for microwave access
WISP	wireless Internet service provider
WLAN	wireless local area network
WLL	wireless local loop
WMAN	wireless metropolitan area network
WRAN	wireless regional area network
WSS	wide-sense stationary
WSSUS	wide-sense stationary uncorrelated scattering
ZF	zero forcing

Index

COMMUNICATIONS ENGINEERING AND EMERGING TECHNOLOGIES SERIES

FUNDAMENTALS OF WIMAX
ANDREWS / GHOSH / MUHAMED
©2007, Cloth, 432 pages, 0-13-222552-2
This book provides the reader with a solid understanding of the fundamentals of WiMAX. The authors explain the technical foundations of WiMAX in a tutorial-like manner, and provide a comprehensive overview of the standard that demystifies its many features. The book also includes a full section on the expected performance of WiMAX networks based on comprehensive link- and system- level simulations.

ULTRA-WIDEBAND COMMUNICATIONS: FUNDAMENTALS AND APPLICATIONS
NEKOOGAR
©2006, Cloth, 240 pages, 0-13-146326-8
Dr. Faranak Nekoogar explains UWB principles and technologies simply and clearly, addressing key issues such as pulse generation, modulation, multiple access techniques, and interference. In addition, she presents a complete market analysis—identifying the most promising applications, initial and future markets, and regulatory trends.

AN INTRODUCTION TO ULTRA WIDEBAND COMMUNICATION SYSTEMS
REED, Ed.
©2005, Cloth, 672 pages, 0-13-148103-7
Authored by leading-edge experts and researchers, *Introduction to Ultra Wideband Communication Systems* systematically addresses every major issue engineers will face in designing, implementing, testing, and deploying successful systems. The authors cover propagation, antennas, receiver and transmitter implementation, standards and regulations, interference, simulation, modulation and multiple access, networking, applications, and more.

THE CDMA2000 SYSTEM FOR MOBILE COMMUNICATIONS: 3G WIRELESS EVOLUTION
VANGHI / DAMNJANOVIC / VOJCIC
©2004, Cloth, 544 pages, 0-13-141601-4
The book represents a comprehensive description of the system architecture and operation of CDMA2000 third-generation wireless communications networks. The book includes a theoretical description of spread spectrum and CDMA technologies. An overview of the entire CDMA2000 network, including CDMA Radio Access Network, Packet Core Network, Mobile Station and the corresponding reference points are given.

AD HOC WIRELESS NETWORKS: ARCHITECTURES AND PROTOCOLS
MURTHY / MANOJ
©2004, Cloth, 832 pages, 0-13-147023-X
This book first presents the fundamental topics involved with Wireless Networking such as Wireless Communications Technology, such as Utlra Wideband Technology, Wireless LANs and PANs, such as Wi-Fi Systems, Wireless WANs and MANs, Wireless Internet, Wireless Sensors Networks, and Hybrid Wireless Architecture (this is the hot topic of cross-layer wireless networks).

PRINCIPLES OF COMMUNICATION SYSTEMS SIMULATION WITH WIRELESS APPLICATIONS
TRANTER / SHANMUGAN / RAPPAPORT / KOSBAR
©2004, Cloth, 560 pages, 0-13-494790-8
This book is a hands-on, example-rich guide to simulating communications systems. The first book to present complete simulation models built with MATLAB, it offers practical guidance on every key aspect of simulation, as well as detailed models that can serve as laboratories for evaluating the potential impact of changes in system design. For all engineers who design or analyze communications systems, and for all engineers who wish to learn simulation techniques.

WIRELESS COMMUNICATIONS: PRINCIPLES AND PRACTICE, SECOND EDITION
RAPPAPORT
©2002, Cloth, 736 pages, 0-13-042232-0
Wireless Communications, Second Edition is the definitive overview of wireless communications technology. Building on his classic First Edition, which became one of the world's bestselling electrical engineering books. Theodore Rappaport covers new 3G air interface standards based on both CDMA or TDMA, including W-CDMA, cdma2000, UMTS, and UMC 136/EDGE. He presents a complete tutorial on the basics of Wireless Local Area Networks, including IEEE 802.11, HIPERLAN, and other alternatives. Additional coverage includes Bluetooth, Local Multipoint Distribution Service (LMDS), and much more. Dozens of new examples and end-ofchapter problems are provided, making the book even more valuable to working professionals and students alike.

SOFTWARE RADIO: A MODERN APPROACH TO RADIO ENGINEERING
REED
©2002, Cloth, 592 pages, 0-13-081158-0
This book fills the gap, covering every issue and technique DSP engineers must understand to successfully utilize signal processing in their radio systems and subsystems. For all RF engineers who want to leverage the power of signal processing; and for signal processing engineers who want to leverage their skills in building software-based wireless communications systems, including cellular phones, PCS devices, and PDAs.

PRINCIPLES OF WIRELESS NETWORKS: A UNIFIED APPROACH
PAHLAVAN / KRISHNAMURTHY
©2002, Cloth, 608 pages, 0-13-093003-2
This is the first book to present a unified common foundation for understanding and building any contemporary wireless network, voice or data – from PCS to wireless LANs to Bluetooth, and beyond. Using extensive practical examples, the authors illuminate the common features, key differences, and specific implementation issues associated with a wide range of leading wireless systems. They begin by outlining the underlying principles of wireless networks, then review the implementation of legacy cellular telephone and mobile data networks based on CDMA and TDMA.

RF MICROELECTRONICS
RAZAVI
©1998, Cloth, 352 pages, 0-13-887571-5
This book is designed to give electrical engineers the RF microelectronics background they need to design state-of-the-art consumer electronics and communications devices. It reviews modulation and detection theory; multiple access techniques, and current wireless standards – including CDMA, TDMA, AMPS and GSM. For electrical engineers working in the communications fields, especially those involved with wireless technology.

COMMUNICATIONS ENGINEERING AND EMERGING TECHNOLOGIES SERIES

PRENTICE HALL PROFESSIONAL

For More Information visit: www.prenhallprofessional.com

PRENTICE
HALL

BOOKS ONLINE

ENABLED

THIS BOOK IS SAFARI ENABLED

INCLUDES FREE 45-DAY ACCESS TO THE ONLINE EDITION

The Safari® Enabled icon on the cover of your favorite technology book means the book is available through Safari Bookshelf. When you buy this book, you get free access to the online edition for 45 days.

Safari Bookshelf is an electronic reference library that lets you easily search thousands of technical books, find code samples, download chapters, and access technical information whenever and wherever you need it.

TO GAIN 45-DAY SAFARI ENABLED ACCESS TO THIS BOOK:

● Go to **http://www.prenhallprofessional.com/safarienabled**

● Complete the brief registration form

● Enter the coupon code found in the front of this book on the "Copyright" page

PRENTICE HALL

If you have difficulty registering on Safari Bookshelf or accessing the online edition, please e-mail customer-service@safaribooksonline.com.